Information Structure

Information Structure

Theoretical, Typological, and Experimental Perspectives

Edited by
MALTE ZIMMERMANN AND CAROLINE FÉRY

OXFORD
UNIVERSITY PRESS

OXFORD
UNIVERSITY PRESS

Great Clarendon Street, Oxford, OX2 6DP,
United Kingdom

Oxford University Press is a department of the University of Oxford.
It furthers the University's objective of excellence in research, scholarship,
and education by publishing worldwide. Oxford is a registered trade mark of
Oxford University Press in the UK and in certain other countries

First Edition published in 2010

Published in the United States of America by Oxford University Press
198 Madison Avenue, New York, NY 10016, United States of America

British Library Cataloguing in Publication Data
Data available

Library of Congress Cataloging in Publication Data
Data available

ISBN 978–0–19–957095–9

Contents

Notes on contributors

Nicholas Asher received his Ph.D. in Philosophy at Yale University. He is currently Director of Research in the CNRS of France at the Institut de Recherche en Informatique de Toulouse and has been for many years Professor of Philosophy and of Linguistics at the University of Texas in Austin. He is the author of *Reference to Abstract Objects in Discourse* (Kluwer, 1993) and *Logics of Conversation* (Cambridge University Press, 2003) and works on issues in semantics and pragmatics, with a special focus on discourse interpretation and on how prosodic information contributes to the interpretation of sentences in a discourse context.

Daniel Büring received his Ph.D. from Tübingen University in 1997 with a dissertation on contrastive topic in German and English. He has since published numerous works on focus, topic and intonation, the syntax and semantic of binding theory, and various topics in formal semantics and the syntax–semantic interface. In 2000 he joined the faculty at the University of California, Los Angeles (UCLA), where he is now a full professor.

Cornelia Ebert (née Endriss) has been Researcher at the Institute of Cognitive Science, University of Osnabrück since March 2007. Before that she was researcher in project A2 "Quantification and Information Structure" of the SFB 632 at the Linguistics Department, University of Potsdam. She received her Ph.D. in Linguistics from the University of Potsdam in 2006.

Gisbert Fanselow is a professor of syntax at the University of Potsdam. He has published in the fields of generative syntax (specializing in German), nominal compounding, gradience in grammar, and several issues in sentence processing.

Caroline Féry has been Professor of Theory of Grammar/Phonology at the Department of Linguistics at the University of Potsdam since 1999. Her interests lie in the prosodic and tonal structure of languages, in the interface between syntax and intonation, and in the theory of grammar. She is the director of the SFB 632 "Information Structure".

Ines Fiedler works as a post-doctoral researcher at the Humboldt University Berlin, in the SFB 632 on "Information Structure". Her main research interests are the interaction of grammar and information structure, mainly in Kwa and Gur languages. She has worked for many years on different aspects of the grammar of Gbe languages, including a morpho-phonological comparison of three of the five Gbe dialect clusters.

Katharina Hartmann received her Ph.D. in 1998 from the University of Frankfurt. Since 2001 she has been a lecturer at the Germanic department of the Humboldt University Berlin. She has published a number of articles on syntactic and semantic aspects of German as well as on the information structure of some Chadic languages.

Roland Hinterhölzl is an assistant professor at the German department of Humboldt University in Berlin. He did his graduate work at the University of Southern California, which he finished in 1999 with a comparative dissertation on restructuring verbs in West Germanic. An extended and revised version of it was published by Oxford University Press in the series Studies in Comparative Syntax in 2006. His current interests concern the interface between syntax, prosody, and information structure, and issues of language change.

Stefan Hinterwimmer is Researcher at the Department of German Studies and Linguistics, Humboldt-Universität zu Berlin and in the project A2 "Semantic and Pragmatic Effects of Topicality" of the SFB 632 "Information Structure". Before that he was Researcher in the project A2 "Quantification and Information Structure" of the SFB 632 at the Department of German Studies and Linguistics, Humboldt-Universität zu Berlin and at the Department of Theoretical Computer Linguistics, University of Potsdam. He received his Ph.D. in Linguistics from the Humboldt-Universität zu Berlin in 2006.

Robin Hörnig is a cognitive psychologist with special interests in linguistics. He finished his Ph.D. in Cognitive Science on the context-(in)variant reference of proper names in 2000. Since then, he has been a member of the University of Potsdam, working in the SFB on "Information Structure" from 2003 on. Together with Thomas Weskott, Reinhold Kliegl, and Gisbert Fanselow he is studying the contextual licensing of marked word order.

Þorbjörg Hróarsdóttir defended her dissertation in 1999, a revised version of which was published as a monograph (*Word Order Change in Icelandic: From OV to VO*). From 2000 to 2005, Hróarsdóttir held a postdoctoral position at the University of Tromsø, focusing on learnability and language change. In the past years, she has also written papers on verb movement in the Scandinavian languages. She is currently leading the project *Syntactic Architecture* (*ArcSyn*).

Larry Hyman, Professor of Linguistics at the University of California, Berkeley, has worked extensively on phonological theory and other aspects of language structure, particularly as concerns the Niger-Congo languages of Africa. His current interests center around phonological typology, tone systems, and the comparative and historical study of the Bantu language family, for which he founded the Comparative Bantu On-Line Dictionary (CBOLD), with funding from the National Science Foundation and in collaboration with the Laboratoire Dynamique du Langage (CNRS/Université Lyon2).

SHINICHIRO ISHIHARA is a researcher of the SFB 632 "Information Structure" at University of Potsdam. He received his Ph.D. at the Massachusetts Institute of Technology in 2003, and has been working on various issues on syntax–prosody interface and its interaction with information structure, from both theoretical and experimental perspectives. Recent publications include "Major Phrase, Focus Intonation, Multiple Spell-Out (MaP, FI, MSO)", which appeared in *The Linguistic Review* in 2007.

ELSI KAISER received her Ph.D. from the University of Pennsylvania in 2003. She spent two years as a post-doctoral researcher at the University of Rochester before moving to the University of Southern California in 2005. In her research, she uses a range of psycholinguistic methods, including visual-world eye-tracking, to investigate adult language processing. Reference resolution, in particular how it is influenced by structural as well as information-structural information, is one of her main research interests.

KATALIN É. KISS is Professor of Linguistics at the Research Institute for Linguistics of the Hungarian Academy of Sciences, and at the Péter Pázmány Catholic University, Piliscsaba. She is the author of six books and about 150 papers, dealing mainly with Hungarian syntax, information structure, the syntax of quantification, and adverbial modification.

MARIA POLINSKY is Professor of Linguistics at Harvard University. Her main interests are in syntax and syntax–information-structure interface. She has done extensive work on Austronesian and Caucasian languages. She is also interested in experimental approaches to information structure, in particular D-linking.

BRIAN REESE received his Ph.D. from the University of Texas at Austin in 2007. He is currently Visiting Assistant Professor in the Linguistics Program at the University of Minnesota. His research interests center around issues in semantics and pragmatics, in particular, the interpretation of questions in discourse and dialog.

BRIGITTE REINEKE held the chair of African languages at the Humboldt University Berlin until her retirement in 2007. In her research work, Kwa languages (Ewe, Akan, Togo mountain languages and Nkonya) and Gur languages were and are at the center of her interest. Concerning the Gur family, she is engaged in the documentation and description of languages spoken in the North West of Benin. She focuses her efforts especially on the noun-class system and the verbal structure, as well as the information structure of these languages.

MATS ROOTH received his Ph.D. in 1985 from the University of Massachusetts in Amherst and is currently Professor of Linguistics and Computing and Information Science at Cornell University (NY). His main research interests lie in computational linguistics, where he has worked extensively on mixed symbolic/probabilistic models of syntax and the lexicon, and natural language semantics, where he has worked and published on contrastive intonation (or *focus*) and related phenomena such as ellipsis and presupposition.

ANNE SCHWARZ currently works as a post-doctoral researcher within the SFB 632 "Information Structure" at the Humboldt University Berlin, where she investigates several Gur and Kwa languages. She has gained considerable Gur expertise from previous research work at the Humboldt University and wrote her dissertation on aspects of the morphosyntax and tonology of Buli. Her current research interests focus on typological and genealogical links among the pragmatic systems of different (Gur and Kwa) languages.

STAVROS SKOPETEAS is a researcher at the University of Potsdam (SFB 632 "Information Structure"). His interests include the development of empirical methods for linguistic field work, the study of the interaction between syntactic operations and information structure in cross-linguistic perspective, and the study of Meso-American and Caucasian languages.

SATOSHI TOMIOKA received his Ph.D. in Linguistics from University of Massachusetts at Amherst in 1997. His past professional experience includes visiting lecturer at Cornell University and the University of California at San Diego and post-doctoral researcher at the University of Tübingen, Germany. Since 1999 he has been teaching at the University of Delaware and is now Associate Professor of Linguistics and Cognitive Science. He has published papers on various topics, such as ellipsis, anaphora, *wh*-interrogatives, plurality, and information structure.

THOMAS WESKOTT joined the SFB 632 on "Information Structure" after finishing his Ph.D. on the processing of German topicalizations in context in 2003. He is working with Robin Hörnig, Reinhold Kliegl, and Gisbert Fanselow on contextual licensing of marked word order. His other interests include the methodological contribution of acceptability judgments to linguistic theory.

MALTE ZIMMERMANN received his Ph.D. from Amsterdam University in 2002. He has since published on quantification in European and non-European languages, on information structure and the realization of focus in Chadic, and on a variety of issues pertaining to the syntax–semantics interface. In 2006 he joined the faculty at the University of Potsdam as junior professor of "Semantics/Theory of Grammar", where he leads the DFG-sponsored project "Focus realization, focus interpretation and focus use from a cross-linguistic perspective" as part of the SFB 632 "Information Structure".

Abbreviations and symbols

Abbreviations

1	first person
2	second person
3	third person
A	argument, class A (person affix)
ACC	accusative
AD	accent domain
ADV	adverbializer
AGT	agentive
ANIM	animate
Asp	aspectual head
ASS	assertion marker
AUX	auxiliary
B	class B (person affix)
CL	class marker
CLF	classifier
CNJ	conjunction
COMP	complementizer
CP	Complementizer Phrase
CSA	complex speech act
CT	contrastive topic
D	definite
D-quantifiers	determiner quantifiers
DAT	dativc
DEF	definite marker
DEM	demonstrative
DET	determiner
DF	domain of focus
DIM	diminutive
DJ	disjunctive
DP	Determiner Phrase
EIOp	Exhaustive Identification Operator
EIP	Exhaustive Identification Phrase
EME	Early Middle English
ES	expletive subject
EVID	evidential
Exh	exhaustive

F	feminine, focus
FEM	feminine
FI	Focus Intonation
FinP	Finite Phrase
FM	focus marker
FOC	focus
FocP	Focus Phrase, Focus Projection
FOF	First Occurrence Focus
FUT	future
GEN	genitive
HAB	habitual
HTLD	Hanging Topic Left Dislocation
IAV	immediately after the verb
IBV	immediately before the verb
IMP	imperative
INCMPL	incompletive
INDEF	indefinite
INS	instrumental
INV	inverse voice
IO	Indirect Object
IP	intonational phrase
IPF	imperfective
IS	Information Structure
LD	left-dislocation
LF	Logical Form
LOC	locative
M	masculine or modifier
MaP	Major Phrase
MASC	masculine
ME	Middle English
NEG	negative
NEUT	neuter
NHG	New High German
N.INV	non-involvemental (evidential)
NOM	nominative
NPI	negative polarity item
NSF	non-subject term focus
NV	neutral version
O	object
OBJ	object
OBL	oblique
OBV	obviative

OE	Old English
OHG	Old High German
OI	Older Icelandic
ON	Old Nordic
OV	object version
P_1	today past tense
P_2	general past
PA	pitch accent
PASS	passive voice
PAST	past tense
P	predicate
PAT	patient
PERF	perfective
PF	Phonetic Form
PL	plural
PLD	primary linguistic data
POS	possessive
poset	partially ordered set
pP	phonological phrase
PP	Prepositional Phrase
PPI	positive polarity item
QM	question marker
PredP	Predicate Phrase
PRES	present
PROG	progressive
PRS	present
PRT	particle
PTF	Prominence Theory of Focus Realization
PV	preverb
PWd	prosodic word
Q	Quantifier
QAP	question–answer pair
QR	Quantifier Raising
QRG	question-related goals
QVE	Quantificational Variability Effect
RAH	Rich Agreement Hypothesis
REL	relative form, relationalizer
Relp	relative pronoun
RFL	reflexive
RP	resumptive pronoun
S	subject

SAAR Sentence Accent Assignment Rule
SARG speech-act related goals
SDRS segmented discourse representation structure
SDRT Segmented Discourse Representation Theory
SG singular
SOF second occurrence focus
SPEC specifier
SUBJ subject
TAM tense-aspect-modality
THM thematic suffix
ToBI tone and break indices
TOP topic
TP Tense phrase
TT Thematic Topic
UBH Universal Base Hypothesis
UG Universal Grammar
V verb
V2 verb second
VE vowel second
vP small verb phrase
VP Verb Phrase
XP Maximal Projection

Symbols

$\wedge$ logical "and"
$\vee$ logical "or"
$'$ prime
$\rightarrow$ implication
$\triangleright$ notation range restriction
$\times$ multiplication
$\leq$ smaller than or equal to
$\exists$ existential quantifier; "there exists"
$\forall$ universal quantifier; "for all"
$\subset$ strict inclusion
$\subseteq$ inclusion
$\sqsubset$ partial ordering
$\in$ belongs to; is a member of
$\ni$ contains as member
$\vdash$ logical consequence
$\models$ true
$\top$ interpolation marker

$\sim$ similar to; asymptotically equal to
$\approx$ approximately equal to
$\succ$ has a higher rank or order
$\neq$ not equal
$\wp$ Weierstrass elliptic function

1

Introduction

MALTE ZIMMERMANN AND CAROLINE FÉRY

The present volume presents a comprehensive overview of different approaches to the study of information structure and its formal expression in natural language. *Information structure* is that cognitive domain that mediates between the modules of linguistic competence in the narrow sense, such as syntax, phonology, and morphology, and other cognitive faculties which serve the central purpose of the fixation of belief by way of information update, pragmatic reasoning, and general inference processes. According to Fodor's (1983: 112) classification, information structure thus constitutes the domain of central general-purpose cognitive processes, as opposed to the modular linguistic systems, whose characteristic function is input analysis. This view on the status and function of information structure is compatible with Krifka's (2008) characterization of information structure as contributing to content-management. Undisputedly, the formal marking of the information-structural status of a given constituent by grammatical means, for instance as given vs. new, topic, or focus, facilitates information update and the actualization of belief states. In addition, the use of a marked linguistic structure over an unmarked, more economical one often triggers pragmatic and other cognitive effects on the side of the hearer, who attempts to find a rationale for the speaker's preference for an uneconomic way of communicating a thought. Such pragmatic and cognitive effects consist, for instance, in the identification of special speech acts, the triggering of implicatures, and an increase in the salience of a given discourse referent, which increases its potential to serve as an antecedent for anaphor resolution. It should also be noted that information structure is presumably universal, while its formal reflexes in the grammatical systems of natural languages are subject to cross-linguistic variation.

There are three basic concepts of information structure (IS) that are frequently encoded in natural language, namely *focus* vs. *background, topic* vs. *comment*, and *given* vs. *new*. In confining our attention to these three concepts, we follow the contributors of the present volume, as well as Krifka (2008). *Focus* may be understood as a classical semantic notion expressing that a focused linguistic constituent is selected from a set of alternatives (Rooth 1985, 1992). *Givenness* is interpreted in the sense of Schwarzschild (1999) as being existentially entailed by the context. *Topichood* may be more difficult to describe in a simple way. The notion of *aboutness topic* (Reinhart 1981) is accompanied by the notion of *contrastive topic*, as well as by a concept of *salience* (E. F. Prince 1981; Lambrecht 1994). All the chapters of this book define these concepts from different perspectives, so that we do not need to further develop them in this introduction; but see e.g. Jackendoff (1972), Rochemont (1986), Erteshik-Shir (1997, 2007), Féry, Fanselow, and Krifka (2008) for background information on the relevant notions. It is our conviction that the study of information structure is necessarily multi-modular and can be successful only when several aspects of grammar and the interacting cognitive domains, such as pragmatic reasoning, the fixation of belief, or the update of information states, are considered simultaneously. On the theoretical side, semantic models need input from research in phonology, syntax, and morphology. On the empirical side, we are going through an era of typological and experimental investigation which promises to be highly profitable for the research on information structure, as it sheds light on different aspects of typological research, psycholinguistics, and language acquisition. In order to comply to this need, the present volume combines theoretical studies of information-structural phenomena with empirical and typological studies, on the one hand, and with experimental studies on language production and language perception (psycholinguistics), on the other. The present collection of chapters thus combines to form a coherent whole.

The book is organized as follows. The first part, called "Topic and Focus", contains six theoretically oriented chapters that belong together in providing interpretations and formal representations of IS features that are directly implemented in the grammar. In this respect, the chapters constitute a framework for the remaining chapters. Some of them develop a semantic framework within which to interpret information structure. Some others introduce prosodic, morphological, syntactic, and even semantic features that are of particular importance for implementing information structure in the grammar. Semantic research has been crucial in the past to settle the issue of the best way of interpreting correlates of focus and givenness, and Chapter 2 "Second occurrence focus and Relativized *Stress F*," goes a step further in this direction. In it, Mats Rooth points out a tight relation between information structure prominence, in terms of focus, and prosodic prominence, in terms of accent. This is done by way of a detailed analysis of *second occurrence focus* (SOF), a phenomenon that has drawn the attention of semanticists and phonologists only in recent years. There are parallels between the semantics and phonology which can be captured by means of two generalizations. The first one is *Stress F*, which posits that a pitch accent must be placed somewhere in the prosodic domain of the focus constituent. The second one, called *Relativized Stress F*, establishes a hierarchy between two bound foci with overlapping or embedding domains. The focus with higher semantic scope is prosodically more prominent than the focus with lower scope. The effects of both constraints can be captured in a strictly local fashion by the introduction of operators that operate as modifiers on semantic combination functions, such as Functional Application. These modifiers pass the information of semantic and prosodic prominence up the syntactic tree. Thus, Chapter 2, even if it has a strong semantic bias, relates the semantic interpretation of focus to the prosody, mediated by the syntax.

Chapter 3 "How focus and givenness shape prosody" by Caroline Féry and Shinichiro Ishihara takes the same direction. It also examines the relationship between information structure and prosody, but this time with an emphasis on the prosodic structures involved. The authors propose that, at least for syntactically unmarked sentences in German and Japanese, it is important to keep apart the import of syntax, expressed in prosodic phrasing, and the import of information structure, reflected in the F0 scaling of pitch accents and boundary tones. Some empirical evidence for this analysis comes from the phonological realization of Second Occurrence Focus in German and embedded *wh*-questions in Japanese. In particular, information structure affects the tonal height of focused and given constituents only by register raising of focused constituents and deaccenting or compression of given material. This chapter complements the preceding paper in refining the phonological part of the semantics–prosody interactions identified in Chapter 2.

A recurrent question in many approaches to information structure is the role played by syntax in expressing information-structural distinctions. Brody (1990) and Rizzi (1997), among others, have been extremely influential in proposing that focused constituents occupy special syntactic positions, in which they can be interpreted. Katalin É. Kiss's work has argued for a preverbal focus position in Hungarian which is interpreted exhaustively. In her Chapter 4, "Structural focus and exhaustivity", she discusses the now classical interpretation of syntactic focus marking in Hungarian. After an informative overview of the large body of the relevant literature, É. Kiss exposes her own analysis of the exhaustivity effect found with focus constructions. Semantically, exhaustive interpretation is the result of an equivalence relation between a set denoted by the focused constituent overtly moved to a designated syntactic position and a set denoted by the background predicate. Drawing on new facts, É. Kiss shows how the distribution of a certain class of elements in Hungarian, namely downward entailing expressions, follows directly from the assumption that exhaustivity is structurally encoded.

The next two chapters complement the discussion of focus in concentrating on the second important information-structural notion of *topic*. Both chapters touch upon the question of how syntax or morphology interact with prosody in the formal expression of this category, and both chapters put forward a unified analysis of aboutness and contrastive topics. Simple topic marking on a constituent expresses that this element must be interpreted outside its immediate speech act. Grammatical topic marking itself is subject to cross-linguistic variation, with German exhibiting syntactic topic marking and Japanese morphological topic marking. The contrastive topic effect can then be understood as the result of prosodic focus marking on a constituent that is already syntactically or morphologically marked as (aboutness) topic. Tomioka points out that such contrastive topic marking gives rise to, for example, scalar implicatures since focus marking invokes a set of alternatives against which the current utterance must be evaluated.

Chapter 5, "The interpretation of topical indefinites as direct and indirect aboutness topics", by Cornelia Ebert and Stefan Hinterwimmer, discusses the interpretation of left-dislocated topics in German. Based on a dynamic conception of left-dislocated *aboutness topics* as addresses in the common ground where information can be stored, the authors show that the marked peripheral topic position in German is not restricted to proper names and definite descriptions, as often assumed in the literature, but that it can also be occupied by a semantically well-defined subset of indefinite quantifying DPs. This restricted subset of quantifying DPs can be reinterpreted as denoting a uniquely identifiable set (of situations), for which reason they can occur as topics.

Chapter 6, "Contrastive topics operate on speech acts", by Satoshi Tomioka, touches upon a classical domain in the study of information structure, namely the role played by the particle *wa* in Japanese, which is often interpreted as a topic particle. The chapter puts forward a convincing solution to three puzzles surrounding contrastive topic (CT) *wa*-marking in Japanese, namely the occurrence of CTs in the absence of additional focus constituents, unlike in intonation languages, the occurrence of CTs in non-assertive speech acts, and the identical morphology of CTs and ordinary aboutness topics. The proposed analysis is minimal in attributing the same basic function to the (contrastive) topic morpheme *wa*. The uniform function of all instances of *wa* is thus to indicate that the constituent it attaches to must be interpreted outside the speech act constituted by the rest of the clause, as is the case with all topics (see above). The contrastive flavour of CT-*wa* is the result of additional prosodic focus marking, which introduces focus alternatives in the Roothian sense. Presence of -*wa* guarantees that these alternatives are not closed off until the speech act level, such that CT-wa indicates the presence of *alternative speech acts*. The assumption that focus alternatives can project to the level of speech act creates a tight link between sentence-based information-structural effects, which consist in the generation of alternatives, and the illocutionary force of an utterance.

A further pragmatic effect of prosodic focus marking is discussed in Chapter 7 by Nicholas Asher and Brian Reese, "Biased questions, intonation and discourse", where it is shown that one and the same prosodic factor, namely contrastive (or emphatic) focus accent, can have the double function of evoking alternatives and the construal of a complex speech act type of *assertion+question*. The question is investigated of which phonological and structural factors have to do with the coding of IS-structure in the narrow sense, and which factors determine other discourse-related notions, such as illocutionary force or speech act. The authors assume a tight relation between sentence-based semantics, consisting, for instance, in the generation of alternatives triggered by focus-accent, and discourse-based pragmatics in terms of speech-act theory. The chapter provides a unified account of so-called *biased questions*—tag questions, negative polarity questions, and emphatic focus questions—which have long been observed to express a certain bias of a speaker towards the possible answers to the question.

The six chapters of Part I thus provide an overview of some theoretical approaches to information structure by discussing the question of how semantics, prosody, syntax, and morphology must be combined in the analysis of information-structural factors and the ways in which they are grammatically encoded. They give us tools for typological and diachronic comparison, as well as for a better understanding of language production and processing.

The second part of the book, "Cross-Linguistic Variation and Diachronic Change", combines five chapters that concentrate on the grammatical expression (or the absence thereof) of focus, topic, and givenness, on the one hand, and on the influence of information structure on diachronic change in grammar, on the other. The typological and diachronic studies complement each other in that the former concentrate on the synchronic variation across different languages, whereas the latter investigate diachronic variation across different stages of one and the same language. Not trivially, some of the chapters show that certain information-structural categories, such as information focus, need not trigger an overt grammatical reflex in all languages of the world. These chapters thus fulfil a control function for the theoretical approaches from the first part. The empirical findings allow for a re-evaluation of the cross-linguistic validity of theoretical claims that have been made almost exclusively on the basis of intonation languages, such as the conviction that focus (contrastive or information) is always formally expressed.

Chapter 8, "Towards a typology of focus realization", by Daniel Büring, sums up a larger typological review of grammatical focus realization in the languages of the world. Büring shows that the reduction of information structure to prosodic effects, as captured by applying Truckenbrodt's (1995) principle of *FocusProminence* "Focus needs to be maximally prominent", goes a long way to this aim. The concept of FocusProminence was originally introduced to account for the addition of pitch accents on a focused constituent, or for the insertion and deletion of prosodic phrase boundaries in information structurally marked patterns. Büring shows in his chapter that FocusProminence can also explain other strategies of focus marking, such as syntactic reordering and the use of particles on the focused constituent. The main idea of the chapter is that focus has to be more prominent than non-focused constituents, and that languages have at their disposal a whole series of strategies to fulfil this need.

In Chapter 9, "Focus in Aghem", Larry Hyman and Maria Polinsky argue against a unified analysis of the immediately postverbal focus in Aghem as a designated position for focus, say FocP. Rather than inflating the inventory of functional projections from FocP(high) to FocP(low), the proposed analysis does away with focus projections altogether. Hyman and Polinsky argue that what appears to be a single designated position at the surface actually corresponds to a set of structurally different positions that are base-generated inside the vP. The authors, then, question the existence of designated syntactic focus positions (in Aghem) and thus come to the opposite conclusion of Chapter 4, which argued for the existence of a focus position in Hungarian. It may be that Hungarian and Aghem are truly different in this respect, but it may also be that syntactic focus positions are dispensable in all languages, including Hungarian. The latter view would support claims in Szendrői (2001) to the effect that Hungarian focus movement is driven by prosodic needs rather than syntactic ones (see also Chapter 8 and Zubizarreta 1998). This position is compatible with a view of information structure that sees grammatical components such as prosody, syntax, and morphology as supporting the needs of information structure in forcing the emergence of marked structures, which are available in a grammar independently of information-structural considerations.

The discussion of focus marking in West African languages is taken up in Chapter 10, "Subject focus in West African languages" by Ines Fiedler, Katharina Hartmann, Brigitte Reineke, Anne Schwarz, and Malte Zimmermann, which discusses focus marking in Kwa and Gur from the Niger-Congo family, and in West Chadic from the Afro-Asiatic family. An extensive empirical survey of 23 languages from these language groups reveals that there is wide variation in the grammatical means employed for marking focus. More strikingly, though, all the languages discussed exhibit an asymmetry between subjects and other arguments when it comes to focus marking. While information focus on non-subjects is generally unmarked in the majority of languages, subject focus must be grammatically marked, no matter whether it is used contrastively or as new information focus. The special status of subjects with respect to focus marking is attributed to the fact that unmarked subjects in their canonical preverbal position are prototypically interpreted as aboutness topics in West African languages. This provides strong evidence against assigning a particular focus interpretation, such as contrast or exhaustiveness, to a particular structural configuration. Quite to the contrary, it shows that an independently available position, say Spec–TP, is associated with a default interpretation as topic.

The syntactic encoding of information structure, possibly triggered by prosodic factors, also plays a central role in the two diachronic chapters, which propose a correlation between the syntactic position of a constituent and its information-structural interpretation. The observed variation of word order in earlier stages of Icelandic and German is reduced to the fact that constituents with different information-structural status (given–new, contrast) are realized in different syntactic positions: Expressions denoting given or contrastive information precede the verb, whereas expressions denoting new information follow it. An interesting result of the two chapters, taken together, is that Icelandic and German developed from a similar starting point with mixed OV/VO word order into opposite directions, with modern Icelandic being strictly VO and modern German being strictly OV.

Þorbjörg Hróarsdóttir's Chapter 11, "Information structure and OV order", addresses the existence of OV orders in historical stages of Icelandic, especially Older Icelandic, and the subsequent loss of these orders. Based on a quantitative analysis of a number of Older Icelandic texts, the author proposes that the loss of OV was triggered by changes affecting the syntactic implementation of information structural distinctions. In older texts, OV order was predominant with DPs representing old information, as for instance pronouns, while VO order was primarily confined to heavier DPs linked to focus and new information. This difference is accounted for by assuming that

DPs representing old information moved to a high syntactic position, while heavy and focused DPs underwent only short movement to a lower syntactic position and thus remained in the VP domain. Extraction of VPs prior to vP movement was instrumental in the realization of the relevant word order differences (OV vs. VO) that were linked to this information structural difference. The chapter also argues that during the seventeenth and eighteenth centuries, a set of sociolinguistic factors gave rise to a high frequency of VO orders. Eventually, a parameter change—loss of VP movement—made the OV orders disappear.

In Chapter 12, "Information structure and unmarked word order in (Older) Germanic", Roland Hinterhölzl argues that the diachronic change in word order from Old High German OV/VO to modern German OV was effected by a diachronic change in the patterns of prosodic markedness, which in turn reflect the IS-structure of an utterance. The chapter accounts for observed word order variations in the Old High German Tatian translation by arguing that this variation is primarily determined by information-structural restrictions. The most important empirical result is that arguments and predicates in Old High German, which cannot be extraposed in modern German, could occur in the postverbal position in embedded clauses if they were part of the focus of the clause. The joint result of this and the preceding chapter is that it is possible to detect effects of information structure in older stages of a language, which in turn allows for the formulation of hypotheses concerning its effects on prosody and word order as a whole.

While the first and second part of the book provide the theoretical foundation and introduce cross-linguistic and diachronic data that can be used in testing for the adequacy of different grammatical models of information-structure, the third part of the book, "Experimental and Psycholinguistic Approaches", turns to a relatively new, but thriving, field of linguistic study. Based on a variety of experiments, the three chapters in this part address the production and processing of information structure and thus constitute the third component of a truly integrative theory of information structure. The emphasis of the three chapters is on production experiments and psycholinguistic online experiments that probe for different aspects of information structure, with special attention being paid to the third information-structural distinction of given vs. new: the one chapter on production investigates the syntactic effects of givenness. The two chapters on perception experiments investigate the resolution of pronominal reference and the parsing of spatial–relational assertions, respectively. Only a deeper understanding of the general processes and mechanisms involved in the production and processing of grammatical structures will lead to a better understanding of what motivates the choice of a marked structure (or reference) over an unmarked canonical structure, as is frequently the case in the expression of focus and topic.

Chapter 13, "Effects of givenness and constraints on free word order", by Stavros Skopeteas and Gisbert Fanselow, presents the results of a production study in which the authors investigate the syntactic effects of givenness in German, Georgian, and a range of additional, typologically unrelated languages. Methodologically, the use of a production experiment sets a new standard for the collection of reliable, and cross-linguistically comparable, empirical data on the interaction of information structure and grammar. Empirically, the results are threefold. First, it shows that givenness manifests itself in natural language in a surprisingly uniform way. A given noun phrase *can* be displaced (even though it need not be so), such that it precedes the new material in the clause. This finding thus confirms the well-known cross-linguistic tendency for given constituents to occur before new ones. Second, the syntactic displacement of given material always involves the more local A-type of movement in the sense of Chomsky (1981). Third, the concrete type of A-movement involved in the expression of givenness is subject to cross-linguistic variation. If a language has A-movement affecting the grammatical functions of the arguments at its disposal, such as passivization, it is this movement type that is induced by givenness. If a language has no operation affecting grammatical functions, or if it does not use it under the relevant circumstances, simple reordering, A-scrambling that is, can serve the expression of givenness. If a language has neither A-scrambling nor grammatical function-changing A-movement, givenness is not reflected in the syntax.

In Chapter 14, "Investigating effects of structural and information-structural factors on pronoun resolution", Elsi Kaiser applies classical psycholinguistic methods in the investigation of two claims concerning the relation of anaphora resolution and information structure. As is well known, weak pronouns like *s/he* anaphorically pick up salient referents; see, for instance, Givón (1983), Ariel (1990), and Gundel, Hedberg, and Zacharski (1993), among others. This restriction has the welcome effect of restricting the search space for potential antecedents. Using a sentence completion experiment, Kaiser investigates which role in anaphora resolution is played by topichood and focus, and also by subjecthood. The somewhat surprising result of this experiment is that anaphora resolution depends not only on the availability of an appropriate antecedent, but also on material that follows the pronoun. Kaiser concludes that anaphora resolution is determined by multiple factors. Subjecthood plays an important role in making an entity a good antecedent for a subsequent pronoun, but the effects of subjecthood are modulated by weaker effects of pronominalization and focusing. The discussion is couched in terms of Centering Theory (Ariel 1990), an influential computationally based model of the local-level component of attentional state in discourse, which assumes that maximally coherent transitions are easiest to process.

Chapter 15, "Given and new information in spatial statements" also reports the results of psycholinguistic experiments on the nature of information structure. Robin Hörnig and Thomas Weskott discuss a number of experiments that investigate the effects of word-order variation on the comprehension of localizing statements, paying special attention to the information-structural status of the spatially related entities as given or new. For instance, given a premise sentence that introduces two spatially related entities (*The A is to the left of the B*), the authors investigate whether the mental localization of a third entity (introduced in a subsequent sentence) relative to the two given entities proceeds faster when its denotandum precedes the localizing expression, or when it follows it. There are two central hypotheses. First, a localizing statement is processed faster when the given entity has the status of *relatum*, that is the entity relative to which the new entity, the *locatum*, is localized. Second, processing is even quicker when the relatum is introduced before the locatum in the sentence. The results confirm both Chafe's (1970) notion of information packaging of new before given objects, and Clark and Haviland's (1977) contrast between *given* and *new*.

Well-conducted production and perception experiments are indispensable tools for testing the validity of claims about the inventory of information-structural categories and their grammatical reflexes. For this reason, the studies on word order, anaphora, and localizing statements discussed in the third part of the book are extremely important for obtaining a better view of how information-structural dimensions can affect the structural design of language-specific grammars. This is because basic cognitive categories, such as familiar/new, the perception of space, and deixis can be understood as universals, even though their grammatical expression in the form of concrete linguistic utterances is not. In disentangling the formal aspects of a linguistic utterance that have their origin on the cognitive side from those that originate in the grammar itself, we will come to a better understanding of the functioning of both individual and universal grammars.

Summing up, despite differences in the immediate object of investigation and methodology, the fourteen chapters in the book converge on the same analytical bottom line, namely the analysis of marked structures—which are typically associated with an informational surplus in one way or other—over unmarked canonical structures. This informational surplus helps the information update of the discourse. Grammatically, it can be expressed in different ways, but it appears from this book that prosodic, syntactic, and morphological strategies are equally employed to this purpose in the languages of the world. All the chapters address—in one way or another—the overwhelming question of how adequately theoretical models can account for real data. Even if the emphasis of the various chapters lies on different aspects such as theoretical modelling, typological comparison, diachronic development, production, or processing, they all show that the study of information structure cannot be ignored in the development of models of the grammar of human language(s). The central questions dealt with in the following chapters are: the phonological, morphological, and syntactic encoding of information-structurally prominent constituents in the grammar; the semantic interpretation of such constituents; interrelations and correlations between the two processes; additional pragmatic effects of grammatical marking; the cross-linguistic variation of such markedness; the diachronic change of patterns of grammatical markedness; and the production and processing of marked structures. The book brings us a step closer to answering these fundamental questions.

Part I

Topic and Focus

2

Second occurrence focus and *Relativized Stress F*

MATS ROOTH

2.1 The semantic and phonological scope of focus

In "anaphoric" or "givenness" theories of the semantics and pragmatics of intonational focus, the first sentence in (1) is in a certain sense the antecedent for the focus in the second sentence. The representation (2) makes the anaphora explicit using an operator "$\sim$", which marks the scope and the antecedent of the focus. The focused phrase is marked with the feature F.

(1) You boil your vegetables?
I MICROwave my vegetables.

(2) $[\text{You boil your vegetables}]_8$

$$\underbrace{[\text{I } \underbrace{\text{microwave}_F}_{\beta} \text{ my vegetables}]}_{\phi} \sim 8$$

The semantic part of anaphoric theories jointly constrains the denotation of the antecedent (here the proposition "John boils John's vegetables", assuming the second speaker is John) and the focus semantic value of the scope of the focus (here the set of propositions of the form "John *R*'s John's vegetables", where *R* is an alternative to "boil"). A couple of versions of the constraint have been given (Rooth, 1992, 1996a; Schwarzschild, 1999). Two options are stated in (3), restricting attention to the clauses which apply to (2).[1] In order for the structure to be licensed, the focus constraint has to be satisfied. In this case, the first one is satisfied because the antecedent "John boils John's vegetables" is an element of the focus semantic value of the scope of the focus. The second version is satisfied because the union of the focus semantic

[1] The constraints include several subclauses or type accommodation principles which are conditioned by semantic type.

value is a disjunction of propositions, and in this case one of them is the antecedent "John boils John's vegetables". Since a proposition p_i entails the disjunction $p_1 \vee \ldots \vee p_i \vee \ldots$, the antecedent entails the union of the focus semantic value.

(3) a. Where ϕ is the scope of the focus, the denotation of the antecedent is an element of $[\![\phi]\!]^f$. (Rooth 1992)

b. Where ϕ is the scope of the focus, the denotation of the antecedent entails the union of $[\![\phi]\!]^f$. (Schwarzschild 1999)

In combination, the notation (2) and its interpretation (3) define a semantic notion of the scope of a focus.[2] Truckenbrodt (1995) argued that this agrees with a phonological notion of the scope of focus, namely the domain of prominence of the focus. To capture the correlation, he proposed a *Stress F* constraint which in my discussion will be formulated as (4).[3]

(4) *Stress F*

Let β be an F-marked phrase with scope ϕ. Then the strongest stress in the phonological realization of ϕ falls within the realization of β.

Suppose stress is represented in the metrical grid formalism (A. Prince 1983), and that the grid representation of (2) is along the lines of (5). Then *Stress F* is satisfied because the strongest stress in the phonology of the whole sentence, namely the stress on the first syllable of *micro*, falls within the phonological interval corresponding to the F-marked element *microwave*. *Stress F* would not be satisfied if the first syllable of *vegetables* had greater grid prominence than the first syllable of *micro*.

(5)

	x						
	x		x		x		
x	x	x	x	x	x	x	x
I	mi	cro	wave	my	vege	ta	bles

(β = mi cro wave; ϕ = I mi cro wave my vege ta bles)

Truckenbrodt's argument for *Stress F* is based on examples like (6), where there is arguably a focus on *American* whose scope is the nominal [an American farmer], rather than the whole sentence.

(6) [$_S$[$_{NP}$ an American$_F$ farmer]$_1$~2 told a [[$_{NP}$ a Canadian$_F$ farmer]$_2$~1] a joke]

2 In Schwarzschild (1999), the scope of an F is marked by the next F up, rather than the ~ operator.

3 See also Jackendoff (1972), Rooth (1996a), Büring (2008), Selkirk (2006a).

In the anaphoric theory, the scope of the foci on *American* and *Canadian* can not be the whole sentence, because the sentence can be used in discourse contexts where there is no available antecedent which has the form "a *P* farmer told a *Q* farmer a joke", or which entails that a farmer of some nationality told a farmer of some nationality a joke. In the representation (6) with a narrower scope for the focus on *American*, the constraint (3a) is satisfied for the focus on *American*, because as long as being Canadian is an alternative to being American, the denotation of the antecedent [a Canadian farmer] is an element of the focus semantic value for the scope of the focus.[4]

The point now is that the semantically motivated scope of focus in (6) agrees with the phonological domain of prominence. Structure (7) is a plausible grid representation. The second syllable in *American* has greater stress than anything else in the phonological interval corresponding to [an American farmer], but it does not have greater stress than *joke* or the second syllable in *Canadian*.[5] But *Stress F* is obeyed, because the semantic scope of the F on *American* (the ϕ in the constraint) is the nominal [an American farmer], rather than the whole sentence.

(7)

```
                                                                  x
        x                              x                          x
        x        x        x            x         x                x
 x   x  x  x  x  x   x    x    x   x   x   x  x  x    x    x      x
an   A  me ri can far mer told a  Ca  na  di an far  mer  a    joke
     \_____β_____/
\__________φ__________/
```

The argument supports the hypothesis that the semantic scope of focus matches the phonological domain of focus prominence. This is a substantial empirical and theoretical claim, and one that is attractive because it postulates a kind of homomorphy between phonology and semantics. However, the homomorphy is arguably poorly articulated in (4), because there is reference to information that is scattered in the syntactic tree and in the phonology,

4 The semantic side of the story about the focus on *Canadian* is symmetric. However, Féry and Samek-Lodovici (2006) point out that there is an asymmetry on the phonological side. In their example (i), each occurrence of *Chevrolet* can be within the scope of the focus on the occurrence of *farmer* to the left. But the first occurrence of *Chevrolet* is realized with an accent.

i An American farmer with a purple Chevrolet was talking to a Canadian farmer with a purple Chevrolet.

5 Evidence for this is that the nuclear accent falls on *joke*. Truckenbrodt notes, "If the phonological domain of a focus would be the clause, regardless of the semantic domain the clause-final default-stress in these examples would not be derived. Instead, one of the foci in each of these examples would attract the nuclear stress of the clause."

and to a correspondence between focused phrases, their scopes, and intervals in a phonological representation. This is explained in Section 2.2. As a corrective, an architecture is proposed with local operators that have a local phonological interpretation of relative prominence, and a local semantic interpretation in terms of the scope of focus. Section 2.3 widens the empirical domain to include second occurrence focus configurations, which are complex configurations where two focused phrases take different, nested scopes. The scope relationships in some second occurrence configurations contradict the simple phonology–semantics correspondence in *Stress F*. This problem is addressed by a "relativized" statement of *Stress F*, which allows a single phrase to contain focused subconstituents that take different scopes outside the phrase. Section 2.4 takes up the technical problem of providing a semantics for the local operators proposed in Section 2.2, and showing how the complex configurations discussed in Section 2.3 are analyzed. Section 2.5 sums up.

2.2 The local operator architecture

Figure 2.1 illustrates a structure of application for *Stress F*. In an embedded position, there is an F-marked phrase *Ashley*, which is the β of (4). Its scope is the entire clause, which is the ϕ of (4). Applying *Stress F* requires (i) locating the phonological interval ϕ' that corresponds to ϕ, (ii) locating the phonological interval β' that corresponds to β, and (iii) checking whether the syllable

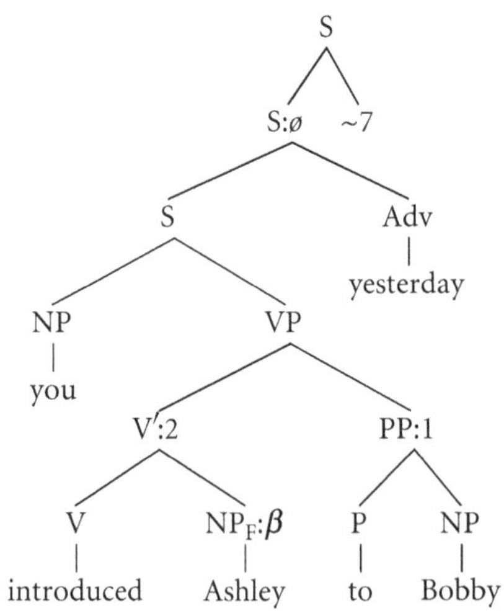

FIGURE 2.1 An application of *Stress F*. β is the focus and ϕ is its scope

σ in ϕ' with maximal prominence in ϕ' falls within β'. This procedure looks at two levels of the syntactic tree that are separated from each other by three nodes, and it refers at two levels to a correspondence between syntactic phrases and phonological intervals. Although a statement of the syntax–phonology interface defines a map between syntax and phonology, determining that a phrase ϕ has phonological realization ϕ', it is a different matter for a grammatical constraint to refer to multiple pairs that stand in the realization relation, as *Stress F* does when it refers to the syntax–phonology pairs $\langle\phi, \phi'\rangle$ and $\langle\beta, \beta'\rangle$. In these respects *Stress F*, rather than stating an elegant homomorphy between syntax and phonology, is an implausibly complex global constraint. Section 2.3 will show that this problem becomes even more severe in configurations involving several Fs with different scopes. But even in basic applications like the one in Figure 2.1, *Stress F* has the character of a descriptive generalization which is not a plausible candidate for a constraint linking linguistic levels, because it refers to so many things at once.

The alternative is to state a local correlation between phonological and semantic interpretation. Rather than stating a global constraint referring to the syntax–phonology correspondence at two tree levels, one would like to state a local correspondence at each tree level. The phonological side of this correspondence in the approach developed here is the familiar one of relative metrical prominence in a binary-branching tree. Phonologically, marking the scope of a focus involves local demarkations of the relative prominence of the two children of binary-branching nodes. In a binary-branching configuration $[\gamma\ \psi]$, the scope of a focus in γ is expanded only if γ is phonologically more prominent than ψ. This is reminiscent of the *sw* architecture for metrical trees proposed in Liberman and Prince (1977), where a binary-branching node $[\gamma\ \psi]$ in which γ is more prominent is represented as $[\gamma_s\ \psi_w]$. For reasons related to the semantics, this prominence relation is represented here as $[\gamma\ \psi]_l$, with *l* indicating prominence on the left. Prominence on the right is indicated with $[\gamma\ \psi]_r$.

This results in the revised representation in Figure 2.2 for the example where *Stress F* was applied in Figure 2.1. On the phonological side, it should be apparent how prominence relations have been localized. Instead of referring globally to the phonological domain of prominence ϕ' of the prominence in β', we have local operators *l* and *r* which represent that β' is maximally prominent within ϕ'. Semantically, the operator *l* should expand the scope of a focus in the left child of the node that bears it, while *r* expands the scope of a focus from the right child. The effect should be that the scope of the focus on *Ashley* in Figure 2.2 is expanded to the clausal level. This semantic side of the architecture is worked out in Section 2.4.

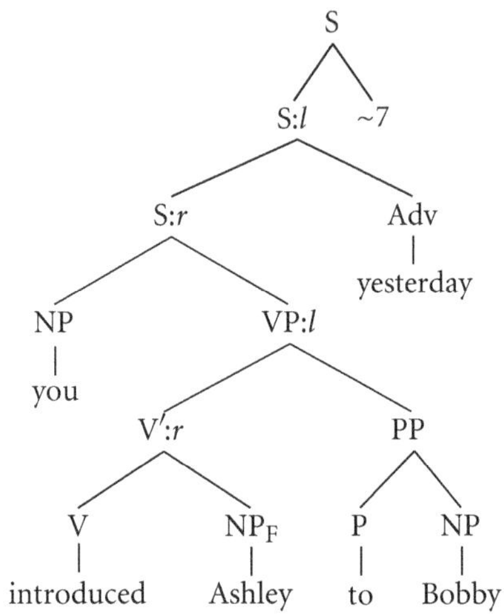

FIGURE 2.2 The scope of focus represented with local prominence operators

2.3 *Stress F* in second occurrence focus configurations

Second occurrence focus (SOF) is an intonational pattern found in certain examples with multiple motivations for focus:[6]

(8) a. Eva only gave xerox copies to the GRADUATE$_F$ students.
No, PETR$_F$ only gave xerox copies to the graduate$_{SOF}$ students.

b. We only introduced Marilyn to JOHN$_F$ Kennedy.
(i.e. not to Bobby and Edward Kennedy)
We also only introduced SUE$_F$ to John$_{SOF}$ Kennedy.

c. Mary only STEAMS vegetables, and even JOHN$_F$ only steams$_{SOF}$ vegetables.

The notation SOF marks a phrase where on semantic grounds one might expect there to be a focus, but where apparently there is no focus marked by a pitch movement. In early literature, there was disagreement about phonetic/phonological status of SOF. Either SOF is not phonetically marked at all (Partee 1991; Krifka 2004a) or SOF is phonologically prominent, though not marked with a pitch accent (Rooth 1992; von Fintel 1994; Rooth 1996b). Recent experimental studies are interpreted as supporting the second position, though the magnitide of the phonetic reflexes is small and not consistent at the token level (Beaver, Clark, Flemming, Jaeger, and Wolters 2007; Howell 2007). There is additional evidence from

[6] These examples are from Partee (1991), Rooth (1993), and Krifka (2004a).

weak pronouns (von Fintel 1994; Rooth 1996b; Beaver *et al.* 2007). English pronouns can be reduced to various degrees; at the extreme, they can lose their onset and be prosodically incorporated into the preceding word as in (9). This process is blocked in SOF configurations, as illustrated in (10).

(9) a. I likim (= him).
b. He likser (= her).

(10) Mary's boyfriend only likes HER.
#Even her BOSS only likser.

The SOF data are interpreted as indicating that the phonological correlate of F is metrical prominence, as in *Stress F* theories, rather than pitch accent (Rooth 1996b; Selkirk 2006a; Beaver *et al.* 2007). Some prominences derived from F surface with pitch accents; these are the ordinary Fs. Others do not; these are the SOFs. Apparently, all examples of SOF in the literature have a special configuration of relative scope of focus, with the scope of the SOF embedded in the scope of the primary F (Rooth 1996b; Büring 2008; Selkirk 2006a). The configuration is schematized in (11).

(11) $[[\ldots \quad \alpha_F \quad \ldots[[\ldots \quad \beta_F \quad \ldots] \sim j]\ldots] \sim k]$

	α_F		β_F
	↓		↓
	ordinary F phonology		SOF phonology

An additional line of evidence is provided in Féry and Ishihara (2005). They look at examples in German where the narrower-scope focus precedes the wider-scope focus. They find that in such configurations the narrower-scope focus is realized with a pitch accent, but one which has a compressed pitch range relative to the wider-scope focus.

We are left with the following picture. In a configuration with two Fs whose scopes are nested, the narrower-scope focus is realized with less phonological prominence than the wide-scope focus. This provides the interface between semantics and phonology/phonetics. The prominence relation results in different realizations, depending on linear order. The narrower-scope focus, if it follows the wider-scope one, is realized with SOF phonology and phonetics, without a pitch movement. If the narrower-scope focus precedes the wider-scope one, it is realized with a compressed pitch range.

The phonology–semantic correlation seen in second occurrence focus—with the wider-scope focus being realized with greater prominence—is reminiscent of the phonology–semantics homomorphy expressed by *Stress F.* Are the prominence relations seen in SOF in fact a *consequence* of *Stress F*, or

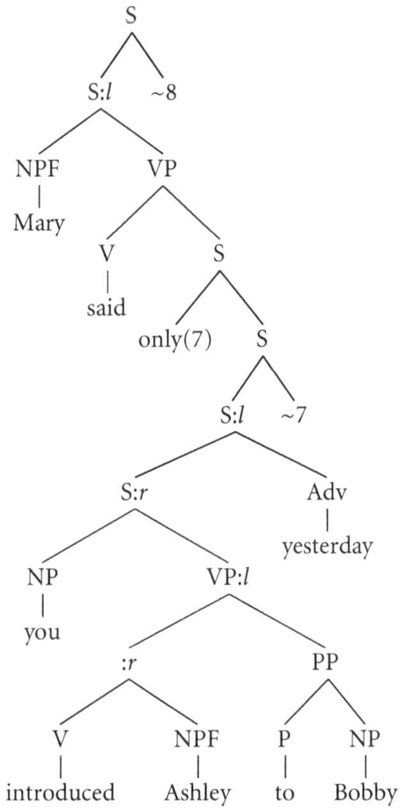

FIGURE 2.3 Local prominence operators in an SOF configuration

of the reformulation of *Stress F* in terms of local operators? Consider the nested-focus example (12), where in the second sentence the focus on *Ashley* is realized with SOF phonology.

(12) John said you only introduced Ashley$_F$ to Bobby yesterday.
Wrong. Mary$_F$ said you only introduced Ashley$_F$ to Bobby yesterday.

The embedded clause in the second sentence of (12) should have a representation similar to Figure 2.2, with a focus on *Ashley* taking scope at the level of the embedded clause. The focus on the matrix subject *F* has scope at the matrix S. Using local prominence operators to represent the scopes, this results in Figure 2.3. The operator *l* at the top has the semantic function of extending the scope of the focus on *Mary* to the matrix. At the same time, it has a phonological interpretation of relative prominence, indicating that the matrix subject is more prominent than the matrix VP. Since the focus on *Ashley* is embedded in the matrix VP, it follows that the the matrix subject, with the wide-scope focus, is more prominent than the embedded object *Ashley*, with the narrow-scope focus. In other words, the prominence relations seen in SOF fall out of the semantically motivated representation of scope of focus, once scope of focus is represented by local operators with both semantic and phonological interpretations.

The same result is obtained with *Stress F* in the formulation in (4). Starting with the more embedded focus, $[\text{Ashley}]_F$ is β_1 and its scope ϕ_1 is the embedded clause. *Stress F* requires that the most prominent syllable in the phonological interval ϕ_1', the phonological realization of the embedded clause, falls within β_1', the phonological realization of the focused phrase *Ashley*. Looking at the higher focus, *Stress F* requires that the most prominent syllable within the realization ϕ_2' of the entire sentence fall within the realization β_2' of the focused phrase *Mary*. Given the subinterval relations in (13), it follows that the most prominent syllable in *Mary* is more prominent than the most prominent syllable in *Ashley*. This is as desired.

(13) $$\underbrace{\ldots\ \underbrace{\text{Mary}_F}_{\beta_2}\ \ldots\ \underbrace{\ldots\ \underbrace{\text{Ashley}_F}_{\beta_1}\ \ldots}_{\phi_1}}_{\phi_2}$$

The logic in this form unfortunately falls apart in SOF examples where there is a single phrase ψ that dominates both Fs, and is dominated by the scopes of both Fs. Consider example (14), where there is an F on *Ashley* with ordinary realization, and an F on *Bobby* with SOF realization. Example (15) is similar, with F on *Gouda and caviar*, and SOF on *New Jersey*.

(14) You know what? You only introduced Mona to Bobby_F yesterday.
You also only introduced Ashley_F to Bobby_{SOF} yesterday.

(15) What foods did you only find in New Jersey last year?
I only found Gouda and caviar_F in New Jersey_{SOF} last year.

Figure 2.4 represents the scopes of focus features in (14). The indexing on F is informal notation for the scope of focus: F1 takes scope at the first c-commanding $\sim$, while F2 takes scope at the second c-commanding $\sim$. The hope is that when the scope relations are expressed with local *lr* operators, the phonological prominence relations will fall out. Instead, there is a contradiction. The tree on the left in Figure 2.5 has operators in the embedded clause that represent the scope of F1. The tree in the right has the operators that represent the scope of F2. The contradiction is seen at the VP node, which is

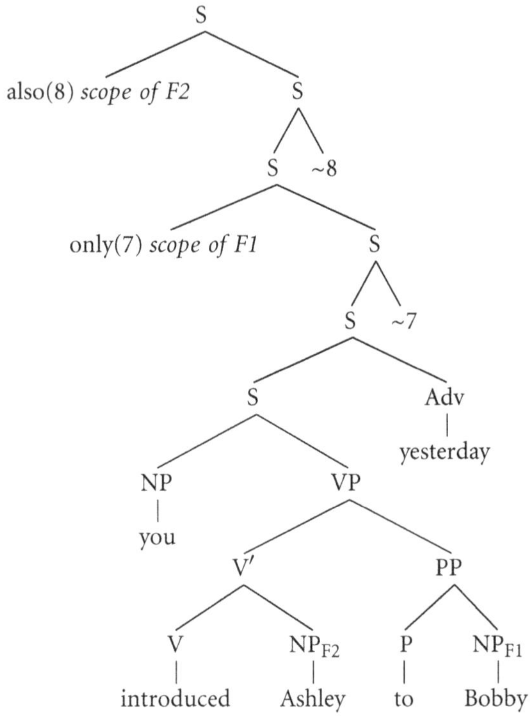

FIGURE 2.4 Scopes of F in example (14). F2 informally annotates the wider-scope focus, and F1 the narrower-scope focus

labeled *r* in the tree representing the scope of F1, and *l* in the tree representing the scope of F2.

Another way of seeing the problem is to apply *Stress F* as formulated in (4) to the tree in Figure 2.4. Starting with F1, *Bobby* is the focused element β_1, and [you introduced Ashley to Bobby] is the scope ϕ_1 of the focus. *Stress F* tells us that the highest stress in ϕ_1 falls within β_1. Second, let us apply *Stress F* to F2, with *Ashley* as the focused element β_2, and [you only introduced Ashley to Bobby] as the scope ϕ_2 of the focus. *Stress F* tells us that the highest stress within ϕ_2 must fall within β_2. These constraints are inconsistent, because β_1 and β_2 are disjoint subintervals of ϕ_1, while ϕ_1 is a subinterval of ϕ_2. Under these circumstances, β_1 cannot contain the greatest prominence in ϕ_1 while β_2 contains the greatest prominence in ϕ_2. This would be comparable to the tallest mountain in North America being in Alaska, while the tallest mountain in the Americas is in Colorado. This logic was discovered by Daniel Büring (Büring 2008).

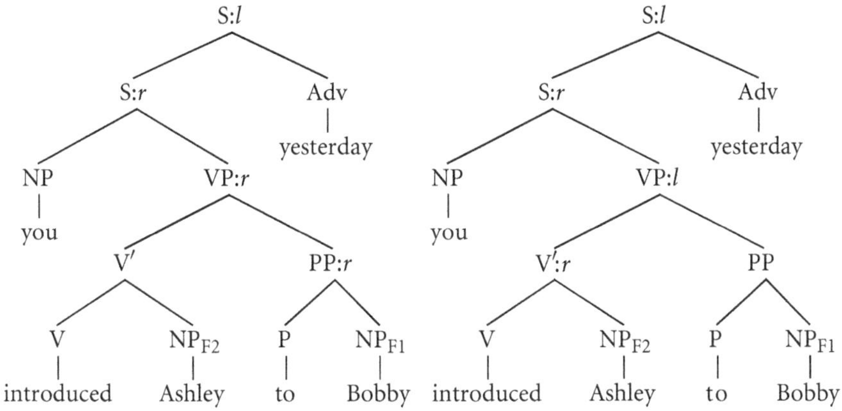

FIGURE 2.5 On the left, local operators in the embedded clause express the scope of F1. On the right, local operators express the scope of F2

To deal with the problem in a rule similar to Truckenbrodt's *Stress F*, one has to "relativize" the constraint to allow for two Fs to take different scopes out of the same phrase. This can be done by stipulating that wider-scope Fs should be ignored in checking *Stress F*. This results in (16).

(16) *Relativized Stress F*
Let β be an F-marked phrase with scope ϕ. Then the strongest stress in the phonological interval corresponding to β is strictly stronger than any stress in the phonological interval corresponding to ϕ which is not contained in the phonological interval corresponding to an F-marked subconstituent of ϕ whose scope is at least ϕ.

In Figure 2.4, *Relativized Stress F* as applied to F1 says that we should ignore prominences within β_2 (which is $[\text{Ashley}]_{F2}$) while checking whether the greatest prominence within ϕ_1 falls within β_1 (which is $[\text{Bobby}]_{F1}$). Suppose the grid structure for ϕ_1 is along the lines of (17). The first syllable of *Ashley* has the greatest prominence. But in checking *Relativized Stress F* for the focus on *Bobby*, any prominences within *Ashley* are ignored, because *Ashley* is contained in a wider-scope focus. *Relativized Stress F* applied to F2 on *Ashley* requires that the first syllable on *Ashley* has greater prominence than the first syllable on *Bobby*, which it does in this representation—and so the contradiction is removed. In general, relativization allows two Fs to take different scopes out of the same constituent, something which is not possible under *Stress F*. At the same time, it entails that the wider-scope focus have greater stress prominence.

(17)

					x							
					x			x				
x		x			x			x		x		
x	x	x	x	x	x	x	x	x	x	x	x	x
on	ly	in	tro	duced	Ash	ley	to	Bob	by	yes	ter	day

If, as maintained in Section 2.2, the original *Stress F* is a complicated global statement that refers to too many pieces of representation at once, then *Relativized Stress F* is much worse—it obviously has the status of a descriptive generalization which should be derived from simpler primitives. This is the project of Section 2.4. Before proceeding with this, I would like to reconsider the empirical status of examples (14) and (15). Schwarzschild (2004) and Büring (2008) discussed examples like (18), which in the scope of Fs are isomorphic with (14) and (15). According to them, there is no way of pronouncing (18b) such that it fits in with the question context (18a). On these intuitions, placing the nuclear accent on *crepes* (18c) or on *Paris* (18d) does not work.

(18) a. What food will Renee only eat in Paris?
b. She'll only eat crepes in Paris.
c. # She'll only eat CREpes in Paris.
d. # She'll only eat crepes in PARis.

I think it is not advisable to draw any deep conclusions from (18), because when it is adjusted by adding some material after the last focus as in (19), by adding some syllables after the stressed syllable marking the first focus as in (20), or both (21), they become good. Presumably they help because they add non-prominent material relative to which the SOF can be perceived as prominent. These adjustments do not affect the scope of focus. If (19) to (21) are good, they are counter-examples to the simple *Stress F* rule and are problematic for the local operator architecture in the way explained in connection with Figures 2.4 and 2.5.

(19) What foods will Renee only eat in Paris next year?
She'll only eat CREPES in Paris next near.

(20) What foods will Paul only eat in Paris?
He'll only eat UDO noodles in Paris.

(21) What foods will Renee only eat in Paris next year?
She'll only eat UDO noodles in Paris next near.

But what about (18)? Maybe some speakers perceive it as a garden path, where, because there is no phonetic evidence for any focus on *Paris*, the only reading which is recovered is one where *only* is associated with focus on *crepes*.

Another way of improving (18c) is to put a rising intonation indicating a partial answer on the first focus *crepes*. With this intonation and pragmatics, the example seems perfect. The rising intonation perhaps removes the garden path effect, because the focus with a rising intonation in this example cannot be read as associated with *only*.

A slightly different hypothesis is that a purely phonological constraint on good metrical configurations is responsible for the judgment that (18) is bad. As pointed out to me by Roger Higgins, there is an unpronounceability effect in (22). Again, the example is saved by adding unstressed syllables after the stressed syllable marking the first focus as in (23). The relation between (22) and (23) is similar to the one between (18) and (20).

(22) Who does only John like?
??Only John_{F} likes Mary_{F}.

(23) Who does only Abernathy like?
Only $\text{Abernathy}_{\text{F}}$//likes Mary_{F}.

Of course, if the data are different, the theory should be different. Büring (2008) has it that SOF examples with the scope pattern like the one in (18) are bad, and attributes this to the syntax–phonology interface in the form of ordinary *Stress F*. The logic is that in a represention like Figure 2.4, unrelativized *Stress F* can not be satisfied, as reviewed above.

I will tentatively assume that the scope configuration in Figure 2.4 is possible—though, any claims about English prosody need to be tested in a laboratory setting and documented in corpus data. As far as I know, no laboratory work on SOF bears on the configuration in Figure 2.4.

2.4 Local operators generalized

Here is an excercise in labeling trees. Starting with the tree in Figure 2.4, mechanically percolate the F-indices 1 and 2 to the level of their scopes. The indices are written in descending order at each node. This produces the tree on the left in Figure 2.6. Next, mechanically add an additional annotation which indicates whether each index came from the left or the right in the binary tree. In the sequence 2, 1 which labels [$_{\text{VP}}$introduced Ashley to Bobby], the index 2 came from the left child (from [introduced Ashley]), while the 1 came from the right child (from [to Bobby]). This results in the label *lr*, with *l* indicating a source on the left, *r* indicating a source on the right, and the linear order in *lr* corresponding to the linear order in "2, 1". In the right tree in Figure 2.6, all nodes are labeled according to this principle. It can be observed that the *l*s and *r*s capture all the information about the scope of the Fs, so that if we

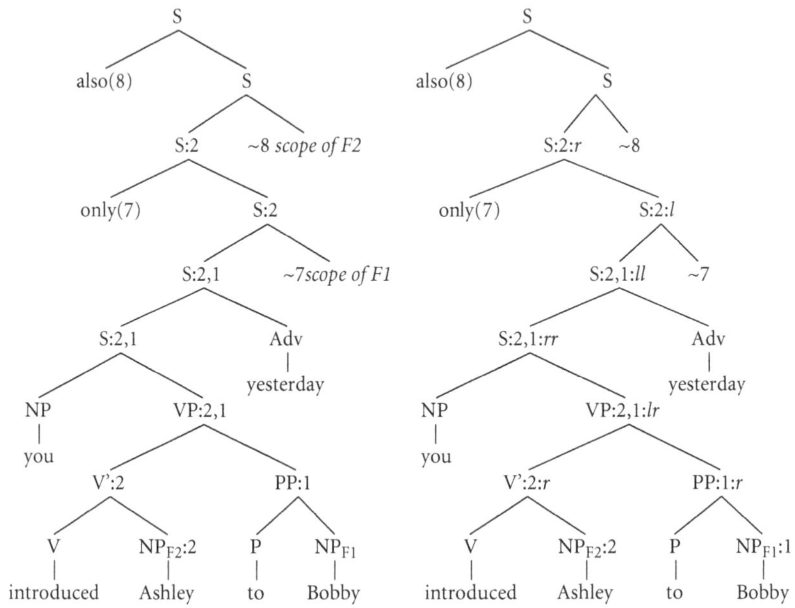

FIGURE 2.6 *lr* annotations added in Figure 2.4

include *l* and *r* in the representation, the F indices (here 1 and 2) are no longer required.

This procedure represents the scope of focus using a *sequence* of *lr* operators at each node, rather than with single operators as in Section 2.2. This allows the scope of multiple Fs to be represented. As before, *l* and *r* are operators which have both a phonological interpretation and a semantic one.[7] Semantically, a sequence lr indicates that there is a widest-scope focus in the left child, and an additional focus in the right child. A sequence rr indicates that there is a widest-scope F in the right child, and an additional F also in the right child. And so forth.

Turning to phonology, the phonological interpretation of any sequence $l\alpha$ which begins with an *l* is that the left child of the labeled node must be metrically stronger than the right child. The phonological interpretation of a sequence $r\alpha$ which begins with *r* is that the right child must be stronger than the left one. So in a sequence with more than one operator, only the first one counts phonologically. This resolves the contradiction in Figure 2.5.

7 This is somehow similar to Wagner's (2006) local-alternative semantics, which uses a local operator on a pair of sisters which has a semantic and a phonological interpretation.

The VP node has the label *lr*, and so the left daughter of the VP, which embeds the primary focus, is more prominent.

My starting point for defining the semantic part of the local-operator architecture is the compositional structured-meaning analysis of Krifka (1991).[8] As in all structured-meaning approaches to focus, the idea is that focus has the effect of structuring the semantic value into a tuple (von Stechow 1989).[9] The denotation of (24a) is normally (24b).[10] It is structured into the tuple (24c). The first element is a version of the normal denotation with a bound variable in the position of the focused phrase, the second element is its denotation. I will call the first elements in structured meanings "focus skeletons". For another example, the structured interpretation of the focused phrase (24d) is (24e), where the skeleton is the identity function. As in Krifka's compositional approach, structuring happens also at compositional levels, not just at the level of the scope of the focus.

(24) a. $[_{\mathrm{VP}}$ introduce $\mathrm{Ashley}_{\mathrm{F}}]$

b. (**introduce ashley**)

c. $\langle \lambda z(\mathrm{introduce} z),\ \mathrm{ashley}\rangle$

d. $\mathrm{Ashley}_{\mathrm{F}}$

e. $\langle \lambda y.y,\ \mathbf{Ashley}\rangle$

On the approach to be developed here, a phrase containing several unbound foci denotes a tuple whose first element is a focus skeleton with the focus positions bound by lambda, whose second element is the denotation of the widest-scope focus, whose third element is the denotation of the next-widest-scope focus, and so forth. *l* and *r* are construed as operators which manipulate structured meanings of this kind, determining the scope of Fs, while also having their phonological effect. Consider the following scheme. All syntactic branching is binary. Each non-terminal node is annotated with a type-raised version of the ordinary semantic operation for the node, which is usually function application (either in the right or left direction). (25) is the semantic derivation tree for [introduced Ashley] with no focus. At terminal

[8] I don't contemplate using structured meanings as a semantics for focus in place of anaphoric/givenness semantics, just as a compositional device.

[9] An option closer to Wold (1996) would be to use a pair of denotations, one of which is a focus skeleton with lambda-bound variables in the position of focused phrases. As in alternative semantics, in a phrase with one focus the other denotation is the ordinary semantic value. In the focus skeleton, the argument order corresponds to the scope of focus. I believe the technical development would be similar to the one in the text.

[10] In this notation a function application term is written fa or (fa) rather than $f(a)$. Application terms are interpreted left-associatively, so that fxy is $((fx)y)$.

nodes, a type-raising operator **u** introduces a unit list to initialize structured meanings. The operator **a** is a type-raised version of the rightward function application operator $\lambda f \lambda a. fa$, and c is an additional operator involved in propagating structured meanings. To obtain the interpretation of the phrase, the lambda term annotating the parent is applied to the lambda terms coming from the two children, in their linear order. In this example, the propagation of structured meanings in the resulting term c**a**(**u introduce**)(**u Ashley**) is trivial, producing the unit set of the ordinary semantic value (**introduce Ashley**).

(25)

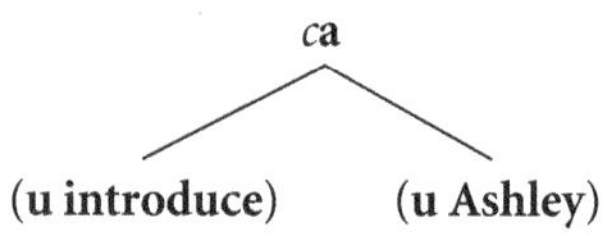

The operators l and r are treated as type-raising operators modifying the semantic operation. They provide for the appropriate projection of structured meanings, in accord with the constraint that l projects a widest-scope focus from the left, while r projects a widest-scope focus from the right. The tree in (26) is the compositional structure for (24a), which has a focus on the right. The operator $c(r\mathbf{a})$ which annotates the parent is a type-raised version of the rightward function application operator which propagates a focus from the right. **f** is the semantic focusing operator. The denotation of the whole phrase, which can be written $c(r\mathbf{a})$(**u introduce**)(**f**(**u Ashley**)), is the non-trivial structured meaning (24c).

(26)

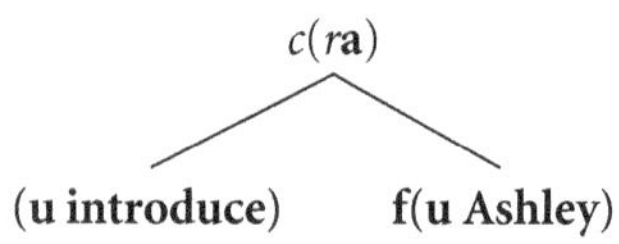

In Figure 2.7, the scheme is applied to the tree on the right in Figure 2.6, restricting attention to the part below the focus interpretation operators. The scope of the Fs is fixed by the operator $l(r\mathbf{a})$, which propagates the widest-scope focus from the left child [introduce Ashley$_{\mathrm{F}}$], and the narrower-scope focus from the right child [to Bobby$_{\mathrm{F}}$]. Figure 2.8 is the version with the wider-scope focus on [$_{\mathrm{NP}}$Bobby]. Everything is the same, except that $r(l\mathbf{a})$ replaces $l(r\mathbf{a})$ in the semantic operation for the phrase [$_{\mathrm{VP}}$introduce Ashley$_{\mathrm{F}}$ to Bobby$_{\mathrm{F}}$]. $r(l\mathbf{a})$ propagates the widest-scope focus from the right child [$_{\mathrm{PP}}$to Bobby$_{\mathrm{F}}$], and the other focus from the left child [introduced Ashley$_{\mathrm{F}}$]. The difference in focus scope is localized at the point where the denotation of [introduce Ashley$_{\mathrm{F}}$] combines with the denotion of [to Bobby$_{\mathrm{F}}$]. Here it has a

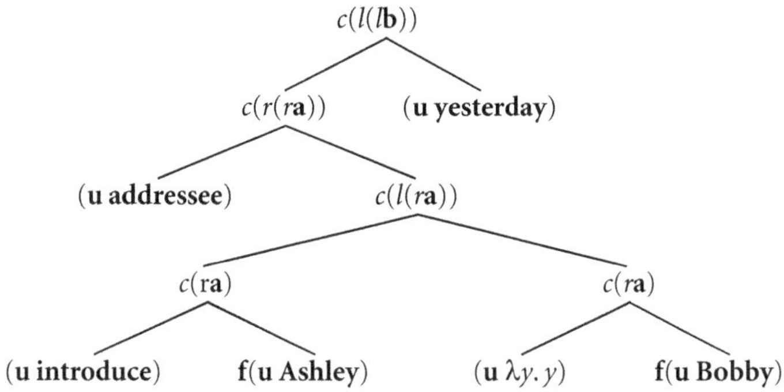

FIGURE 2.7 Compositional structure with wider scope for the focus on *Ashley*

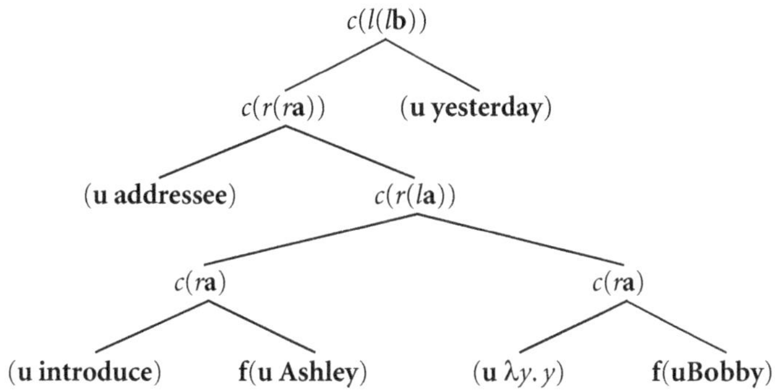

FIGURE 2.8 Compositional structure with wider scope for the focus on *Bobby*

local phonological effect (on the relative prominence of the two phrases) and a local semantic one (on the scopes of the two Fs).

Structured meanings are formalized as lists. To be systematic, even phrases without any Fs have list denotations. This is illustrated by (25), in which each phrase denotes the unit set of its standard semantics. Semantic operations have to manipulate lists, and also, it will turn out, ordered pairs. This is done with the lambda calculus encodings of list and pairing operations stated in Figure 2.9.[11] Using these definitions, the list denoted by $[_{\mathrm{NP}}\mathrm{Ashley}]_{\mathrm{F}}$ is obtained as **cns** $(\lambda z.z)$ (**u Ashley**). Where L is that list, the first element can be extracted as **hd** L, and the second element as **hd** (**tl** L). The non-focused verb *introduced* denotes a unit list, which can be written as **cns introduce nil**. To interpret the complex phrase [introduced $\mathrm{Ashley}_{\mathrm{F}}$], the two structured

[11] These definitions are as in Paulson (1996: 386–7). I checked semantic derivations using an implementation in SML of normal-order lambda reduction (Sestoft 1996).

pr	$= \lambda x.\lambda y.\lambda s.sxy$	(**pr** xy) is the ordered pair of x and y
fst	$= \lambda p.p\ (\lambda x.\lambda y.y)$	**fst** p is the first element of the ordered pair p
snd	$= \lambda p.p\ (\lambda x.\lambda y.y)$	**snd** p is the second element of the ordered pair p
nil	$= \lambda x.x$	empty list
hd	$= \lambda x.$ **fst** (**snd** x)	**hd** x is the head (first element) of the list x
tl	$= \lambda x.$ **snd** (**snd** x)	**tl** x is the tail (list of remaining elements) of the list x
cns	$= \lambda x.\lambda y.$**pr** $(\lambda x.\lambda y.y)$ (**pr** xy)	**cns** yx is the list with head y and tail x
u	$= \lambda x.$**cns** x **nil**	**u** x is the unit list with element x

FIGURE 2.9 Pairing and list operators

meanings have to be combined. If there was no focus or structuring, the denotations of *introduce* and *Ashley* would be combined with the rightward function application operator $\lambda f.\lambda a.fa$. In the structured-meaning scheme, they are instead combined using cr**a**. This is a type-raised rightward function application operator, modified by an operator r which projects a focus from the second (right) argument.

Operators which project structured meanings are defined in two parts. First, there are operators l_1 and r_1, which modify the semantic operation which applies to the focus skeletons. Where o_1 is such an operation, $l_1(o_1)$ accommodates an additional focus from the left child, and $r_1(o_1)$ accommodates an additional focus from the right child.

Second, there are operators l_2 and r_2, which manipulate the additional meaning components corresponding to focused phrases. Lists of focused-phrase denotations have to be merged together, observing the constraint that l projects the widest-scope focus from the left argument, and r from the right argument. Where o_2 is an operator which manipulates focus lists, $l_2(o_2)$ accommodates an additional focus coming from the left child (corresponding to the first argument of the operator) and $r_2(o_2)$ accommodates an additional focus coming from the right child (corresponding to the second argument of the operator).

Operators such as **a**, l**a**, $r(l$**a**$)$, and **b** (which is the type-raised backward function application operator) have the type of ordered pairs. l is defined in terms of l_1 and l_2 to map an ordered pair of this kind to another ordered pair. c acts as an application operator which combines an operator like $r(l$**a**$)$ with the two structured meanings denoted by the child phrases.

The operators l, r, and c are defined in Figure 2.10, using the list operations of Figure 2.9. Figure 2.11 defines type-raised function application operators.

$$
\begin{aligned}
l_1 &= \lambda f \lambda a \lambda b \lambda x. f(ax)b \\
r_1 &= \lambda f \lambda a \lambda b \lambda x. fa\,(bx) \\
l_2 &= \lambda f \lambda a \lambda b.\mathbf{cns}\,(\mathbf{hd}\,a)(f(\mathbf{tl}\,a)b) \\
r_2 &= \lambda f \lambda a \lambda b.\mathbf{cns}\,(\mathbf{hd}\,b)(fa\,(\mathbf{tl}\,b)) \\
r &= \lambda o.(r_1(\mathbf{fst}\,o))(r_2(\mathbf{snd}\,o)) \\
l &= \lambda o.(l_1(\mathbf{fst}\,o))(l_2(\mathbf{snd}\,o)) \\
c &= \lambda o.\lambda a.\lambda b.\mathbf{cns}\,((\mathbf{fst}\,o)(\mathbf{hd}\,a)(\mathbf{hd}\,b))((\mathbf{snd}\,o))(\mathbf{tl}\,a)(\mathbf{tl}\,b))
\end{aligned}
$$

FIGURE 2.10 Operators projecting structured meanings. *ro* applies the semantic operation o, while projecting an additional focus from the second (right) argument. *lo* applies the semantic operation o, while projecting an additional focus from the first (left) argument

It remains to define the focusing operator. As already exemplified above, this has a simple definition which adds the identify function to the front of the list denoted by the argument. This operation is defined in (27).

(27) $\mathbf{f} = \lambda y.\mathbf{cns}(\lambda x.x)y$

In sum, the ls and rs in representations such as Figure 2.6 are given a semantic interpretation as modifiers of semantic operations. When l modifies a given operation o, the combination lo projects an extra widest-scope focus from the left child. When r modifies an operation o, the combination ro projects an extra widest-scope focus from the right child. A composite operation $l(r(o))$ projects a widest-scope focus from the left child, and a narrower-scope focus from the right child.

There is another part of the semantic analysis which I will only sketch. The lr operators have a systematic correspondence with focusing operators. For instance, we would not want a phrase whose right child contains no F to be annotated with the semantic operator $c(r\mathbf{a})$, since r projects a focus from the right child. This kind of issue comes up in a couple of places in variable-free and structured approaches to compositional semantics. Pieces of meaning of functional type have certain intended interpretations (with one argument position corresponding to a free pronoun, another argument

$$
\begin{aligned}
\mathbf{ar} &= \lambda a \lambda b.ab \\
\mathbf{al} &= \lambda a \lambda b.ba \\
\mathbf{fnl} &= \lambda a \lambda b.\mathbf{nil} \\
\mathbf{a} &= \mathbf{pr\ ar\ fnl} \\
\mathbf{b} &= \mathbf{pr\ al\ fnl}
\end{aligned}
$$

FIGURE 2.11 **a** is the type-raised rightward function application operator. **b** is the type-raised leftward function application operator. **b** and **a** can be modified by l and r, e.g. $r(l(\mathbf{a}))$

position corresponding to a free trace, another corresponding to a focus, and another corresponding to an ordinary compositional argument). These interpretations constrain how pieces of meaning can combine, and what type-raising operators are inserted in semantic composition. One way of controlling such a system is to use an enriched system of semantic-type labels to regulate semantic composition. For instance, consider a type system where a VP has the type *et*, and a subject NP has type *e*. Given the phrase order [NP VP], these are combined with a leftward function composition operator of type $e(et)t$. Suppose an NP with a free focus of type *e* has a type label $e \rhd e$. This is not an appropriate argument for a leftward function application operator of type $e(et)t$. The mismatch is fixed with a type-raising operator l with type $(e(et)t)(e \rhd e)(et)(e \rhd t)$, which modifies rightward function application to propagate a focus from the first (left) argument. The ultimate value type $(e \rhd t)$ represents a standard type t with a free focus of type e. On this approach, regulating the distribution of ls and rs is reduced to checking type consistency.

This completes the semantics of *lr* sequences. The phonology is stated in (28). The constraint says that a node labeled with a sequence which starts with l must have a strong left child, while a node labeled with a sequence that starts with r must have a strong right child.

(28) Phonology of *lr* labels

(i) The left child of any node labeled $l\alpha$ is stronger than the right child.

(ii) The right child of any node labeled $r\alpha$ is stronger than the left child.

2.5 Conclusion

The structure in (29) is the new representation of the SOF configuration. The operators l and r fix the scope of the Fs and have a phonological interpretation in terms of metrical prominence.

(29)
$$[[\ldots[\ldots[[\ldots \underset{\substack{\downarrow\\ \text{ordinary F}\\ \text{phonology}}}{\alpha_F} \ldots][\ldots \underset{\substack{\downarrow\\ \text{SOF}\\ \text{phonology}}}{\beta_F} \ldots]]_{lr}\ldots] \sim j \ldots] \sim k]$$

To complete the phonological part of the theory, the interpretation of the focus-projecting operators should be embedded in an account of prosodic phonology, including matters such as pitch-accent assignment and the correspondence among metrical structure, phrasing, and syntactic structure, such as the one discussed in Féry and Ishihara (this volume).

The semantic solution in Section 2.4 is of a technical character. While it can be used in conjunction with a constrained architecture for focus interpretation such as alternative semantics—because alternatives can be introduced at the top level—the recursive operations do not employ alternative semantics. Structured meanings are a semantic device which have also been used for other semantic problems, such as a *de re* interpretation (Cresswell and von Stechow 1982) and the exceptional scope of indefinites (Abusch 1994). Therefore they should be regarded as a general device for constrained non-compositionality. It is hard to dismiss the employment of structured-meaning devices on general grounds, especially if they are used in the interpretation of universal features. If the semantics of the feature in question is universal, it has to be stipulated only once in linguistic theory, and applying an a priori criterion of simplicity might lead in the direction of an incorrect theory of universal grammar. Still, as a methodological strategy, it seems reasonable to use structured meanings only as a last resort.

Once we see the technical solution outlined in Section 2.4, we could try to do the same thing in recursive alternative semantics (Hamblin 1973; Rooth 1985; Shan 2004). Perhaps the semantics of SOF configurations could be expressed with a tuple of meanings, consisting of an ordinary denotation, an alternative set representing the widest-scope focus, and an alternative set representing the narrower-scope focus. This is reminiscent of Büring's architecture for topic interpretation (Büring 1997) and of Wagner's attempt to recast it in terms of foci with nested scopes (Wagner 2008). My hunches about whether this strategy could work are mixed.

3

How focus and givenness shape prosody

CAROLINE FÉRY AND SHINICHIRO ISHIHARA

3.1 Introduction

This chapter proposes a representation of prosody that distinguishes formally between the imports of syntax, which acts on the formation of prosodic phrases, and those of information structure (focus and givenness), which affects F0 register scaling. It is proposed that the representation of the syntax–prosody interaction has to be able to express both phenomena separately in order to keep their prosodic effects apart. It is also proposed that the resulting prosody is partly language-dependent.

Section 3.2 introduces the model with German and Japanese, and, to a lesser extent, with English. Starting with the relationship between syntax and prosodic phrasing, we can lean on a vast amount of literature on the subject, which agrees that prosodic phrasing is mapped from syntax. How exactly this happens and which constituency originates from the mapping have been the subject of extensive discussions, and we will not at this point present an all-ready formalization of syntax-based phrasing. In Section 3.2.1 we will assume that prosodic phrasing involves recursive constituents.

The second component of our model is the effect of information structure on prosody, discussed in Section 3.2.2. Focus and givenness, the two information-structural properties considered in this chapter, change the F0 registers that a speaker uses at a certain point in a sentence. Focus enlarges it, givenness compresses it, and the height of pitch accents and boundary tones

We would like to express our gratitude to Lisa Selkirk, Elsi Kaiser, and Malte Zimmermann, whose comments have helped to improve the model presented here. Our work is part of project A1 of the SFB 632 on Information Structure in Potsdam and Berlin, financed by the DFG. Many thanks also to our colleagues Gisbert Fanselow and Ingo Feldhausen.

are changed accordingly. Crucially, these effects are limited to F0 scaling and do not affect prosodic phrasing, at least in the languages considered here. Section 3.3 illustrates the proposal with empirical studies from German and Japanese. Section 3.4 sums up the main aspects of our proposal in comparison with earlier ones.

3.2 The model

3.2.1 *The impact of syntax on prosody*

Our main claim is that the prosodic effects of syntactic structure and information structure should be kept apart. For the first part, we assume that syntactic structure is mapped onto prosody as prosodic phrasing, and that prosodic phrasing is recursive. In an all-new sentence, that is, in a sentence without focus-new-given partition (see Section 3.2.2. for definitions), the formation of prosodic phrases as well as the tonal pattern and scaling depend entirely on the morpho-syntactic structure. Prosodic phrases have heads in the form of abstract grid positions, which may be realized as pitch accents, as is the case in English and German. The prosodic heads are not necessarily realized with pitch accents, but can be expressed by duration, intensity or even by completely different criteria, like tones or segmental modifications. The prosodic heads correlate with metrical or hierarchical prosodic structure (see for instance Halle and Vergnaud 1980, Selkirk 1984, Nespor and Vogel 1986). An example of such a metrical structure appears in (1) for English. In most studies, relevant levels of phrasing have been assigned a variety of names, such as Minor and Major Phrases, phonological phrases and intermediate phrases, accent domains and rhythmic groups, among others (Selkirk 1984; Nespor and Vogel 1986; Ladd 1990; Hayes 1995). However, as soon as the syntax–prosody mapping suggests additional levels of phrasing, such models have to add new names for domains or allow recursivity of at least some of the levels. In the following, we give up the distinction between smaller and larger domains, and prefer to consider prosodic phrasing as a recursive structure. To keep the terminology as theory-neutral as possible, we call the levels mapped from syntactic constituents "p-phrases". Following proposals by Wagner (2005) and Ito and Mester (2006), p-phrases can be embedded into each other. All levels of phrasing shown in (1), except for the lower one—which is the level of the prosodic word (PWd)—and the upper one—which is the level of the intonation phrase—are p-phrases. It is easy to see that p-phrases can be added when the individual constituents are extended.

(1)

```
(                                                   x ) IP
(                      x ) (                        x ) p
(          x ) (       x ) (            x)(         x ) p
( x ) ( x ) ( x ) ( x ) ( x ) (         x)(   x ) (  x ) PW
Princess Diana's sudden death has been the source of many speculations
```

In an unmarked case like (1), in other words, in the case of an all-new sentence without any focused or given constituents, the p-phrases of a given level are scaled relative to one another in a completely predictable way (see Kratzer and Selkirk 2007, Truckenbrodt 2002). In the following, we make use of the phonetic abstraction of the phrasal reference line introduced by van den Berg, Gussenhoven, and Rietveld (1992), a register line of constant height during an intonation phrase (such as a matrix clause), running at the height of the domain-initial peak. To account for downstep among smaller domains, the phrasal reference line is progressively downstepped, as illustrated in Figure 3.1. In other words, smaller downstep (among accents) is embedded in larger downstep (modeled by the phrasal reference line). Every phrase is downstepped as compared to the preceding one. This downstep takes place at all levels of phrasing. Each prosodic level defines its own downstep pattern, in which the reference top line of every phrase is lower than the reference top line of the preceding phrase of the same category (see also Bruce 1977, Ladd 1990, Truckenbrodt 2002, and Féry and Truckenbrodt 2005 for similar models). In

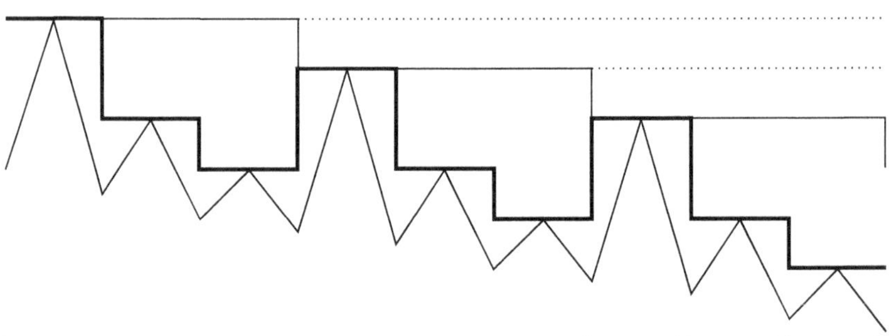

{(x)(x)(x)}{(x)(x)(x)}{(x)(x)(x)}

FIGURE 3.1 Embedded downstep pattern

a complex sentence, downstep is thus recursive and a property of embedded prosodic phrases.

Applied to (1), such a model delivers an intricate pattern of downstep relationship between the different phrases at all levels of phrasing, part of which is shown in (2). The highest reference line starts with the first high tone of the intonation phrase and remains available until the end of its domain, here the end of the sentence. In the illustration it is the dotted grey line. In other words, the top lines do not disappear when following ones become more pervasive. They survive until the end of the relevant domain, and can determine subsequent scaling. This accounts for embedding of register domains inside one another, like the one described below for Japanese.

The next lower level of phrasing introduces the first downstep. It separates the subject of the sentence from the VP. The first sister of this downstep relation is at the same height as the beginning of the intonation phrase, but the second part is one step lower. Again, (2) indicates the domain of every reference line with thinner lines. In the first part of this relationship, the grey line stands for the domain of the first sister, and the thin black line for the domain of the second sister. Within these constituents, smaller sister constituents are again inducing a downstep relationship, and so on. Féry and Truckenbrodt (2005) found for German that at a higher level of prosodic constituency, when clauses are coordinated, a deeper level of embedding induces a steeper downstep.[1]

(2)

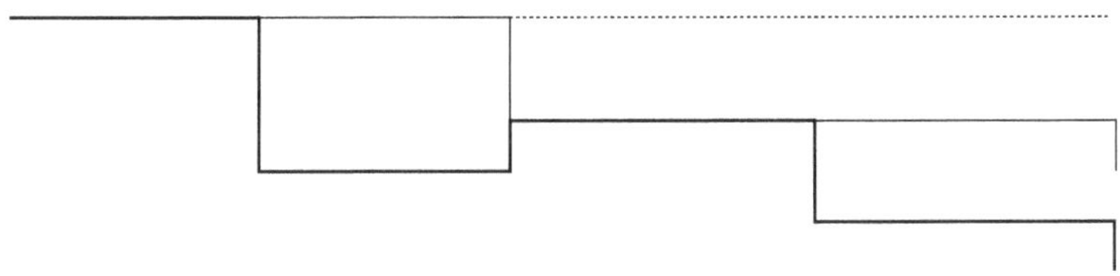

```
(                                                              x )

(                        x ) (                                 x )

(          x ) (         x ) (              x )(               x )

( x   )( x ) ( x   )( x ) (  x  )(     x )(  x   ) (           x )
```
Princess Diana's sudden death has been the source of many speculations

[1] This result is tentatively replicated for smaller prosodic domains in Figure. 3.1, though no empirical results are available so far that would confirm this hypothesis.

F0 registers are calculated relative to the preceding phrases of the same level of prosodic phrasing (see Pierrehumbert 1980, van den Berg *et al.* 1992, Truckenbrodt 2004, Féry and Truckenbrodt 2005). The most evident cue for registers is the height of initial pitch accents in each prosodic domain, but the boundary tones at the end of prosodic domains can also give indication as to the relevant top lines. The boundary tones can be subject to reset (see Truckenbrodt 2002), and the height reached by a reset boundary tone is determined by the top lines. The calculation of register relationship can be complicated by embedding of register downstep, upstep, final lowering, as well as by purely tonal assimilations and dissimilations (as those found by Féry and Kügler 2008 for German). But once these effects are factored out, the pattern is straightforward. Though we cannot exclude that some languages do not present a pattern of downstep between prosodic phrases of the same level, at least in Japanese and in German, downstep seems to be a pervasise phenomenon, which is crucial in the perception of emphasis between accents.

3.2.2 *Information structure*

In this chapter we are mainly interested in the effect of focus and givenness on the prosodic structure of sentences from a theoretical perspective. Our main claim is that focus and givenness affect the F0 scaling of certain prosodic domains, but do not directly affect prosodic phrasing. Before illustrating our model, we start the discussion with some definitions of relevant information structural categories—namely, focus and givenness. Other information structural categories such as topic are left out of consideration.

We assume that focus is realized by prosodic prominence, as formulated in (3) (Jackendoff 1972; Truckenbrodt 1995; Büring 2001), though we do not try to define prominence, assuming that it can be realized in different ways (see also Büring, this volume).

(3) *Focus Prominence*
Focus is realized by prominence in its focus domain.

Throughout the chapter, we use the term "focus" in the sense of Rooth (1985, 1992), an element that singles out referents from a set of alternatives. Focused elements are bound by the focus operator ~. We do not distinguish between what has been called "narrow focus", "contrastive focus", and "identificational focus." However, we clearly distinguish it from what has been called "broad focus" and "information focus", in other words, elements that are new in the discourse. Following Selkirk (2008) we assume that discourse-new elements are in the "default" state, not bearing foci or being given.

The prosodic domain of focus corresponds to the semantic scope of focus, and (3) implies that focus is interpreted and gets its prosodic prominence in this domain. The domain contains the focused phrase and identifies the background information relevant to the semantic denotation of focus (see Chapter 4 in Truckenbrodt 1995 for an explanation of the focus domain). The focus domain does not necessarily coincide with a single prosodic constituent, as will be illustrated in Section 3.3. In our examples, focus will be formally indicated with a feature F, giving its scope. As has been shown by Féry and Samek-Lodovici (2006) for embedded foci and by Féry and Ishihara (2009) for Second Occurrence Foci, (3) can be violated when prosodic considerations are overriding the need for a focus to be prominent in its domain. The other important terminological concept is "givenness", for which a modified version of Schwarzschild's (1999) definition is adopted. A given constituent is entailed from the context in a precise semantic sense. Schwarzschild's proposal amounts to freely assigning F-marks instead of letting them percolate through the syntactic tree, as in Selkirk's (1995) proposal. But the two constraints in (4) restrict the occurrence of these marks. Rule (4a) entails that no F-marking indicates givenness, (4b) keeps the number of F-marking to a minimum.

(4) a. GIVENness: A constituent that is not F-marked is given.
   b. AvoidF: Do not F-Mark.

We depart from Schwarzschild's analysis in that instead of assigning F-marks to all non-given elements, we assume that all given elements are G-marked. Féry and Samek-Lodovici (2006) show that G-marks for givenness are needed as well. Unless they are focused, given constituents are not accented, a state which is expressed by (5). This constraint directly addresses the prosodic pattern of discourse informed constituents.

(5) Destress–Given
   A given phrase is prosodically non-prominent.

The present chapter proposes a more precise articulation of the relationship between new, focus, and givenness, on the one hand, and prosodic structure, on the other, than (3) and (5) alone are able to achieve. It is proposed that in German and Japanese the influence of information structure is mediated through F0 registers corresponding to focus and givenness domains.

With these definitions, we now illustrate our model. As mentioned above, information structure is reflected in changes in register scaling of prosodic

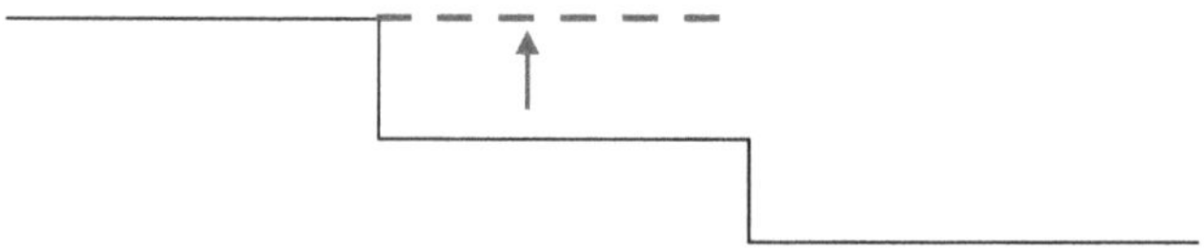

FIGURE 3.2 Boosting of the top line of the F-marked

domains and/or focused element.[2] In other words, information structure does not manipulate the boundaries of prosodic phrases as they have been defined by syntax, but instead changes their pitch registers by widening or narrowing them. The most immediate effect of this manipulation is that pitch accents can be higher or lower than in the unmarked situation, according to their focus or given status. If a sentence contains a focus, the F0 register of the focus is affected such that its reference top line is raised, provoking a sudden boost of the pitch accent correlating with the focused word or exponent. And when given material appears in a sentence, on the other hand, this part of the sentence (if any) is compressed.[3] We assume that the higher pitch accent because of focus is the consequence of the register change within the focus domain, and that lower F0 realization for given material is the result of the compression of the prosodic domain associated with givenness. The height of the individual pitch accents—generally understood as heads of prosodic phrases, but in some other cases just the most prominent part of the focus domain—is the result of the transformation of the reference top lines rather than the result of directly boosting or lowering the individual tones associated with accents. One important difference with models directly manipulating pitch accents is that the relationship between different parts of the sentence is changed. We claim that information structure changes the scaling of the entire sentence, instead of targeting only the most prominent pitch accents. We return to this point below.

When a narrow focus disturbs the unmarked and regular downstep pattern, the top line of the domain corresponding to the focus is raised. This is illustrated in Figure 3.2. In contrast, givenness lowers the top line of the given domain, as shown in Figure 3.3.

Pitch accents are scaled relative to these top lines. An individual accent will be higher or lower than in the unmarked case, depending on its status as focused or given, because it is constrained by the top line of its domain.

[2] We are aware that some languages reflect information structures by other prosodic ways, or even by other grammatical components. See for instance Fiedler *et al.* (this volume) and Hyman (this volume).

[3] As Michael Wagner observed (pers. comm.) in a sentence ending in a continuation rise, the register can also be narrower because the bottom line is raised.

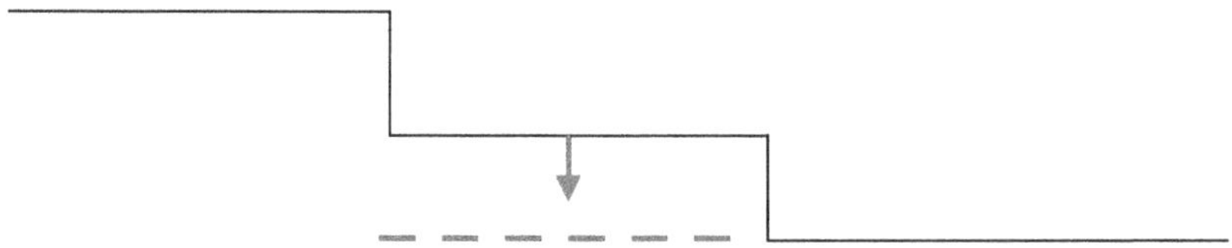

FIGURE 3.3 Lowering of the top line of the given domain

Different cases are illustrated in Figure 3.4. Figure 3.4a shows the default pattern, in which two prosodic domains are downstepped relative to each other. In Figure 3.4b, narrow focus raises the top line of the domain in which it appears. In Figure 3.4c, givenness lowers the top line. Finally, in Figure 3.4d both effects are present. This can be observed in Second Occurrence Focus (see Section 3.3).

An additional input to the model is the prenuclearity and postnuclearity of the relevant domains. In Japanese, German, and English, at least, there is a difference between prenuclear and postnuclear effects of givenness. Prenuclearly—that is before the last accent of a sentence—the effects of syntax on prosody are pervasive. All prosodic domains are kept more or less intact, even if givenness may lower the relevant top lines. Postnuclearly, however,

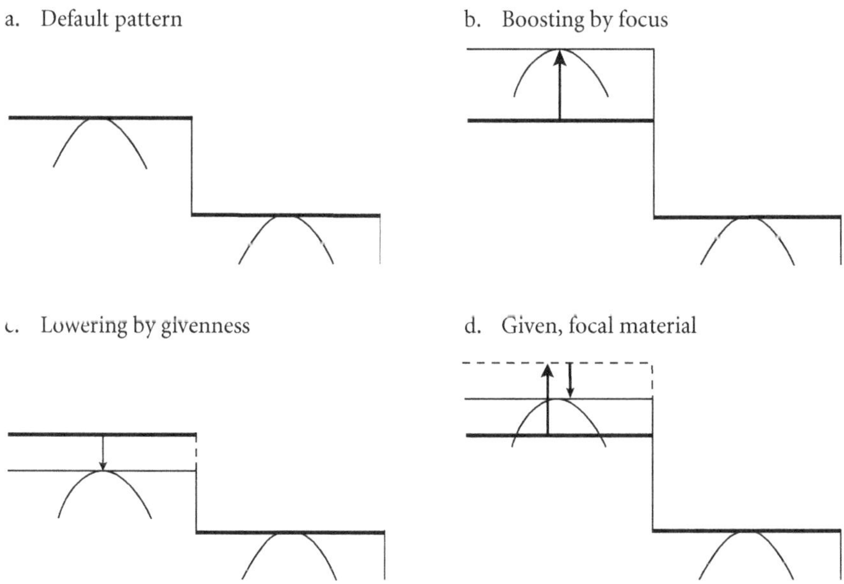

FIGURE 3.4 Pitch accents constrained by the top lines of their respective prosodic domains

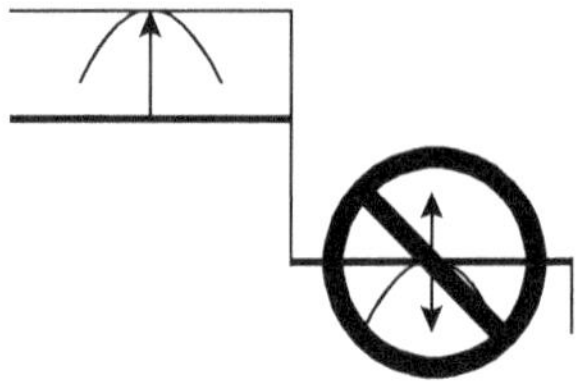

FIGURE 3.5 Postnuclear suppression of register

the top lines are lowered to a minimum, as shown in Figure 3.5, and are confounded with the baseline. There is no room anymore for the realization of pitch accents. In German, deletion of postnuclear accents is due to the extreme compression or reduction of register. This is especially true when the last accent corresponds to a contrastive narrow focus and when the material following this focus is given. In Japanese, a similar effect can be observed, but to a lesser extent; the compression is not so radical as in German (see Ishihara and Féry, in preparation, for a comparison between these two languages and Hungarian).

We have assumed until now that prosodic phrases are raised or lowered. It will be shown below that focus and givenness domains do not need to be isomorphic to prosodic domains. Though we have results only for Japanese so far (see below), we assume that the domain of manipulation of reference lines can be smaller or larger than prosodic phrases in German, as well.

To sum up the proposed model, phrase formation is accompanied by prosodic heads, often in the form of pitch accents, whose heights are adjusted to default reference lines. When the sentence has no focus-new-given partition—that is, when it is all-new—all prosodic phrases are downstepped relative to each other, and all accents are realized as predicted by the syntax, modulo some variations addressed in the next section. But as soon as information structure comes into play, the reference lines are changed and pitch accents may become more prominent because the register of their domain is extended. Alternatively, they may become nearly undetectable because their register lines are compressed.

3.3 Empirical results

This section provides empirical evidence for our model. The first set of data, discussed in Section 3.3.1, supports the view that only syntax is active in the formation of prosodic phrases. It is in sharp contradiction with models claiming that syntax and information structure have a similar power to shape phrasing (see Gussenhoven 1983, 1992; Truckenbrodt 2006). The second set of data, discussed in Section 3.3.2, addresses pitch scaling. Our model assumes that pitch accents are scaled to reference lines, which are coextensive with focus domains and givenness domains. It can be contrasted with models taking pitch accents to be directly manipulated. The latter kinds of model do not make any claim as to the scaling of other accents in the sentences, or to boundary tones, which should be blind to information structure. Our model, by contrast, assumes that pitch accents and boundary tones, being scaled to reference lines which are in relationship with other reference lines of the same sentence, are radically transformed by a changed information structure.

3.3.1 *Prosodic phrasing*

It is usually assumed in the literature that, in German, a p-phrase requires a head, which is realized in the form of a phrasal pitch accent, and in Japanese, a p-phrase is the domain of downstep between lexical pitch accents. We show that both criteria are not conclusive, and that a more abstract concept of prosodic phrasing, relying entirely on syntactic structures, is to be preferred.

3.3.1.1 *Prosodic phrasing in German* A number of researchers link the role of information structure directly or indirectly to the formation of prosodic phrases. It has been proposed a number of times that focus and givenness change the prosodic phrasing by adding or deleting prosodic constituents. A well-known example for Germanic languages is Gussenhoven (1983, 1992), who formulates the Sentence Accent Assignment Rule (SAAR), reproduced in (6), see also Truckenbrodt (2006).

(6) *SAAR* (Gussenhoven 1992)
If focused, every predicate, argument, and modifier must be accented, with the exception of a predicate that, discounting unfocused constituents, is adjacent to an argument.

As can be seen from the examples in (7), Gussenhoven relates the presence of a pitch accent strictly to the formation of a prosodic phrase, taking the accent as the head of the phrase. Narrow focus may increase or reduce the number of prosodic phrases. In (7b), the predicate (P) is not adjacent to an argument (A). As a result, argument, modifier (M), and verb are phrased individually, whereas in (7a) predicate and argument are phrased together by virtue of being adjacent. In (7c), the modifier is given, but both argument and verb are focused, and, as a result, argument, modifier, and verb are all phrased together.

(7) a. (Any news?)
[Our DOG disappeared]$_F$ [ÁP]

b. (What happened?)
[Our DOG MYSTERIOUSLY DISAPPEARED]$_F$ [Á] [Ḿ] [Ṕ]

c. (Talking about mysteries...)
[Our DOG]$_F$ mysteriously [disappeared]$_F$ [ÁMP]

The generalizations expressed in SAAR are intuitively adequate and go a long way to explain sentence accent assignment. Still, it may be questioned whether the strict link between prosodic phrases and pitch accents is desirable and necessary. We think that it is not. Even though the prosodic structure is mapped to the syntactic structure (see Cinque 1993, Selkirk 1995, Kratzer and Selkirk 2007, Féry and Samek-Lodovici 2006 for different approaches of how this is done), the assumptions that pitch accents are necessarily heads of prosodic phrases, or that every prosodic phrase is necessarily headed by a pitch accent, cause problems in certain cases because of the variation observed in pitch accent distribution. Not only sentences with a marked information structure, but also all-new sentences display a certain amount of variation in their pitch accents. Moreover, it remains to be demonstrated that prosodic phrases are truly deleted because of deaccenting of accents.

In a production study (Féry and Herbst 2004) it was shown that in the case of a modifier (adverbial or PP) sandwiched between the object and a transitive verb, the verb is often not accented, though the deaccenting of the verb happened less regularly than when the direct object immediately preceded the verb. The accent properties of all-new sentences like the one in (8) were compared with (9), containing a modifier which had already been introduced in the preceding question, and which was thus considered as given.

(8) Argument–Modifier–Verb (the VP is new)
{Melina is a real entertainer! How did she entertain you this time?}
Melina hat eine Arie auf der Wanderung gesungen.
Melina has an aria on the walk sung
'During the walk, Melina sang an aria.'

(9) Argument–Modifier–Verb (the modifier is given)
{I heard that you had a lot of fun on the walk with Melina. What did she do?}
Melina hat eine Arie auf der Wanderung gesungen.

If accents serve as indicators of a prosodic phrase by virtue of being heads, a single accent on the modifier or on the preceding object correlates with a

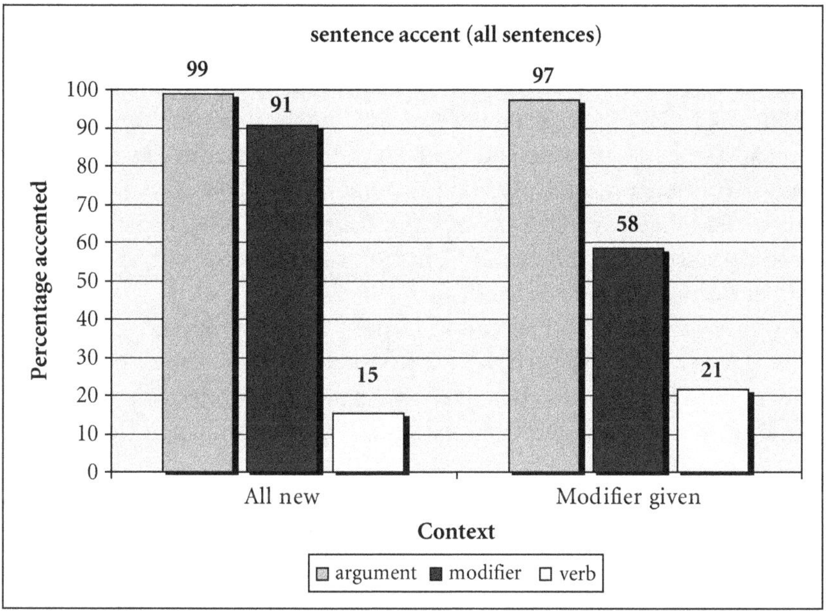

FIGURE 3.6 Percentage of accented phrases (from Féry and Herbst 2004)

single prosodic phrase, whereas two accents, one on the verb and one on the constituent preceding it, reveal the presence of two prosodic phrases.

A comparison of (8) and (9), displayed in Figure 3.6, clearly shows that the modifier was accented more often when it was part of an all-new sentence (8) than when it was given (9), namely in 91 percent vs. 58 percent of all cases, respectively. This can be seen by comparing the middle bars in Figure 3.6. Again, the argument was clearly accented in nearly all cases (see the leftmost bars).

A striking discrepancy between the predictions of SAAR and the results obtained in the experiment concerns the accenting of the verb. The percentage of accented verbs does not change much across the conditions. The verb was accented in 15 percent of the cases when the modifier is new, and in 21 percent of the cases when the modifier is given, a result going into the opposite direction from the one posited by Gussenhoven. We conclude that there is no correlation between accenting of the verb and accenting of the modifier.[4]

[4] In VPs consisting of just a modifier and a verb (and no object), the verb is accented in the majority of cases (unpublished data). In no more than 31 sentences out of 210 (15%), the modifier is the only accented element. SAAR does not predict such a change in the accent structure just because the object is absent.

On the basis of these results, it is reasonable to assume that pitch-accent assignment is an indicator of phrasing only up to a certain point. The fact that the object was (nearly) always accented confirms the assumption of a metrical grid in which the object is the head of the sentence. But the optionality of the pitch accents on the verb and on the modifier leads us to assume that the verb is part of the prosodic phrase of the preceding constituent, argument or modifier, and that its variable accenting is the result of other constraints than prosodic phrasing, like different information content of the verb and/or of the modifier for instance.

A recursive phrasing of such an example, as was proposed in Féry and Herbst (2004), is to be preferred to a structure predicted by SAAR. The p-phrase of the modifier is embedded in the p-phrase projected by the whole VP and comprising the object and the verb. This is illustrated in (10).

(10) Melina hat [eine Arie [auf der Wanderung]$_{P}$ gesungen]$_{P}$.

The object is the head of the larger p-phrase, and is assigned a pitch accent because of this property. But both the verb and the modifier can also get a pitch accent, the modifier because it is also a head of a p-phrase, albeit one of a lower, embedded kind, and the verb because it is separated from its head, and also form a prosodic unit of some kind.

The upshot is that an account of phrasing which relies entirely on the physical presence of pitch accents is fragile when it comes to variation. Moreover, it assumes that prosodic phrasing is contingent on phonetic realization, and that each uttered sentence comes with its own prosodic phrasing. The alternative proposed here is that prosodic phrasing is part of the syntactic derivation. It is an abstract structure submitted to phonological and phonetic realization. As such, it is not yet realized, and can not have physical pitch accents. It determines potential locations of accent realizations, which are represented as grid positions, but the actual realization of pitch accents depends on a number of factors, like adjacency to other accents, givenness, and so on.

3.3.1.2 *Prosodic Phrasing in Japanese* In Japanese, pitch accent is part of lexical information and can thus not be taken as a defining characteristic of p-phrases in the same way as in German. Instead of pitch accents, tonal scaling (downstep, reset) is used as evidence for the existence of p-phrases. It has long been assumed that p-phrase (i.e. what has been called *Major Phrase* or *intermediate phrase* in the literature)[5] is the domain of downstep (Poser 1984; Pierrehumbert and Beckmann 1988; Kubozono 1993; among many others).

[5] We no longer use these terms because once we adopt recursive p-phrasing, as proposed in Ito and Mester (2006), we no longer need the distinction between major and minor phrase, or intermediate and accentual phrase.

That is, if a sequence of two pitch accents is subject to downstep, they belong to the same p-phrase. If, by contrast, the second pitch accent is reset to the height of the first one, there is a p-phrase boundary between them. This definition of p-phrase, together with some standard assumptions, however, raises several problems. In this section we discuss some of them.

One of the widely accepted analyses of p-phrasing in Japanese is Selkirk and Tateishi's (1991) end-based model. They claim that the left edge of a syntactic maximal projection (XP) corresponds to the left edge of a p-phrase. In this analysis, the existence or absence of XP-boundaries is the deciding factor for the existence or absence of p-phrase boundaries, and for the corresponding downstep effect. The analysis predicts that in a left-branching structure like (11a), both N2 and N3 show downstep, while in a right-branching structure like (11b), only N3 does.

(11) a. Left-branching structure: [[N1 N2] N3]
Major Phrase structure: (N1 N2 N3)

b. Right-branching structure: [N1 [N2 N3]]
Major Phrase structure: (N1) (N2 N3)

This prediction is not completely borne out. Kubozono's (1993) experimental data show that N2 in (11b) shows downstep as well, even though the amount of lowering is significantly smaller than in (11a).[6,7] Furthermore, recent

[6] For the sake of fairness, it should be mentioned that in Selkirk and Tateishi's (1991) data, there was no downstep on N2 in a structure like (b), just as predicted by their analysis. We believe, however, that their results are influenced by an (unwanted) effect of focus. The sentences used in their experiment, for instance, the one in (i), are structurally ambiguous between a left-branching and a right-branching parse:

(i) a. Left-branching:
[[Aoyama-no Yamaguchi-no] aniyome-ga] inai
-GEN -GEN sister-in-law-NOM not.there
'We cannot find the sister-in-law of Yamaguchi from Aoyama.'

b. Right-branching:
[Aoyama-no [Yamaguchi-no aniyome-ga]] inai
'We cannot find Yamaguchi's sister-in-law from Aoyama.'

(Selkirk and Tateishi 1991: 523)

It is highly plausible that the speakers, who read both sentences and hence were fully aware of the difference in syntactic structure as well as meaning, purposefully tried to disambiguate the two sentences by placing focus on N2 in the (b)-sentence. If that is the case, Selkirk and Tateishi's results do not show the pure effect of p-phrasing.

[7] Kubozono (1993) tries to account for the difference by proposing a phonetic boosting effect called *metrical boost*, which applies to the phrase on the left side of XP (i.e. where Selkirk and Tateishi's analysis predicts there to be a p-phrase boundary). The crucial difference between the two analyses is that the metrical boost is a phonetic effect of expanding the pitch range on the XP-leftmost phrase,

studies show that downstep is not limited within what the earlier analyses claimed to be p-phrases. Kubozono (2007) shows that the downstep can be observed after a syntactic XP-boundary, which is considered to correspond to a p-phrase boundary. Experimental results reported by Ishihara (2008) also confirm that pitch reset effect at the XP-boundary is only partial, and a smaller amount of downstep effect remains after the XP-boundary. If so, we need to reconsider the definition of either p-phrase or the domain of downstep.

In our analysis, we adopt recursive p-phrasing. By allowing p-phrase embedding, not only can we maintain the assumption that p-phrase is the domain of downstep, but we can also explain the smaller amount of downstep across XP-boundaries. First, we assume that XPs in the left- and right-branching structure are mapped as embedding p-phrases, as shown in (12). In the left-branching structure, N1 and N2 form a single p-phrase, which forms a larger p-phrase with N3, while in the right-branching structure, N1 forms a large p-phrase with a smaller p-phrase containing N2 and N3. Second, maintaining the standard assumption that p-phrase is the domain of downstep, we expect a pitch reset at each left edge of p-phrases.

In the left-branching structure in (12a), all left edges are aligned at the beginning of the entire phrase. Therefore no reset is expected within the phrase. There are two p-phrases, the smaller p-phrase containing N1 and N2, and the bigger one containing the smaller p-phrase and N3. Inside the former, downstep is expected on N2, and in the latter, on N3. As a result, we observe a successive downstep.

In the right-branching structure in (12b), there are also two p-phrases. The smaller one containing N2 and N3, and the bigger one containing N1 and the smaller p-phrase. In this case, we expect a pitch reset at N2, because a left edge of the smaller p-phrase appears here. However, we also expect a downstep effect at this position, because downstep is expected at the larger p-phrase, between N1 and the smaller p-phrase. As a result, we observe a small amount of downstep effect at N2.

Japanese p-phrase further supports the view that p-phrase is formed recursively, according to syntax. Branching structure is mapped onto p-phrasing, and the different amount of downstep observed in the left- and right-branching structure can be explained by recursive p-structure.

In German, pitch-accent distribution does not speak for a one-to-one mapping between accents and p-phrases. A recursive structure, allowing variable pitch-accent assignment, makes better predictions.

and can apply multiple times, whereas p-phrase boundary insertion is a phonological operation that can be applied only once.

(12) a. Left-branching structure: [[N1 N2] N3]

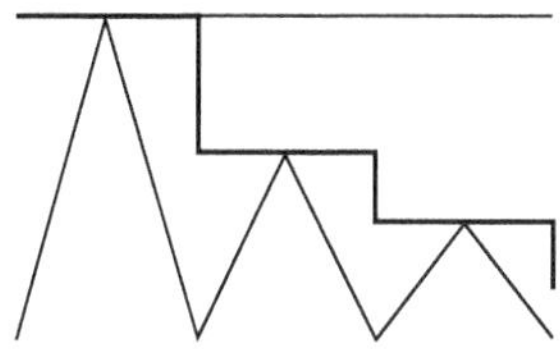

P-phrase structure: ((N1 N2) N3)

b. Right-branching structure: [N1 [N2 N3]]

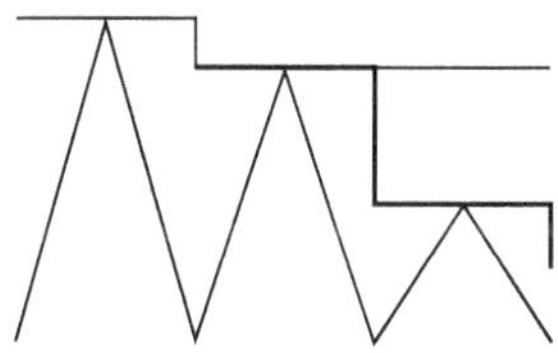

P-phrase structure: (N1 (N2 N3))

3.3.2 *Pitch range and pitch scaling*

In this section, results on pitch scaling are summed up which show how tones are scaled differently as a result of varying information structure. As before, results are first shown for German, and in a second step for Japanese.

3.3.2.1 *Pitch scaling in German* Two experiments on German are summarized, one on Second Occurrence Focus (SOF) (Féry and Ishihara 2009), and another one on the scaling of pitch accents in sentences with a simple syntactic structure under varying information structure conditions (Féry and Kügler 2008).

In Féry and Ishihara (2009) it is shown that the pitch height of a word depends on its status as First Occurrence Focus (FOF, a focused expression that appears in the discourse for the first time), Second Occurrence Focus (a focused expression that has already appeared in the discourse and is repeated) or just given (Non-Focus). SOF is illustrated in (13) and (14) with one of the sentences from the object set used in our experiment.[8] The study was a follow-up on studies by Rooth (2004), Bartels (2004), and Beaver *et al.* (2007), who looked at SOF in postnuclear position in English. In our study, prenuclear position was also considered, which led to clearer results concerning

[8] The experimental sentences were divided into two sets: in one of them the relevant expressions were the subject of the sentence, and in the other one, they were the object.

the realization of pitch accents (see also Rooth, this volume, for relevant discussion).

(13) Second Occurrence Focus: prenuclear
{Many women have invited several relatives to the village fair.
FOF: Aber Eva hat nur ihren Bruder eingeladen.
'But Eva has invited only her brother.'}

SOF: *Nur ihren Bruder hat auch Maria eingeladen.*
'Only her brother.ACC has invited also Maria.'

(14) Second Occurrence Focus: postnuclear
{Many women have invited several relatives to the village fair.
FOF: Aber Eva hat nur ihren Bruder eingeladen.
'But Eva has invited only her brother.'}

SOF: *Auch Maria hat nur ihren Bruder eingeladen.*
'Also Maria has invited only her brother.'

As far as pitch accents are concerned, we showed that there is a hierarchy in terms of pitch height among FOF, SOF, and non-focus: an FOF is realized higher than an SOF, which in turn is higher than a given constituent (or non-focus). This hierarchy is best observed in the sentence initial position (dark bars in Figure 3.7, corresponding to the subject set). In postnuclear position, pitch accent differences were cancelled between SOF and non-focus, since a postnuclear deaccenting had taken place in this position, and suppressed any potential differences. This can be seen in comparing the final two light bars which stand for the object set.

We analyzed SOF as being simultaneously given and focused, and posited that the intermediate height of a SOF is the reflex of this double status. We also found a difference in duration, reproducing Beaver *et al.*'s results. More precisely, a focused expression, be it FOF or SOF, is always longer in duration than a non-focused one.

The pitch range is thus sensitive to information structure, and assignment of accents is not just a yes or no matter.

The second relevant study for German was reported in Féry and Kügler (2008), who investigated the production of sentences of the following type.

(15) Subject **Verb** (focus on the verb)
{The animals don't like to fight. Why are they angry with the sheep?}
Weil der Hammel [angefangen hat]$_F$
'Because the sheep started (a fight).'

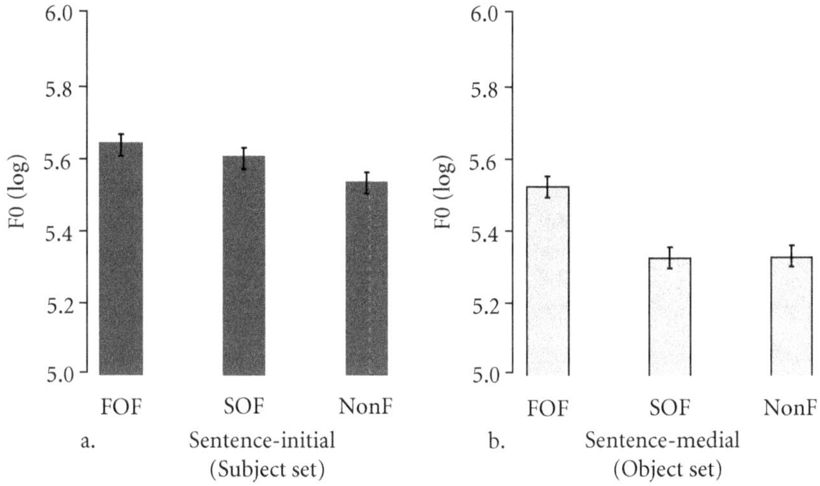

FIGURE 3.7 Mean log-transformed F0 for FOF/SOF/Non-Focus in sentence-initial/medial conditions (with 95% Confidence Interval) (Féry and Ishihara 2009: 297)

(16) **Subject Object Verb** (all-new condition)
{'Why were the animals happy?'}
[Weil der Hummer den Rammler eingeladen hat]$_F$.
'Because the lobster has invited the buck.'

(17) Subject Object **Indirect Object** Verb (focus on the dative complement)
{The buck wanted to introduce the sheep to the lion. Why didn't he do that?}
Weil der Hammel den Rammler [dem Hummer]$_F$ vorgestellt hat.
'Because the sheep introduced the buck to the lobster.'

Three parameters were systematically varied. First, the number of arguments, which varied between one and three (subject, direct object, and indirect object). Second, the word order, and third the given-new status of the arguments and the verb. Altogether 26 conditions were created. In total 2,340 (130 sentences × 18 speakers) were recorded, and 2,277 were analyzed.

The aim of this experiment was to quantify the downstep coming from the syntax, the tone-boosting effect of focus, and the tone-lowering effect of givenness. Figure 3.8 shows how downstep regulates the high tones of each argument and the verb in the four sentence lengths.[9] This pattern is thus the

[9] It must be observed that only 34 percent of the all-new sentences had this pattern. The other realizations had upstep on the preverbal arguments, or the verb was unaccented.

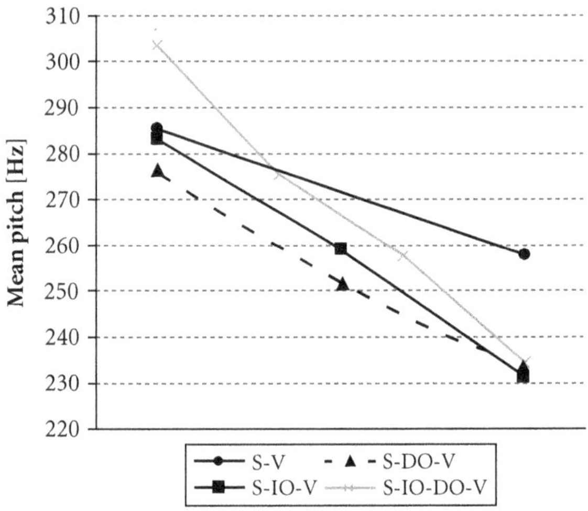

FIGURE 3.8 Downstep pattern in all-new sentences Féry and Kügler 2008: 688)

result of the syntactic structure which predicts a lowering on each high tone in an intonation phrase.

The effect of focus and givenness appears in Figure 3.9 for sentences with narrow focus on the first (a), the second (b), the third argument (c), and the verb (d). It is easy to see that the focused argument causes an F0 raising of the high tone, and that the other given arguments are lower than in the default case shown in Figure 3.9.

This experiment confirmed that the downstep and lowering effects caused by syntax and by information structure are distinct. There was no indication that the prosodic phrasing could be distinguished in the same way.

Tonal scaling is thus remarkably sensitive to changes in information structure, and varies with the height of the accented constituents in a very fine-grained way. Postulating only prosodic phrases for these results is not sufficient. When the status of one constituent is changed to focused, SOF or given, not only is the F0 height of its accent affected, but also the height of the other constituents in the same sentence. There is no indication that the relationship between syntax and prosodic structure is changed as well, and even if we wanted to relate the changes in F0 height to prosodic domains, it would be a delicate matter to decide which prosodic level is affected by being neighbor of a focus, and to distinguish between pre- or post-focality. The solution we offer to separate prosodic phrasing and pitch scaling, and treat them as two independent factors entering the prosodic structure of a sentence, can account for these results.

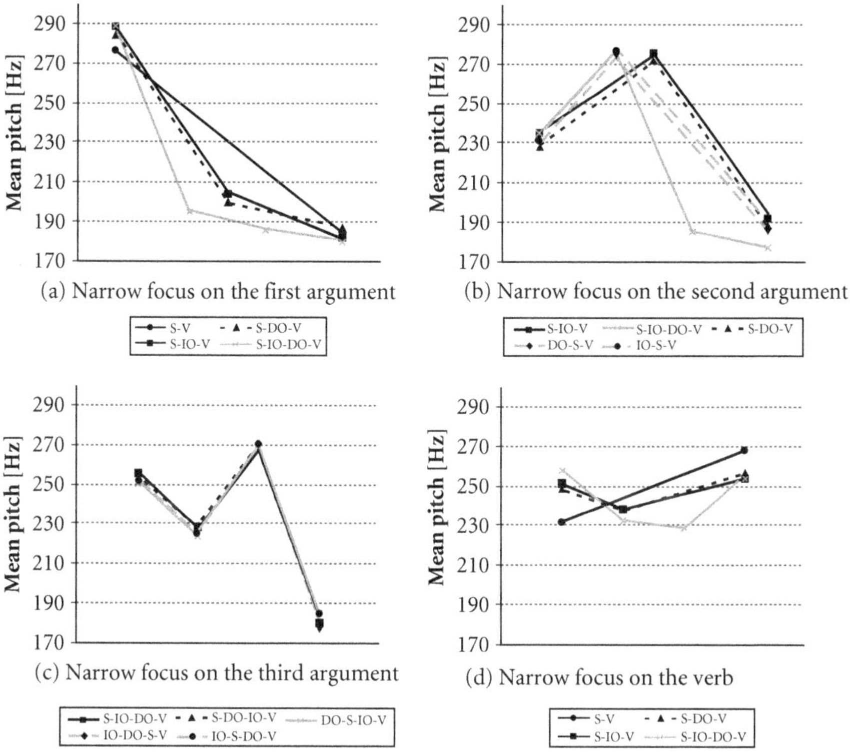

FIGURE 3.9 Representation of pitch peak realization of narrow focus on different constituents. (Féry and Kügler 2008: 690).

3.3.2.2 *Pitch scaling in Japanese*

In the case of Japanese, too, phonetic effects of focus have often been explained in terms of prosodic phrasing. It has been claimed that focus influences p-phrase structure. Pierrehumbert and Beckmann (1988) proposed that focus creates a p-phrase boundary on its left, hence blocking downstep effects. In addition, the F0 downtrend observed on post-focal material has often been analyzed as downstep as a result of the elimination of all p-phrases after focus (Nagahara 1994).[10] Under this line of analysis, focus directly affects prosodic phrasing, by adding or deleting p-phrase boundaries.[11]

[10] See, however, Shinya (1999), who argues against this line of analysis, and supports Poser's (1984) claim that the F0 boosting effect of focus is independent of MaP boundaries. See also Sugahara (2003), who experimentally showed that MaP boundaries can be observed in the post-focal area when it is discourse new.

[11] Although some authors claim that focus creates an Intonation Phrase prominence (Truckenbrodt 1995; Selkirk 2008), the basic concept remains the same in that focus modifies prosodic phrasing. See Sections 3.2 and 3.4 for relevant discussion.

In Ishihara (2007) it is claimed on the basis of several experimental results that p-phrase and Focus Intonation (FI) domains are independent of each other. The former is purely syntax-based, while the latter is created by changing the pitch register of the focused phrase (by F0 boosting) and the post-focal material (by F0 compression). The experimental results show that downstep, which lowers the pitch realization of non-initial pitch accents within a p-phrase, and post-focal compression, which lowers the pitch realization of post-focal material, are independent phenomena. That is, a pitch reset after downstep at the following p-phrase boundary, and a pitch reset after an FI can be observed separately. In (18), where there is no focus, the embedded clause verb *nonda* "drank" and the preceding PP *nomiya-de* "at the bar" form a single p-phrase. Accordingly, downstep is observed on the verb, followed by a pitch reset on the matrix adverbial phrase *imademo* "still".

(18) Declarative sentence (Default pitch contour)

Náoya-wa	[$_{CP}$ Mári-ga	wain-o$_i$	[$_{VP}$ nomíya-de t_i	nónda]
Naoya-TOP	Mari-NOM	wine-ACC	bar-LOC	drank

to]	ímademo	omótteru
that	still	think

'Naoya still thought that Mari drank something at the bar.'

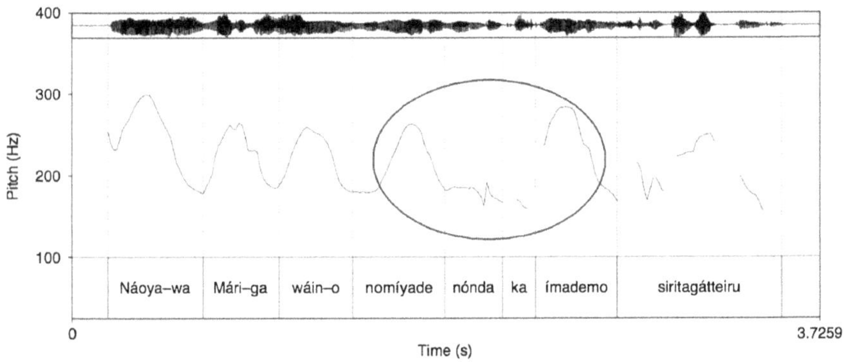

In a *wh*-question like (19), where the *wh*-phrase behaves prosodically as a focused phrase and hence triggers a FI, the F0 peak on the *wh*-phrase is raised, and the post-focal material is compressed.

(19) *Wh*-question (FI starting from the *wh*-phrase)

dáre-ga	[$_{CP}$ Mári-ga	wáin-o$_i$	[$_{VP}$ nomíya-de t_i	nónda]
who-NOM	Mari-NOM	wine-ACC	bar-LOC	drank

to]	ímademo	omótteru	no?
that	still	think	Q

'Who still thinks that Mari drank wine at the bar?'

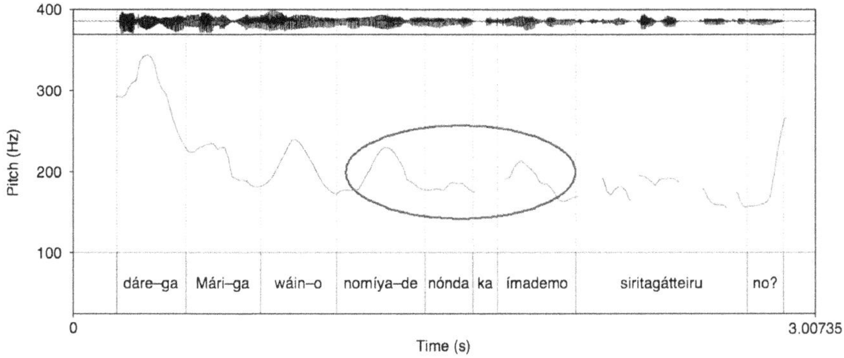

It should be noted here that we can still observe a downstep on the embedded verb *nonda*, which is realized lower than the preceding PP *nomíya-de*, and the subsequent pitch reset on the matrix adverbial phrase, which is realized roughly as high as *nomíya-de*, even though the entire pitch contour is compressed due to the focus in sentence-initial position (see also Figure 3.10). If we took the FI to be a large p-phrase created by deletion of p-phrase boundaries, as the earlier accounts claim, then we would expect downstep on all the phrases in the sentence, and no pitch reset would be expected. What we see here instead is that the p-phrase structure is maintained within the (more or less evenly compressed) post-focal domain. The normalized mean F0 peaks and valleys given in Figure 3.10[12] confirm this observation: within the compressed pitch contour in the *wh*-question (condition B, solid line), the F0 peak on the embedded verb shows a larger fall due to downstep, and that of the matrix adverb shows a pitch reset.

Furthermore, another experiment by Ishihara (2008) reveals clear phonetic differences between focal F0 boosting and p-phrase boundary. In (20), the pitch realizations of the first three nouns (N1–N3) were measured. In (20a), there is neither focus nor p-phrase boundary on N3, whereas there is a focus in (20b) and a p-phrase boundary in (20c).

(20) a. No focus, no p-phrase boundary on N3

[$_{VP}$ [$_{DP}$ Náoya-no áni-no wáin-o] waingúrasu-de
Naoya-GEN brother-GEN wine-ACC wineglass-with
nónda]
drank

'(I) drank Naoya's brother's wine with a wineglass.'

[12] The results from 11 subjects are normalized to factor out the individual difference in pitch range. Each condition has 198 samples (six sentences recorded three times per subject, produced by 11 subjects).

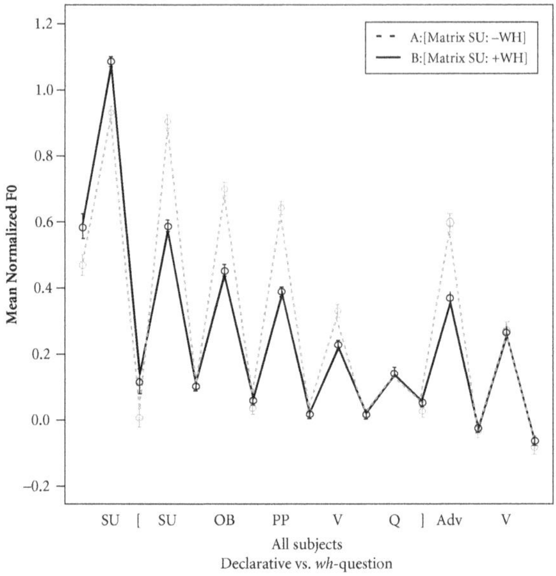

FIGURE 3.10 Normalized mean F0 peaks and valleys of (15) and (16)

b. Focus on N3

[VP [DP	Náoya-no	áni-no	náni-o]	waingúrasu-de
	Naoya-GEN	brother-GEN	what-ACC	wineglass-with

nónda]	no?
drank	Q

'(Lit.) Naoya's brother's what did you drink with a wineglass?'

c. P-phrase boundary before N3

[DP	Náoya-no	áni-ga]	[VP	wáin-o	waingúrasu-de
	Naoya-GEN	brother-NOM		wine-ACC	wineglass-with

nónda]
drank

'Naoya's brother drank wine with a wineglass.'

The results are summarized in Figure 3.11. When N3 bears focus, as shown in the dashed line, the F0 peak on N3 is raised strongly compared to a non-focus. Nothing else is changed. When there is a p-phrase boundary in front of N3, as shown in the dotted line, not only the F0 peak on N3 is raised (but not as strongly as in the case of focus), showing a pitch reset, but also the F0 peak on N2 is lowered. This suggests that focus only affects the F0 peak of the focused

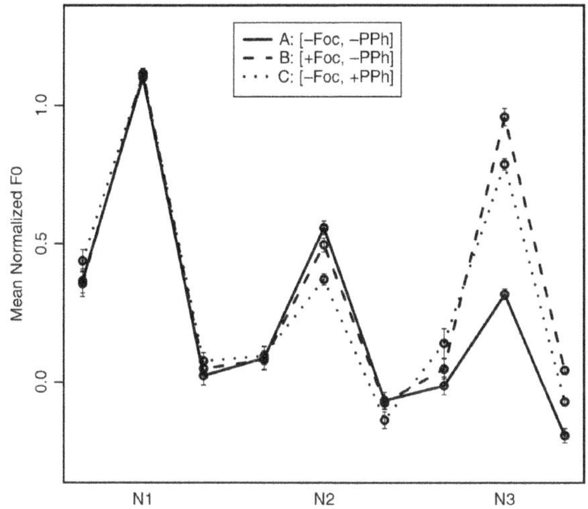

FIGURE 3.11 Normalized mean F0 of the first three words in (19a) (solid line); (19b) (dashed line); and (19c) (dotted line)

phrase, whereas an insertion of a p-phrase boundary affects not only the p-phrase-initial phrase but also the final phrase of the preceding p-phrase. If there is such a clear difference between the phonetic effects of focus and those of p-phrase boundary, they should be treated differently in the phonological analysis.

In sum, we find an extreme sensitivity of pitch scaling to information structure, both in German and in Japanese, and in different domains of information structure. It does not seem desirable, or even possible, to explain all effects by changes in the prosodic phrasing.

3.4 Comparison with earlier approaches

In this section we compare our approach with some of the other models that have been proposed in the literature. The core of our approach is the clear separation of prosodic influence of syntactic structure and those of information structure. This property sharply contrasts with those in many analyses that attempt to capture those effects in a single phonological representation.

Earlier analyses may be divided into two subgroups depending on whether they adopt prosodic phrasing or not. In one group of analyses, an information structural feature is assigned directly to morpho-syntactic elements, and both syntactic and information structural effects are encoded directly into prosody (most often as pitch accents), without any direct reference to prosodic phrasing. In another group of analysis, syntactic and information structural features are encoded onto prosodic phrasing.

3.4.1 *Pitch accents without mediation of prosodic phrasing*

The first type of model which does not make a clear distinction between the effects of syntax and those of information structure posits that pitch accents are directly assigned to morpho-syntactic constituents or to focused constituents. The distribution of default pitch accents is driven by left- or rightmost Nuclear Stress Rules or by projection rules which take the predicate–argument structure of the sentence into account. Pitch accents are usually represented in the phonological representation as positions in metrical grids or metrical trees. In a second step, information structure modifies the default or unmarked pitch accent distribution. Focus constituents get an accent and given constituents lose their default accent (Jackendoff 1972; Schmerling 1976; Ladd 1980; von Stechow and Uhmann 1986; Cinque 1993; Rooth 1985, 1992; Schwarzschild 1999; Steedman 2000; Büring 2006; Beaver *et al.* 2007, among many others)

A famous representative of this group is Selkirk's (1995) Focus Projection Theory and its variants and descendants. F-marked constituents are assigned a pitch accent and the F-feature percolates up in the syntactic structure. In Selkirk's analysis, givenness is derived from the lack of F-marking (see also Schwarzschild 1999 and Büring 2006). As we saw above, however, such direct association of givenness and deaccentuation causes problems in certain cases such as Second Occurrence Focus, where a given element may or may not bear pitch accent depending on its relative location to the First Occurrence Focus. In our model, on the other hand, syntactic effects are mapped to prosodic phrasing while the information structural effects are encoded onto pitch register. This allows interaction of the two different effects.

A serious drawback of such an approach is that pitch accents are always modified individually. Focus just changes one pitch accent, for instance, and leaves the remaining of the representation untouched. In other words, no global effect in the scaling of accents can be accounted for by such approaches. If all pitch accents of a sentence have to be modified, this can be done only by addressing every tone one by one. Furthermore, boundary tones or phrasal tones are either completely left out of such representations, or assumed to be anchored in a different part of the phonology, namely in the intonational phonology (Pierrehumbert 1980).

3.4.2 *Pitch accents are mediated by prosodic phrasing*

In the second group of analyses, prosodic phrasing mediates the syntactic and information structural effects (Gussenhoven 1983; Beckman and Pierrehumbert 1986, 1992; Selkirk 1986; Truckenbrodt 2006). There are two main assumptions in this line of analysis. The first general assumption is that *all* prosodic effects are explained in terms of prosodic phrasing. This means that prosodic effects of syntactic structure and those of information structure are both encoded onto prosodic phrasing. The second prevalent assumption (at least for intonation languages such as English and German) that comes with this analysis is that heads of prosodic phrases are marked by pitch accents. With the combination of the two assumptions, syntactic and information-structural effects are explained in terms of insertion or deletion of prosodic boundaries.[13]

According to some versions of syntax–prosody mapping principle (e.g. Selkirk 1986; Truckenbrodt 1995), syntax determines the location of prosodic boundaries. Once the prosodic structure is determined according to syntax, the head of each prosodic phrase bears a pitch accent, under the second assumption mentioned above. Information structure (focus and givenness), on the other hand, affects the location of pitch accent. Focus requires a pitch accent, and givenness the lack of it. Again, under the second assumption, this amounts to saying that when there is a focus, it must bear a pitch accent, and consequently, become the head of a prosodic phrasing. This forces the modification of prosodic phrasing. A pitch accent assigned to focus forces insertion of a prosodic boundary. The opposite phenomenon takes place for given material. Givenness requires the absence of pitch accent, and hence, may eliminate pitch accents assigned according to the syntax–prosody mapping. Under the second assumption, elimination of pitch accent directly means the elimination of prosodic phrase boundary, because no prosodic phrase may exist without its head. All in all, both syntax and information structure affect the realization of prosodic phrasing, and nothing else. A similar view explains earlier approaches to Japanese: an additional upstep associated with focus forces the insertion of a prosodic boundary, and compression of F0 means that prosodic phrases have been deleted.

But such an analysis cannot explain some of the data discussed in this chapter. Japanese data presented in Figure 3.11, for example, clearly show that a prosodic boundary inserted at the syntactic boundary and the F0 boosting

[13] Selkirk (2006a) proposed for Bengali that a morphemic tone $[H]_{FOC}$ necessarily aligns with the end of a syntactic focus, creating a prosodic boundary. In this approach, a tone associated with focus creates a prosodic phrasing, and deletion of post-focal prosodic phrases follows.

triggered by focus behave differently. We need a more elaborate model than insertion and deletion of prosodic phrases if we want to explain the information-structural effects on prosody.[14]

3.4.3 *Comparison with our model*

Our model integrates several main components of the tradition from which it originates. A crucial ingredient is that prosodic phrases are mapped from the syntax. In the case of German and English, the location of a default pitch accent correlates with the location of heads of p-phrases. In the case of Japanese, each p-phrase serves as a domain of downstep. The distribution of pitch accents is thus regulated by constraints on the syntax and on the accent projection, aspects which we did not evoke in the paper. In the case of Japanese, each p-phrase serves as the domain of downstep.

Our model differs from the earlier accounts in two respects: (i) it does not assume the Strict Layer Hypothesis and adopts a recursive p-phrasing instead; (ii) it denies that a unique prosodic structure is shaped in the same way by syntax and by information structure, and instead proposes that the prosody responds to different influences in different ways. In particular, we claim that the main import of information structure on prosody is its effect on F0 scaling. As shown in detail in Section 3.2, raising, lowering, and deletion of pitch accents are mediated by the effect of information structure on focus and givenness domains. The pitch accents are scaled to top and bottom lines of F0 register of these domains, which can be isomorphic to prosodic domains, but do not have to. These registers are calculated as a function of what precedes and follows. It is this aspect of the model which differs most from the proposals presented above. Many of them consider the changes in F0 as the phonetic correlate of a change in the prosodic phrasing. In another line of thought, an accent standing for narrow focus may be the correlate of a metrical grid position assigned at a different level of phrasing from an accent standing for an information focus (Selkirk 2002). An important consequence of our proposal is that it can account for the global effect that a focus has on the remainder of the sentence. The scaling of all pitch accents is the perceived consequence of the changed structure of the registers as a consequence of the focus–new–given partition. It must not be postulated that every pitch accent is subject to a special rule principle or constraint which raises or decreases it.

[14] However, it has been shown for diferent languages that prosodic phrasing may be sufficient as an indication of focus. Kanerva (1990) shows for Chichewa that a change in phrasing triggered by focus is not necessarily accompanied by prosodic prominence in the form of pitch accents. The focus structure is thus expressed by the phrasing and nothing else (see also Fiedler *et al.*, in this volume).

3.5 Conclusion

In this chapter, we have presented both empirical results and theoretical reflections on the interaction between syntax, information structure, and prosody. It has been shown that the effects from the syntax and those from the information structure should be kept apart in order better to understand how the prosodic structure is obtained at the level of the sentence. We have compared an intonation language with free assignment of pitch accents (German) with a pitch-accent language in which pitch-accent distribution is lexically governed (Japanese). Syntax regulates prosodic phrasing and default pitch scaling, whereas information structure acts on the F0 of entire focus and givenness domains. An additional aspect of our proposal is the systematic recursivity of prosodic phrases.

4

Structural focus and exhaustivity

KATALIN É. KISS

4.1 Introduction

The Hungarian sentence has been known to have a structurally determined focus slot in its left periphery. Most approaches to the syntax and the syntax–semantics interface of Hungarian agree that this preverbal focus position is associated with a [+exhaustive] feature. At the same time, however, theories have also been proposed which ignore the exhaustivity of the Hungarian preverbal slot, concentrating their attention on the universal properties of focus, also shared by the prosodic focus of languages with no invariant focus position. Recently, Wedgwood (2005) has explicitly argued against the [+exhaustive] feature of the Hungarian preverbal focus, claiming that the exhaustivity often associated with foci in languages of both the Hungarian type and the English type is a mere pragmatic implicature. This chapter discusses and refutes Wedgwood's claim, and will present new evidence for the exhaustivity of structural focus.

Section 4.2 introduces the standard arguments for the [+exhaustive] feature of Hungarian structural focus, and the syntactic means of deriving this feature. Section 4.3 briefly discusses focus theories ignoring exhaustivity, whereas Section 4.4 presents Wedgwood's (2005) theory, denying the encoded exhaustivity of structural focus altogether. After refuting an apparent counter-argument of Wedgwood in Section 4.5, Section 4.6 puts forth new evidence supporting the obligatory [+exhaustive] feature of structural focus, based on the interpretation of scalar elements. A scalar element *n*, interpretable as "at least *n*" out of focus, can only mean "exactly *n*" in the preverbal focus slot, which is derived from the exhaustivity of structural focus, involving the

I owe thanks to Werner Frey, Nick Asher, Caroline Féry, Knud Lambrecht, Christopher Piñón, Anna Szabolcsi, Daniel Wedgwood, and the participants of my focus seminar at Potsdam University in the spring term of 2006 for many useful discussions. The writing of this paper was in part supported by Grant TS 49873 of OTKA, the Hungarian National Research Fund.

exclusion of all alternatives but that denoted by the focused constituent. It is shown that the "exactly *n*" interpretation is obligatory in focus position irrespective of the pragmatic conditions. What is more, scalar elements for which the "at least *n*" interpretation would lead to a semantic anomaly (i.e. scalar elements representing a value in the negative domain of a bidirectional scale) are obligatorily focused in Hungarian.

4.2 The standard interpretation of structural focus

In the left periphery of the Hungarian sentence, left-adjacent to the V, there is an optionally filled structural position functioning as a focus slot. In focusless, neutral, sentences, the post-topic sentence part, functioning as the logical predicate, usually begins with a verbal particle (1a) or a bare nominal complement (1b).[1] Recent studies on Hungarian syntax (e.g. Csirmaz 2004; É. Kiss 2008a) analyze the logical predicate of such neutral sentences as a PredP (subsuming a split VP), with the verbal particle or bare nominal (both marked as [+pred]) occupying Spec,PredP.

(1) a. Péter [$_{\text{PredP}}$ meg vette Kertész könyvét].
Peter PRT bought Kertész's book-ACC
'Peter bought Kertész's book.'

b. Péter [$_{\text{PredP}}$ könyvet vett].
Peter book-ACC bought
'Peter bought some book(s).'

In focus constructions, the predicate section of the sentence begins with the focus, to be immediately followed by the V (2a, b). Since Bródy (1990), it has been generally believed that the focus constituent occupies the specifier of a designated functional projection called FocP. In the presence of a focus, the V moves across the verbal particle/bare nominal complement into a head position adjacent to the focus. As shown by Olsvay (2000) and Horvath (2005), this is the head position of a syntactic projection below FocP (called Non-Neutral Phrase, perhaps a version of FinP). In neutral sentences and focus constructions alike, the main stress falls on the left edge of the logical predicate. The post-focus section of focus constructions is destressed.

[1] *Meg* in (1a) is a resultative verbal particle, a telicity marker. Verbal particles appear in accomplishment and achievement sentences containing a [+specific] theme; they predicate the result state of the theme. In (1b) *meg* is missing because the theme is a non-specific noun phrase. The bare nominal of (1b) occupies the same Spec,PredP position as the verbal particle of (1a); after all, it is also a secondary predicate predicated of the incorporated internal argument of the V.

(2) a. Péter [$_{\text{FocP}}$ KERTÉSZ KÖNYVÉT [$_{\text{NNP}}$ vette [$_{\text{PredP}}$ meg t_V]]].
'It was Kertész's book that Peter bought.'

b. [$_{\text{FocP}}$ PÉTER [$_{\text{NNP}}$ vett [$_{\text{PredP}}$ könyvet t_V]]].
'It was Peter who bought some book(s).'[2]

Such a focus position is not unique to Hungarian. A survey carried out in the framework of the EUROTYP project found that half of the European languages have a structural focus position (É. Kiss 1998c). Hinterhölzl (this volume) argues for a preverbal focus position in Old High German. Fiedler *et al.* (this volume) point out a clause-final structural focus position in a number of West African languages. Aghem is known to have a postverbal focus slot (although Hyman and Polinsky (this volume) argue that—unlike left-peripheral focus positions—it is not an A-bar position filled by movement).

Hungarian native speakers share the intuition that (1a) and (2a), as well as (1b) and (2b) have different truth conditions. The focus movement of the object in (2a) and the subject in (2b) adds a special component to the meaning of the sentence. Whereas (1a) is true in every situation in which Peter bought a set of objects (or a set of books) including Kertész's book, (2a) is only true in a situation in which the set of objects/books bought by Péter consists of Kertész's book and nothing else. Similarly, (1b) is true whenever the set of persons who bought books includes Peter; (2b), on the other hand, is only true in a situation in which Peter is the only (relevant) person who bought books.

The formalization of this intuition goes back to the work of Szabolcsi (1980, 1981a, 1981b, 1983, 1985, etc.). In these early studies of Szabolcsi, focus is analyzed to express exhaustive listing, that is, to provide an exhaustive list of the referents for which the statement expressed by the sentence is true. Observe how Szabolcsi (1980, 1981a) paraphrases the meaning of the focus construction in (3a):

(3) a. [$_{\text{FocP}}$ PÉTER [aludt a padlón]].
Peter slept the floor-on

b. 'For every x, x slept on the floor iff x = Péter'

Naturally, the universal quantifier is to be interpreted on a relevant set determined by the given situation or context (e.g. the set of persons staying in a particular apartment at a given time).

Evidence of the [+exhaustive] feature of structural focus is provided by the fact that (3a) and (4) cannot be simultaneously true, that is, (3a) is not a logical consequence of (4) but contradicts it.

[2] In (2a) *könyvét* could also be presupposed and destressed. Irrespective of whether either of its subconstituents is unstressed and given, it is the whole constituent in Spec,FocP, *Kertész könyvét*, which is analyzed as focus.

(4) [$_{\text{FocP}}$ PÉTER és PÁL [aludt a padlón]].
Peter and Paul slept the floor-on
'It was Peter and Paul who slept on the floor.'

The negated version of (3a), spelled out in (5a), on the other hand, is compatible with (4), as shown in (5b):

(5) a. Nem [$_{\text{FocP}}$ PÉTER [aludt a padlón]].
'It wasn't Peter who slept on the floor.'

b. Nem [$_{\text{FocP}}$ PÉTER [aludt a padlón]] hanem [$_{\text{FocP}}$ PÉTER és PÁL ([aludt a padlón])].
'It wasn't Peter who slept on the floor but it was Peter and Paul (who slept on the floor).'

Szabolcsi's formula is also intended to cover structural foci in all-new sentences, for example:

(6) [$_{\text{FocP}}$ PÉTER [aludt a padlón]] vagy [$_{\text{FocP}}$ A VENDÉG
Peter slept the floor-on or the guest
[ment szállodába]]?
went hotel-to
'Was it Peter who slept on the floor, or was it the guest who went to a hotel?'

Sentence (6) can only be asked if in one of the two alternative worlds Peter exhausts the set of persons who slept on the floor, and in the other one the guest exhausts the set of persons who went to a hotel.

Further evidence of the exhaustivity of structural focus presented by Szabolcsi (1980) involves conditionals. A non-focused *if*-clause is a unidirectional conditional; a focused *if*-clause, on the other hand, is understood as a bidirectional, corresponding to an *if and only if* clause. (The focus function of a subordinate clause is indicated by the focus position of the pronominal associated with it. The clausal constituent itself cannot appear in Spec,FocP, presumably owing to a prosodic constraint requiring that the focus constituent form a phonological word with the V.) Compare:

(7) a. Meg-kapod a pénzt akkor, ha el-végzed a munkát.
PRT get-you the money then if PRT do-you the job-ACC
'You get the money in case you do the job.'

b. [$_{\text{FocP}}$ AKKOR$_i$ [kapod meg a pénzt]] [ha elvégzed a munkát]$_i$.
'It is in the case you do the job that you get the money.'

Kenesei (1986: 149), proposing an alternative to the focus interpretation of Szabolcsi (1980, 1981a), did not discard the [+exhaustive] feature, but represented it as something presupposed rather than asserted. In Kenesei's theory, the focus expresses exclusion by identification. This term is based on the intuition that focusing means both the identification of a referent and the exclusion of potential alternatives. (Notice that in Hungarian, universal quantifiers such as *mindenki* "everybody", involving no exclusion, cannot be moved into the focus position.) Kenesei's logical formula for focus contains an iota operator:

(8) a. [$_{\text{FocP}}$ PÉTER [aludt]].
'It was Peter who slept.'

b. $\iota x\, [\text{Slept}\,(x)] = p$

Formula (8b) presupposes that there is a unique individual who slept, and asserts that this individual is Peter.

In her (1994) study, Szabolcsi basically adopts Kenesei's notion of focus, however, she proposes to change the formalism in such a way that it can also handle plurals:

(9) $\lambda z \lambda P\, [z = \iota x\, [P(x)\, \&\, \forall y[P(y) \rightarrow y \leq x]]]$

Modifying Kenesei's notion of "exclusion by identification" to some extent, É. Kiss (1998a) describes the focus function as "exhaustive identification". The article assumes a focus operator which operates on the set of contextually determined elements for which the predicate of the sentence can potentially hold, and exhaustively identifies the proper subset of this set for which the predicate actually holds, excluding the complementary subset. The preverbal focus represents the value of this operator.

As shown by É. Kiss (1998a), a constituent conveying new information, answering an explicit or implicit *wh*-expression, can also appear in the postverbal part of the sentence; however, postverbally it has a non-exhaustive reading. Compare the two alternative ways of answering the *wh*-question in (10a):

(10) a. [$_{\text{FocP}}$ KIKET [hívtál meg ma estére]]?
who-PL-ACC invited-you PRT today evening-for
'Who did you invite for tonight?'

b. [$_{\text{FocP}}$ PÉTERT és PÁLT ([hívtam meg])].
'It is Peter and Paul (that I invited).'

c. Meg hívtam PÉTERT és PÁLT.
'I invited Peter and Paul.'

The answer in (10b) is false if I also invited others than Peter and Paul. The answer in (10c), on the other hand, does not imply or implicate that I invited nobody but Peter and Paul; on the contrary, native speakers would use it to suggest that the answer is partial, non-exhaustive. The structural focus, illustrated in (10b), is called identificational focus. The prosodic focus in (10c) is a mere information focus. The preverbal identificational focus is optionally present; it represents the value of an operator; it is exhaustive; it has scope; it involves distributional restrictions; it is always a major constituent (an argument of the V or an adjunct); it is associated with a fixed structural position; and it is derived by movement. The information focus, on the other hand, is a mere pragmatic–prosodic phenomenon marked by primary stress, expressing new information. Every sentence contains an information focus, the size of which can vary. The prosodic focus of English is information focus. The English equivalents of the Hungarian preverbal identificational focus are the cleft and the pseudo-cleft constituents.

In É. Kiss (2006a, b), I analyzed the preverbal focus as a specificational predicate—adopting Higgins's (1973) analysis of the English pseudo-cleft constituent, and Huber's (2000) analysis of the German and Swedish cleft constituent. In this approach, which I maintain in this chapter, the focus referentially identifies the set determined by the presupposed section of the sentence by listing its members. The meaning of (10b), for example, can be explicated informally as follows:

(10) b′. 'Who I invited are Peter and Paul.'

The focus as a specificational predicate is predicated of the open sentence determined by the post-focus projection. Focus movement serves the purpose of realizing in syntactic structure the specificational predicate–argument (i.e. the focus–presupposition) articulation.[3]

In this framework, the [+exhaustive] feature of focus is a semantic consequence of its specificational predicate role. The focus serves to specify the set

[3] Szabolcsi (1980) was the first to notice the "qualitative", that is, predicative, nature of focus in examples like (i), where the two foci can denote the same individual because of the non-referring, predicative function of the focus constituents. The coreferent reading is impossible in any other structural position; see (ii), (iii):

(i) [$_{\text{FocP}}$ A BARÁTNŐMET$_i$ [hívtam meg]], nem [$_{\text{FocP}}$ A MINISZTER
my friend-ACC invited-I PRT not the minister's
FELESÉGÉT$_{i/j}$].
wife-ACC
'It was my friend that I invited, not the minister's wife.'

(ii) *A barátnőmet*$_i$ meg-hívtam, *a miniszter feleségét*$_{*i/j}$ nem.
'My friend, I invited, the minister's wife, I didn't.'

(iii) Meg-hívtam *a barátnőmet*$_i$, de nem hívtam meg *a miniszter feleségét*$_{*i/j}$.
'I invited my friend, but I didn't invite the minister's wife.'

determined by the presupposed sentence part by listing its members, and it fulfils this function if it enlists the members of the set exhaustively. Thus the exhaustivity of structural focus is neither a lexical property (like the *wh* feature of interrogative expressions), nor a feature assigned by a focus operator in the course of the derivation, but a semantic consequence of the specificational predicate role of the focused constituent. Universal quantifiers cannot be focused because they cannot function as predicate nominals (cf. Giannakidou and Quer 1995).

The most recent theory attributing a [+exhaustive] feature to the Hungarian preverbal focus is the Exhaustive Identification Operator theory of Horvath (2005). According to Horvath, the preverbal focus position in the Hungarian sentence is, in fact, a kind of quantifier position, the specifier of an Exhaustive Identification Phrase (EIP), to be filled by a constituent combined with an invisible Exhaustive Identification Operator (EIOp). The source of exhaustivity is the EIOp, whose function is described by Horvath (2005) as follows:

(11) EI operates on a set of contextually or pragmatically given elements for which the predicate phrase can potentially hold; it identifies the exhaustive proper subset of this set for which the predicate phrase actually holds.

In Horvath's terminology, the term "focus" is restricted to the pragmatic–prosodic information focus of É. Kiss (1998a); it is defined as a syntactically unencoded interface phenomenon, marked prosodically across languages. The relation between exhaustivity and focus is indirect; the EIOp, similar to *even* and *only*, requires association with focus. This is not an absolute requirement though; Horvath also quotes examples in which the constituent associated with the EI operator is presupposed and deaccented; the information focus, bearing primary stress, is a quantifier preceding the EIOpP. For example:

(12) MINDEN fiú [$_{\text{FocP}}$ Marit [kérte fel táncolni]], nemcsak
every boy Mary-ACC asked PRT to.dance not only
a barátja.
her friend
'Every boy asked Mary [and no one else] for a dance, not only her friend.'

4.3 Focus theories ignoring the exhaustivity of structural focus

Though the exhaustive interpretation of the preverbal focus has never been questioned by native Hungarian speakers, including linguists, there have been proposals which, aiming to formulate a universal theory of focus, regard the [+exhaustive] feature of the Hungarian preverbal focus as irrelevant. In the theory of Bródy (Bródy 1990, 1991, 1995), constituents in the preverbal focus position have a [+f] feature, and they have been raised to Spec,FocP to enter into a checking relation with the [+f] feature of the Foc head. Universal quantifiers, which can be adjoined to FocP in visible syntax and be assigned primary stress in Hungarian, are assumed by Bródy to have the same [+f] feature. In Bródy's analysis, Q-raising to FocP is also movement into the checking domain of the Foc head. Bródy also attributes a [+f] feature to postverbal constituents bearing a primary stress, e.g. the postverbal stressed object of (10c), and to the prosodic focus of the English sentence. These constituents are raised to Spec,FocP at LF. If an English prosodic focus, or a Hungarian postverbal information focus also has a [+f] feature, then [+f] cannot encode exhaustive identification. If exhaustive identification means the identification of the proper subset of a relevant set, and the exclusion of the complementary subset, then [+f] cannot involve exhaustive identification in the case of universal quantifiers either. That is, the [+f] feature in the theory of Bródy has no [+exhaustive] component.

Exhaustivity plays no role in the focus theory of Szendrői (2003) either, where focus movement is triggered by the requirement of stress–focus correspondence—as proposed by Reinhart (1995). According to the Stress–Focus Correspondence Principle of Reinhart, the focus of a clause is any constituent containing the main stress of the intonational phrase, as determined by the stress rule. Focus and main stress can be made to coincide in two ways: by syntactic transposition, or by stress shift onto the constituent to be focused. According to Szendrői, Hungarian always adopts the first option. In Hungarian, the Nuclear Stress Rule marks the leftmost position of the VP as the position of main stress, hence focus movement is movement to the left edge of the VP. Examples like (10c), containing a postverbal focus, are claimed by Szendrői to represent VP-focus, and the difference between the interpretation of (10b) and (10c) is claimed to be a difference between narrow focus and wide focus. Exhaustivity plays no role in the analysis.

4.4 A focus theory denying the exhaustivity of structural focus

Whereas Bródy (1990, 1991, 1995) and Szendrői (2003) are merely silent about the Hungarian preverbal focus being obligatorily interpreted as [+exhaustive], Wedgwood explicitly denies that structural focus encodes a [+exhaustive] feature (see Wedgwood 2005). He is aware of the fact that the Hungarian preverbal focus is usually understood to be exhaustive; however, he claims that its exhaustivity is a mere implicature, which is elicited by a narrow focus also in languages with no focus movement, as in English. The exhaustivity implicature can be derived either from Grice's Maxim of Quantity (Grice 1975), or from the general principles of inference of Sperber and Wilson's Relevance Theory (1995). The first sub-maxim of Grice's Maxim of Quantity says "make your contribution as informative as is required (for the current purposes of the exchange)". When new information is presented in a restricted context (in answer to an explicit or implicit question), an exhaustive answer is the optimally relevant answer, therefore, exhaustivity is always expected.

Naturally, an answer is understood to be exhaustive only with respect to the given context—as pointed out by Szabolcsi as early as in (1980, 1981a, b) based on examples like (13):

(13) [$_{\text{FocP}}$ JOSEPH CONRAD [született lengyelnek]].
Joseph Conrad was.born Polish-DAT
'It was Joseph Conrad who was born Polish.'

Joseph Conrad does not exhaust the set of all people born Polish; however, he may exhaust the set of those born Polish when talking about great English writers.

Narrow focus, that is, that congruent with a *wh*-expression, is non-exhaustive only in two cases: (i) if it is mutually manifest to the interlocutors that the speaker does not expect an exhaustive answer, as in:

(14) Honnan tudhatom meg a menetrendet?
'Where can I learn the railway timetable?'
Meg-tudhatod az INTERNETRŐL.
PRT learn-can-you the internet-from
'You can learn it from the internet.'

(ii) the answer is not interpreted as exhaustive if the speaker makes it manifest that he is only giving partial information, as in (15):

(15) a. Kiket hívtál meg szombat estére?
'Who did you invite for Saturday night?'
b. Többek között PÉTERT ÉS ÉVÁT (hívtam meg).
others among Peter-ACC and Eve-ACC invited-I PRT
'Peter and Eve, among others.'

The Hungarian preverbal focus and the English cleft focus are exhaustive in the unmarked case because they usually represent narrow foci. The correlation between narrow focus and exhaustivity is derived from Sperber and Wilson's Principle of Relevance (1995):

(16) *Principle of Relevance*
Hearers seek to optimize the communicative rewards of processing an utterance, relative to the effort this processing demands.

A context elicits an utterance with wide focus when existing assumptions need to be strengthened; the hearer's commitment to their truth needs to be increased. A context eliciting a narrow focus requires a greater processing effort, involving the contradiction and elimination of existing assumptions. Such an effort is worthwhile if the identification of the alternative to be contrasted with the existing assumptions is exhaustive.

If exhaustivity is a mere pragmatic implicature also in the case of the Hungarian preverbal focus and the English cleft focus, then it should be cancellable under appropriate pragmatic conditions. Wedgwood claims that this is indeed the case.

4.5 Apparent arguments against the [+exhaustive] feature of structural focus

An obvious argument allegedly refuting the exhaustivity of the Hungarian preverbal focus is provided by examples like (15b), in which the expression *többek között* "among others" explicitly denies the exhaustivity of the preverbal *Pétert és Évát* "Peter and Eve". This counter-argument, however, is only apparent. The constituent that occupies the structural focus position in (15b) is *többek között Pétert és Évát* "Peter and Eve among others". This can be proved by the constituency test provided by *még...is* "even". *Még...is* is a complex modifier consisting of a proclitic and an enclitic, which must be cliticized to one and the same constituent. Whereas *még...is* cannot surround a string of two arguments, or an argument and an independent adjunct—see (17a, b)—*még többek között Pétert és Évát is* is grammatical; see (17c).

(17) a. **Még* Évának egy könyvet *is* viszek ajándékba.
even Eve-DAT a book-ACC also take-I present-for
'I take even a book to Eve as a present.'

b. **Még* holnap Évát *is* meg látogatom.
even tomorrow Eve-ACC also PRT visit-I
'I visit even Eve tomorrow.'

c. *Még* többek között Pétert és Évát *is* meg
even others among Peter-ACC and Eve-ACC also PRT
látogattam.[4]
visited-I
'I visited even Peter and Eve, among others.'

If the Spec,FocP of (15b) is occupied by *többek között Pétert és Évát* "Peter and Eve, among others", then the focus of (15b) is an exhaustive focus. It specifies the members of the set determined by the presupposed part of the given sentence by naming two of them, and referring to the rest of them as a group with the pronoun *többek* "others". If *többek* "others" were omitted, the set could only include Peter and Eve and no one else.[5]

Wedgwood's main argument against the encoded exhaustivity of structural focus is borrowed from Horn (1981). Horn's argument is based on the following minimal pair (translated into Hungarian by Wedgwood 2005: 137):

(18) a. ??Azt tudtam, hogy Mari meg-evett egy pizzát, de
that knew-I that Mary PRT-ate a pizza but
most vettem észre, hogy egy pizzát evett meg.
now took-I notice that a pizza ate-she PRT
'I knew that Mary ate a pizza, but I have just noticed that it was a pizza that she ate.'

b. Azt tudtam, hogy Mari meg-evett egy pizzát, de most
that knew-I that Mary PRT-ate a pizza but now
vettem észre, hogy csak egy pizzát evett meg.
took-I notice that only a pizza ate-she PRT
'I knew that Mary ate a pizza, but I have just noticed that it was *only* a pizza that she ate.'

Horn and Wedgwood assume that the [+exhaustive] feature is encoded by the particle *csak* "only" in (18b). The two clauses of (18b) can be contrasted

[4] The fact that *többek között* can also be separated from *Pétert és Évát* does not refute their initial constituency. Adjuncts of nominal heads can be freely extraposed in Hungarian. A focused constituent must be head-final, hence the adjunct of a focused nominal must, in fact, either be extraposed or be turned into a premodifier. Compare the following three examples.

(i) *[$_{\text{FocP}}$ A házat a sziklák alatt [vettem meg]].
the house the cliffs below bought-I PRT
'I bought the house below the cliffs.'

(ii) [$_{\text{FocP}}$ [$_{\text{DP}}$ A házat t_i][vettem meg]] [$_{\text{PP}}$ a sziklák alatt]$_i$.

(iii) [$_{\text{FocP}}$ A sziklák alatti házat [vettem meg]].

[5] As an anonymous reviewer pointed out, *többek között* is always unstressed. The reason for this is presumably the pronominal status of *többek* "several [ones]".

because of the presence of exhaustivity in the second clause. In the second clause of (18a), on the other hand, the structural focus itself carries no [+exhaustive] feature, which is why the two clauses cannot be coordinated by the contrastive *de* "but".

In my opinion, this argument is not convincing. The first problem is that (18a) is ungrammatical under the prosody in (19a), and is grammatical under the prosody in (19b). The degraded status, indicated by the double question mark, that native speakers assigned to (18a) presumably indicates that it takes some effort to find a context eliciting (19b).

(19) a. *Azt tudtam, hogy Mari meg-evett egy pizzát, de most vettem észre, hogy [$_{\text{FocP}}$ egy PIZZÁT [evett meg]].
'I knew that Mary ate a pizza, but I have just realized that it was a pizza she ate.'

b. Azt tudtam, hogy Mari meg-evett egy pizzát, de most vettem észre, hogy [$_{\text{FocP}}$ EGY pizzát [evett meg]].
'I knew that Mary ate a pizza, but I have just realized that it was one pizza she ate.'

As for (19a) (i.e. the prosodically more unmarked version of (18a)), if two clauses differ only to the extent that the object has a [+exhaustive] feature in the second one, then the relation between the two clauses is not "contrast"; rather, the second clause makes the first one more precise. The first clause of (19a) does not encode the exhaustivity of the object, but does not exclude it, either. The second clause, imposing a [+exhaustive] reading on the object, merely restricts its interpretative possibilities. Since the meanings of the two clauses are not identical, it should be possible to conjoin them without a tautology arising; however, the conjunction must be appositive instead of contrastive:

(20) Mari meg-evett egy pizzát, pontosabban [$_{\text{FocP}}$ egy PIZZÁT
Mary PRT-ate a pizza more.precisely a pizza-ACC
[evett meg]].
ate-she PRT
'Mary ate a pizza, more precisely, it was a pizza she ate.'

Sentence (18b) is acceptable with a contrastive conjunction because the second clause not only makes the meaning of the first one more precise by fixing the value of the exhaustivity feature of the object, but it also adds a negative evaluation to it by means of *only*. As argued in É. Kiss (1996, 1998a), *only* is an evaluative scalar modifier expressing that the scalar degree it modifies represents a low, or at least non-maximal, degree of the given scale. The synonymy of (21a, b) might seem to suggest that *only* is merely the spelling out of the exhaustivity feature of the focus. That this is not the case is proven by the ungrammaticality of (22b) as opposed to (22a).

(21) a. [$_{\text{FocP}}$ ÉVA [volt Mária lánya]].
Eve was Mary's daughter

b. [$_{\text{FocP}}$ Csak ÉVA [volt Mária lánya]].
only Eve was Mary's daughter

(22) a. [$_{\text{FocP}}$ MÁRIA [volt Éva anyja]].
Mary was Eve's mother

b. *[$_{\text{FocP}}$ Csak MÁRIA [volt Éva anyja]].
only Mary was Eve's mother

A focus modified by *csak* is presented as a scalar value contrasted with higher values of the given scale. A single daughter represents a low value on the scale of potential daughters (consisting of a single daughter, a set of two daughters, a set of three daughters, etc.), a mother, on the other hand, cannot be opposed to higher values on the scale of mothers.

The expression *only a pizza* can generate either a qualitative or a quantitative scale. Compare:

(23) a. Mari csak egy PIZZÁT evett.
'It was only a pizza that Mary ate.'

b. Mari csak EGY pizzát evett.
'It was only one pizza that Mary ate.'

In the case of (23a), the scale of foods can be a qualitative scale extending, say, from pizza to beafsteak. In this case *only* contrasts pizza with the more delicious, more sophisticated, more valuable kinds of food. But (23a) can also evoke a quantitative scale. The degrees of this type of scale are of the kind *a pizza, a pizza and a salad, a pizza, a salad, and an ice cream*, etc. In the case of (23b), the scalar degrees above *one pizza* are *two pizzas, three pizzas*, etc. The latter reading can also be elicited without *only*, given that the numeral *egy* "one" (the stressed equivalent of the indefinite article in Hungarian) represents a low (in fact, the lowest) degree of the scale of positive whole numbers also without the evaluative particle *only*. That is why (19b) is acceptable; its second clause can be contrasted with the first one because the second clause not only has the value of the exhaustivity feature of the object fixed, but it also assigns to the object a low scalar value.

In view of these considerations, Wedgwood's arguments do not prove that the [+exhaustive] feature of structural focus can be absent, hence they do not support the conclusion that the exhaustivity of structural focus is a cancellable pragmatic implicature.

4.6 A further argument for the [+exhaustive] feature of structural focus

In Hungarian, expressions containing a numerical modifier *n* can be interpreted as "at least *n*; *n* or more" in every sentence position but Spec,FocP (see É. Kiss 2006c). Observe how *két gyereket* "two children" is interpreted in VP-internal position (24a), in topic position (24b), and, supplied with the maximizer *is* "even", in distributive quantifier position (24c)—as opposed to Spec,FocP (25):

(24) a. Aki [$_{\text{PredP}}$ fel-nevelt *két* *gyereket*], az 15%
who up brought two children (s)he 15%
nyugdíjemelésre jogosult.
pension-raise-to entitled-is
'Who brought up (at least) two children is entitled to a 15% pension raise.'

b. Aki [$_{\text{TopP}}$ *két gyereket* [$_{\text{PredP}}$ fel-nevelt]], az 15% nyugdíjemelésre jogosult.
'Who brought up (at least) two children is entitled to a 15% pension raise.'

c. Aki [$_{\text{DistP}}$ *két gyereket is* [$_{\text{PredP}}$ fel-nevelt]], az 15% nyugdíjemelésre jogosult.[6]
'Who brought up (at least) two children is entitled to a 15% pension raise.'

(25) Aki [$_{\text{FocP}}$ KÉT GYEREKET [nevelt fel]], az 15% nyugdíjemelésre jogosult.
'Who brought up [exactly] two children is entitled to a 15% pension raise.'

What we attest in (24a–c) is not surprising. According to a well-known hypothesis, the basic meaning of a numerical modifier *n* in natural languages is "at least *n*", "*n* or more"—see for instance Horn (1972, 1981), Levinson (2001), and Kadmon (2001).[7] It is easy to see why. The truth of the statement

[6] *is* in (24c) is a distributive particle. A numeral associated with it cannot be focused; it must land in Spec,DistP, the landing site of distributive quantifiers, for example, universals.

[7] Not everyone shares this view, though. Horn (1992, 1996), Geurts (2006), and Breheny (2008) argue for theories in which the "exactly *n*" interpretation of numerals is primary, and the "at least *n*"

John brought up two children is preserved also in case the number of children brought up by John increases to three, four, or ten (as one cannot bring up ten children without also bringing up two children). What needs to be explained is why a numerical modifier is still mostly interpreted at its face value. The face-value interpretation of numerical modifiers is attributed to a so-called scalar implicature, which fixes the upper bound of "at least *n*" by adding the meaning component "at most *n*" to it. The presence or absence of this implicature is supposed to depend on pragmatic conditions, as follows from Grice's Maxim of Quantity, requiring a speaker to be as informative as is necessary for the current purposes of the exchange (Grice 1975). For example, if we are interested in a person's family status, and he tells us that he has raised two children, we will associate the expression *two children* with a scalar implicature, that is, we will interpret it as "exactly two children"—because a person asked about his family status is sufficiently informative only if he gives us the exact number of his children. If, on the other hand, we are talking about a benefit allotted to people who have brought up a certain number of children, it seems sufficient to fix the lower boundary of the number of children required in order to receive the benefit. That is, in this case no scalar implicature is induced. Accordingly, *két gyereket* "two children" in (24a–c) is understood to mean "at least two children".

In focus position, the interpretation "at least *n*" is impossible—no matter what the pragmatic conditions are. Thus no matter how pragmatically unlikely it may be, (25) can only mean that only the persons who have raised exactly two children (no less and no more) are entitled to the benefit.

The meaning difference between a non-focused and a focused numerically quantified expression becomes even more obvious under negation. Compare:

(26) a. János NEM nevelt fel két gyereket.[8]
John not brought up two children
'John did not bring up two children.'

b. János NEM [$_{\text{FocP}}$ két gyereket [nevelt fel]].
'It wasn't two children that John brought up.'

interpretation is derived. Breheny (2008) derives the "at least" reading from the "exactly" reading by pragmatic reasoning. Geurts (2006) distinguishes quantifier and predicate senses of a numeral, the former associated with an "exactly" interpretation, and the latter associated with an "at least" interpretation. The two readings are related by type-shifting rules.

[8] What triggers the inversion of the V and the particle (more precisely, the extraction of the V across the particle) in (24a) is not a focus but the negative particle.

Sentence (26a) is true only if John brought up less than two children, that is, if he brought up one child or no child at all. Example (26b), on the other hand, is true whether the number of children brought up by John is smaller than two or larger than two. (Example (26b) cannot be used if John brought up no child—because a structural focus is associated with an existential presupposition. The presupposed part of the sentence is understood to determine an existing, non-empty set.)

The fact that a focused numerical modifier *n* cannot be interpreted as "at least *n*; *n* or more" must be a consequence of the [+exhaustive] feature of focus. If focusing means the exhaustive identification of the alternative named by the focused expression from among the set of potential alternatives, then the focusing of an expression containing the numeral *n* will exclude the alternatives containing $n+1$, $n+2$, $n+3$ etc. If the [+exhaustive] feature of focus were a pragmatic implicature, it ought to be sensitive to pragmatic conditions. The fact that it is not affected by pragmatics indicates that it is a semantic feature, an inherent property of structural focus.

Actually, focused noun phrases involving no numerical modifier can also be looked upon as scalar elements in the case of which the upward extending interpretation is blocked, resulting in an "at least *x* and at most *x*" reading. Consider again the examples under (10), repeated here as (27).

(27) a. [$_{\text{FocP}}$ KIKET [hívtál meg ma estére]]?
who-PL-ACC invited-you PRT today night-for
'Who have you invited for tonight?'

b. [$_{\text{FocP}}$ PÉTERT és PÁLT [hívtam meg]].
Peter-ACC and Paul-ACC invited-I PRT
'It is Peter and Paul that I have invited.'

c. [$_{\text{PredP}}$ Meg-hívtam PÉTERT ÉS PÁLT].
'I have invited Peter and Paul.'

d. [$_{\text{TopP}}$ Pétert és Pált [$_{\text{PredP}}$MEG-hívtam]].
'Peter and Paul, I have invited.'

Let us generate all the possible subsets (the power set) of the set of those who I could have invited for tonight, and let us order them along a scale. The one-member subsets, e.g. {Peter} or {Paul}, will represent the lowest degree of the scale, the two-member subsets like {Peter, Paul} will represent the second degree of the scale, the three-member subsets such as {Peter, Paul, Stephen} will represent the third degree of the scale etc. A non-focused *Peter and Paul* means "at least Peter and Paul" because—in addition to forming a two-member subset—it can also represent a subset of larger, three- or four-member, subsets. If, on the other hand, *Peter and Paul* is focused, it can only denote the two-member subset {Peter, Paul}. That is, exhaustivity and the blocking of upward extension along a scale are two sides of the same coin.

The correlation between the "at least *n*" vs. "exactly *n*" interpretation of a numeral and the discourse function of the numerically modified expression has already been noticed—by Szabolcsi (1981a, b), Fretheim (1992), van Kuppevelt (1996), and Wedgwood (2005), among others. What Fretheim observed was that the interpretation of a numerical modifier *n* in Norwegian is related to the stress of the modified expression. If *n* is part of an unstressed, contextually given, salient expression, *n* means "at least *n*". Otherwise, *n* means "exactly *n*". In fact, the so-called scalar implicature, supplementing the meaning "at least *n*" with the upper bound "at most *n*" depending on pragmatic conditions, can only be elicited by salient, topical numerically modified expressions. In the case of foci, the upper bound is obligatory, which suggests that it is not imposed upon the meaning of *n* by a weak pragmatic implicature but is part of the meaning of the utterance.

Van Kuppevelt (1996) extended Fretheim's observation to scalar elements other than numerals. Similar to Fretheim, van Kuppevelt argues that the "exactly *n*" reading of focused scalar expressions is not a pragmatic implicature but a semantic entailment. Actually, van Kuppevelt uses the term *satisfactory comment* instead of *focus*. He establishes the information structure of a sentence by questions. A satisfactory comment is (the non-presupposed part of) a uniquely determining answer to an explicit or implicit question. An answer which leaves open the possibility of alternatives (to be eliminated by a further subquestion) does not count as a satisfactory comment.

Wedgwood (2005, sect. 5.2.4) also discusses English examples corresponding to those in (20) and (21), and identifies the "exactly *n*" reading of numerals as their exhaustive reading. He claims that the "exactly *n*" interpretation of numerals correlates with narrow focus function. In examples like those in (20),

> what is really being questioned is the existence of a set of (someone's) children of a certain cardinality... Any assertion of the existence of a set of a given cardinality is compatible with the existence of sets with higher cardinalities because the smaller set can be simply a proper subset of a larger one... In contrast, when the existence of the set is presupposed but its cardinality is in question... it is precisely the choice of one cardinality from among the set of alternative cardinal values that is at stake, so the assertion that a certain cardinality holds implicitly excludes other cardinalities, whether higher or lower. (Wedgwood 2005: 165)

My point is that what has been observed about stressed numeral expressions by Fretheim, about numeral expressions functioning as satisfactory comments by van Kuppevelt, and about numeral expressions functioning as narrow foci by Wedgwood is grammaticalized in Hungarian. The "exactly *n*" reading of numerals is associated with a particular structural position with an encoded [+exhaustive] feature, and is not sensitive to pragmatic or contextual conditions.

In van Kuppevelt's framework, it is contextually determined what counts as a satisfactory comment. Compare his illustrative minimal pair:

(28) [Harry did a lot of shopping this afternoon.]
How many books did he buy?
#He bought four$_{\text{Comment}}$ books. In fact he bought seven.

(29) [Did Harry get a free book in this shop? If he bought four books, he got one.]
How many books did he buy?
He bought four$_{\text{Comment}}$ books. In fact he bought seven.

In the context of (28), the numeral is a satisfactory answer only under its "exactly four" reading. In the context of (29), on the other hand, a process of "topic weakening" takes place, which renders the numeral a satisfactory answer also under its basic "at least four" interpretation.

In Hungarian, the situation is much clearer. In the two different contexts, both the questions and the answers have different structures, and the interpretation is structurally determined. In the context of (28), a *how many* question is asked, which is answered with a focused numerically modified expression:

(30)

[$_{\text{FocP}}$ HÁNY	KÖNYVET	[vett	Harry]]?
how.many	book-ACC	bought	Harry

'How many books did Harry buy?'

[$_{\text{FocP}}$ NÉGY	könyvet],	#valójában	hetet	is.
four	book-ACC	in.fact	seven-ACC	even

'Four books, in fact, seven.'

The focused numeral phrase is correctly predicted not to allow an upward extending interpretation.

In the context of (29), we have two options. We can ask a yes–no question, with the numerically modified expression in postverbal position. In that case, its meaning is "at least four" both in the question and in the answer (31). Alternatively, we can ask a *wh*-question, with the *how many* phrase and the corresponding numeral phrase in focus position. In that case, the numeral phrase must be interpreted at face value (32).

(31) [Did Harry get a free book in this shop? If he bought four books, he got one.]

VETT Harry négy könyvet?
bought Harry four book-ACC
'Did Harry buy four books?'

Igen, VETT négy könyvet. Valójában hetet is.
yes bought-he four book in.fact seven-ACC even
'Yes, he bought four books. In fact, he bought seven.'

(32) [Did Harry get a free book in this shop? If he bought four books, he got one.]
[$_{\text{FocP}}$ HÁNY KÖNYVET [vett]]?
'How many books did he buy?'

[$_{\text{FocP}}$ HÉT KÖNYVET].
'Seven books.'

#[$_{\text{FocP}}$ NÉGY KÖNYVET].
'Four books.'

That is, in Hungarian the preverbal structural focus position blocks the "at least *n*" reading irrespective of pragmatic conditions.

Wedgwood (2005) would correctly analyze the sentences in (31) to contain a wide focus, with the numeral having an "at least four" reading. What would be problematic for him is the discourses in (33) and (34), containing a non-preverbal narrow information focus.

(33) Kapott Harry ajándékkönyvet? Ha vett négy könyvet, kapott egyet ajándékba.
'Did Harry get a free book? If he bought four books, he got one for free.'
[$_{\text{FocP}}$ HÁNY KÖNYVET [vett]]?
'How many books did he buy?'

Vett négy könyvet. Valójában hetet is.
bought-he four book-ACC in.fact seven-ACC even
'He bought four books. In fact, he bought seven.'

(34) Kapott Harry ajándékkönyvet? Ha vett négy könyvet, kapott egyet ajándékba.
'Did Harry get a free book? If he bought four books, he got one for free.'

Négy könyvet (szerencsére) VETT. Valójában hetet is.
four book-ACC (luckily) bought-he in.fact seven-ACC even
'Four books, luckily, he bought. In fact, he bought seven.'

(*Négy könyvet* in (34) is to be pronounced with a contrastive intonation, i.e. with a rise or fall–rise developing into a hat contour.) If the information focus of a sentence is the constituent congruent with the *wh*-phrase of the question eliciting the sentence, then (33) and (34) contain a narrow focus. This narrow focus is stressed, but its stress is weaker than the main stress falling on the left edge of the logical predicate (in these cases, the V). As shown by these examples, a narrow information focus is not identical with a structural focus! *Négy könyvet* is a narrow information focus left in situ in the VP in (33), and is a contrastive topic in (34). The verbal projection cannot be deleted in these examples because the non-focus position of the numerically modified expression cannot be indicated otherwise. Actually, for a contrastive topic to be identifiable, it is sufficient to spell out an adverb adjoined to the verbal projection:

(35) [$_{\text{PredP}}$ HÁNY KÖNYVET [vett Harry]]?
'How many books did Harry buy?'
[$_{\text{TopP}}$Négyet [$_{\text{PredP}}$ BIZTOSAN/MINDENKÉPPEN [$_{\text{FocP}}$]]].
four-ACC for.sure /by.all.means
'Four, for sure/by all means.'

That is, the obligatory blocking of the "at least" interpretation of numerals is structure-dependent in Hungarian; it is a concomittant of the preverbal focus position; it cannot be accounted for in pragmatic terms.

The claim that exhaustivity is a grammaticalized property of the Hungarian preverbal focus is also supported by examples of different types.

Indefinite numerals like *pár* "couple of", *néhány* "some" have different meanings in focus position and in non-focus position. In focus position, the upward extension of their interpretation is blocked. Compare the following examples:

(36) a. Hogy sikerült az adománygyűjtés?
how succeeded the fund-raising
'How was the fund-raising?'

b. [$_{\text{TopP}}$ *Pár* *forint* [$_{\text{PredP}}$ össze-gyűlt]] így meg tudtuk
couple forint PRT gathered so PRT could-we
venni az ajándékot.
buy the present-ACC
'A few forints were collected, so we could buy the present.'

c. %[$_{\text{FocP}}$ PÁR FORINT [gyűlt össze]] így meg tudtuk venni az ajándékot.
'It was few forints that were collected, so we could buy the present.'

(37) a. Hogy sikerült az adománygyűjtés?
'How was the fund-raising?'

b. %[$_{\text{TopP}}$ *Pár forint* [$_{\text{PredP}}$ össze-gyűlt]] így nem tudtuk megvenni az ajándékot.
'A few forints were collected, so we couldn't buy the present.'

c. [$_{\text{FocP}}$ PÁR FORINT [gyűlt össze]] így nem tudtuk megvenni az ajándékot.
'It was few forints that were collected, so we couldn't buy the present.'

In (36b), the topicalized *pár forint* "a couple of forints" has an upward extending interpretation, meaning "a couple of forints or more", "a couple of forints in the least", which licences a positive result clause. As shown in (37b), the same sentence is incompatible with a negated result clause. In (36c), on the other hand, the focused *pár forint* "a couple of forints" means "a couple of forints" and no more, which is incompatible with a positive result clause, but can licence a negative result clause, as shown in (37c).

Sok "many" is known to be ambiguous between a partitive, or proportional, and an absolute, or "counting" reading (a term of Szabolcsi 1997). The upward extending interpretation is only allowed by the partitive, or proportional *sok* "many". Under this reading, the *sok* phrase denotes a relatively large subset of a set, the size of which can be extended until it becomes identical with the maximal set. Thus the following example is not contradictory:

(38) [$_{\text{TopP}}$ *Sok* *kollégát* [$_{\text{PredP}}$ meg-hívtam]]; tulajdonképpen
many colleague-ACC PRT invited-I actually
mindet.
all-ACC
'I invited many colleagues; actually all of them.'

In focus position, on the other hand, the *sok* phrase is not partitive or proportional; it denotes an indefinite absolute number; it functions as a "counting" quantifier:

(39) [$_{\text{FocP}}$ SOK KOLLÉGÁT [hívtam meg]].
many colleague-ACC invited-I PRT
'It was many colleagues that I invited.'

Sentence (39) means "the number of colleagues I invited is many". *Sok* in (39) evokes no superset, no scale. Szabolcsi (pers. comm.) demonstrates the meaning difference between the partitive *many* and the "counting" *many* with a minimal pair of the following type:

(40) a. [$_{\text{TopP}}$ *Sok kolléga* [meg-jelent]] [$_{\text{TopP}}$ *sok kolléga*
many colleague PRT appeared many colleague
[nem jelent meg]].
not appeared PRT
'Many colleagues appeared, many colleagues did not appear.'

b. %[$_{\text{FocP}}$ SOK KOLLÉGA [jelent meg]] [$_{\text{FocP}}$ SOK KOLLÉGA [nem jelent meg]].
'It was many colleagues that appeared, it was many colleagues that did not appear.'

The second clause of (40a) makes a statement about the complementary subset of the set denoted by the non-focused, partitive *many* phrase. In (40b), on the other hand, the focused, "counting" *many* evokes no superset and no complementary subset, hence the second clause is uninterpretable. The absolute reading of *sok* "many", enforced in focus position, corresponds to the "exactly *n*" reading of focused numerals.

Interestingly, *kevés* "few" phrases are obligatorily focused in Hungarian:

(41) a. *Meg-hívtam *kevés kollégát.*
PRT invited-I few colleague-ACC
'I invited few colleagues.'

b. *[$_{\text{TopP}}$ *Kevés kollégát* [$_{\text{PredP}}$ meg-hívtam]].
'Few colleagues, I invited.'

c. [$_{\text{FocP}}$ KEVÉS KOLLÉGÁT [hívtam meg]].
'It was few colleagues that I invited.'

In the presence of a preverbal focus, a *kevés* phrase can also appear behind the V; however, as shown in É. Kiss (1998b), it occupies a focus position (the specifier of a lower FocP projection) postverbally, too. (Whereas a postverbal numeral can, but need not, be analyzed as a lower, Second Occurrence Focus, for *kevés* phrases, which can only occur in Spec,FocP, this is the only option. Notice that, whereas a numeral can stand postverbally also in lack of a preverbal focus, a postverbal *kevés* phrase is only licensed by a preverbal, First Occurrence Focus.) In multiple focus constructions, such as (42), the V is attracted across the lower Foc to the higher Foc head, that is why the filler of the lower Spec,FocP surfaces postverbally.

(42) [$_{FocP}$ A ROSSZ IDŐ miatt [$_{NNP}$ hívtam [$_{FocP}$ KEVÉS KOLLÉGÁT [$_{NNP}$ t$_V$ [$_{PredP}$ meg t$_V$]]]]].[9,10]

[$_{FocP}$ A	ROSSZ	IDŐ	miatt [$_{NNP}$	hívtam [$_{FocP}$	KEVÉS
the	bad	weather	because	invited-I	few

KOLLÉGÁT [$_{NNP}$ t$_V$ [$_{PredP}$	meg t$_V$]]]]].
colleague-ACC	PRT

'It was because of the bad weather that I invited FEW COLLEAGUES.'

As argued in É. Kiss (2006c), a *kevés* phrase must be moved into Spec,FocP because in other sentence positions it would be associated with an upward extending interpretation, which is inadmissible in the case of a scalar value in the negative domain of a bidirectional scale. Horn (1989: 235), in fact, claims that

> positive and negative quantifiers, modals, and related operators must be represented on distinct, though related, scales. There can be no single scale on which operators like *some* and *not all*, or *possible* and *unlikely*, can be plotted. Rather, there is one scale defined by the positive operators and one by their negative counterparts.

In Horn's theory of quantitative scales, *few* and *many*, *beautiful* and *ugly* cannot belong to the same scale because, if P_j outranks P_i on a Horn scale, then, by definition, a statement containing an instance of the former unilaterally entails the corresponding statement containing the latter. In other words, assuming Horn's (1972, 1989) notion of scale, P_i and P_j can be regarded as values of the same scale iff the interpretation of P_i can potentially be extended up to P_j. In Horn's theory, *kevés kolléga* "few colleagues" and *sok kolléga* "many colleagues" cannot belong to the same scale because the extension of the interpretation of *kevés kolléga* to the scalar value represented by *sok kolléga* is impossible, given

[9] In a more unmarked version of (38), the verbal particle immediately follows the overt V. É. Kiss (2008a) argues at length that V movement frees up postverbal word order, and postverbal constituents are rearranged according to the law of growing constituents of Behaghel (1932).

[10] The interpretation of a postverbal *pár* "couple of" phrase also depends on whether or not the preverbal focus slot is filled. The postverbal *pár* phrase in the focus construction in (i) can be understood either as a (second) focus on a par with (37c), or as a non-focus, on a par with (36b).

(i) [$_{FocP}$ CSAK ESTÉRE gyűlt össze *pár forint*] így nem tudtuk meg-venni az ajándékot/meg tudtuk venni az ajándékot.

[$_{FocP}$ CSAK	ESTÉRE	gyűlt	össze	*pár*	*forint*]	így	nem	tudtuk
	only	evening-by	gathered	together	couple	forint	so	not	could-we

meg-venni	az	ajándékot/meg	tudtuk	venni	az	ajándékot.
PRT buy	the	present /PRT	could-we	buy	the	present

'It was only by the evening that a couple of forints were collected, so we couldn't buy/we could buy the present.'

In the absence of a preverbal focus, the postverbal *pár*-phrase cannot be analyzed as the specifier of a second, lower, focus projection, and, accordingly, it is interpreted as a non-focus:

(ii) Össze-gyűlt estére *pár forint*, így meg tudtuk venni az ajándékot/*nem tudtuk meg-venni az ajándékot.
'A couple of forints were collected by the evening, so we could buy/*we couldn't buy the present.'

that the statement *I invited many colleagues* does not entail the statement *I invited few colleagues.*

Native speakers of Hungarian handle this problem differently. For them, *kevés kolléga* "few colleagues" and *sok kolléga* "many colleagues" clearly represent opposing values of the same scale, with *kevés kolléga* in the negative domain, and *sok kolléga* in the positive domain of the scale. What is ruled out is the upward extending interpretation of *kevés kolléga*—by its obligatory focusing. Upward extending interpretation is generally blocked in the case of scalar expressions in the negative domain of bidirectional scales, and the means of this is obligatory focusing.

As discussed in detail in É. Kiss (2006c), this generalization extends to negative adverbs of degree, manner, and frequency, as well. These types of adverb are obligatorily moved to focus position, unlike their positive counterparts:

(43) a. János [$_{\text{PredP}}$ *gyakran* [$_{\text{PredP}}$ meg-látogat]].
John often PRT visits-me
'John often visits me.'

b. *János [$_{\text{PredP}}$ *ritkán* [$_{\text{PredP}}$ meg-látogat]].
John seldom PRT visits-me

c. János [$_{\text{FocP}}$ RITKÁN [látogat meg]].
'It is seldom that John visits me.'

(44) a. János [$_{\text{PredP}}$ *nagyon* [$_{\text{PredP}}$ el -fáradt]].
John very-much PRT got.tired
'John got tired very much.'

b. *János [$_{\text{PredP}}$ *alig* [$_{\text{PredP}}$ el -fáradt]].
John hardly PRT got.tired

c. János [$_{\text{FocP}}$ ALIG [fáradt el]].
'HARDLY did John get tired.'

(45) a. János [$_{\text{PredP}}$ *ügyesen* [$_{\text{PredP}}$ meg-csinálta a feladatot]].
John skilfully PRT did the job
'John did the job skilfully.'

b. *János [$_{\text{PredP}}$ *ügyetlenül* [$_{\text{PredP}}$ meg-csinálta a feladatot]].
John unskilfully PRT did the job

c. János [$_{\text{FocP}}$ ÜGYETLENÜL [csinálta meg a feladatot]].
'It was unskilfully that John did the job.'

What forces the focusing of negative adverbs is the need to block the upward extension of their interpretation. Since they represent scalar values in the negative domain of bidirectional scales, an upward extending interpretation could lead to a semantic anomaly. The possibility of upward extension being structure-dependent in Hungarian, it can only be blocked by movement into the structural focus position.

4.7 Conclusion

The chapter has argued for the claim that the structural focus of the Hungarian sentence, occupying Spec,FocP, and structural focus (e.g. the English cleft), in general, encodes the feature [+exhaustive]. After reviewing the standard arguments for the exhaustivity of structural focus, a new argument against it, proposed by Wedgwood (2005), has been examined, and has been shown to be based on a false premise. Finally, a new type of evidence has been presented. It has been argued (following van Kuppevelt 1996 and Wedgwood 2005) that the "exactly *n*" reading of a focused numeral (as opposed to its basic, "at least *n*" meaning) is a manifestation of its exhaustivity. It has been shown that the "exactly *n*" meaning of focused numerals is grammaticalized in Hungarian; it is associated with the preverbal focus position, causing obligatory focus movement in the case of certain types of scalar elements, and predictable, structurally determined meaning differences in the case of others.

5

The interpretation of topical indefinites as direct and indirect aboutness topics

CORNELIA EBERT AND STEFAN HINTERWIMMER

5.1 Introduction

In this chapter we deal with the interpretation of sentences that contain indefinite DPs marked as topics. It has often been observed that topical indefinites can be interpreted in either of the following ways: (1) they receive widest scope (in episodic sentences) (cf. Cresti 1995 and Jäger 1996); (2) they are interpreted generically (in sentences with generic tense) (cf. Kuno 1972 and Kuroda 1972); and (3) they induce so-called Quantificational Variability Effects (in the presence of adverbial quantifiers), that is to say that the respective sentence is interpreted as if the quantificational force of the indefinite depended on the quantificational force of the Q-adverb (cf. Partee 1991 and Chierchia 1995). However, so far, these three observations have not been related to each other systematically. Our contribution in this chapter is to show that the three readings that are in principle available to topical indefinites can be based on one underlying principle. In other words, topicality has a truth-conditional effect, which can be seen with topical indefinites. This effect has different shapes dependent on where the topical indefinite appears. However, all these shapes are just instances of the same underlying principle that determines the interpretation of topical items. The chapter is structured as follows: in Section 5.2, we give some background on determiners and adverbial quantifiers. In Section 5.3, we lay out the basic facts we want to account for and in Section 5.4, our analysis is presented in detail. Section 5.5 concludes the chapter.

5.2 Background: Determiner quantifiers and adverbial quantifiers

It is well known that languages such as English and German have two different types of quantificational elements: on the one hand, there are items like *every*, *no*, *most*, and *a*, which syntactically behave like determiners insofar as they combine with NPs, forming DPs that occupy argument positions at the surface. They are therefore called quantificational determiners or determiner quantifiers (D-quantifiers). On the other hand, there are items like *always*, *never*, *usually*, and *sometimes*, which syntactically behave like adverbs insofar as they appear in VP- (or vP-) as well as in TP-adjoined position. We will refer to them as adverbial quantifiers (A-quantifiers).

Concerning the semantics of D-quantifiers, the by-now standard view is that they take two expressions which denote sets of individuals as arguments and map them onto a proposition that is true if the respective sets stand in a certain lexically specified relation to each other (see Barwise and Cooper 1981). The important point for our current purposes is that the arguments of D-quantifiers are strictly determined by (LF-)syntax. While prosodic as well as contextual information might have an influence on the truth conditions of sentences containing two or more D-quantifiers (see below), there is no way for this kind of information to alter the order in which a D-quantifier is combined with its two arguments: the NP-complement of a quantificational determiner is always interpreted as its first argument (the restrictor), while the LF-sister of the entire DP is always interpreted as its second argument (the nuclear scope).[1]

Turning to A-quantifiers, matters are different. Here, syntax does not entirely determine which part of the clause is to be interpreted as the restrictor and which part is to be interpreted as the nuclear scope. Furthermore, the domain of quantification is different in this case. It is by now standard to assume that Q-adverbs quantify over situations or events. Controversies arise solely concerning the question of whether Q-adverbs can additionally quantify over individuals, at least in some cases (see below) (cf. de Swart 1993; von Fintel 1994; and Herburger 2000 for discussion).

[1] With the possible exception of weak quantifiers like *many*, for which precisely this claim has been made (cf. Herburger 2000), in order to explain the fact that sentences like (i) can be interpreted as paraphrased in (ii):

(i) Many ScandiNAvians won the Nobel Prize in literature.

(ii) There is a large number Nobel Prize winners in literature who are Scandinavians.

But see Cohen (2001) for the claim that this effect can also be explained as a consequence of the inherent vagueness of *many*.

Consider example (1a): depending on where the main accent (indicated by capital letters) falls, the sentence gets a different interpretation, as shown in (1b–d).

(1) a. John always goes to the beach with Mary.
    b. John always goes to the beach with MAry =
       'All situations where John goes to the beach with someone are situations where he goes to the beach with Mary.'
    c. John always goes to the BEACH with Mary =
       'All situations where John goes somewhere with Mary are situations where he goes to the beach with Mary.'
    d. JOHN always goes to the beach with Mary =
       'All situations where Mary is accompanied to the beach by someone are situations where she is accompanied by John.'

It is therefore reasonable to assume that intonation plays a decisive role in the interpretation of sentences with A-quantifiers. Rooth (1985, 1992, 1995) has developed a formally precise theory that accounts for these intuitions: he assumes that the entire sentence (minus the quantifier) is interpreted as the nuclear scope of an A-quantifier, while the restrictor is determined on the basis of an algorithm which is sensitive to focus-marking—where focus-marking is indicated by intonational prominence. Simplifying somewhat, Rooth assumes that the semantic effect of focus-marking is the introduction of a (contextually restricted) set of alternatives to the (denotation of the) respective constituent. These alternatives are then composed with the rest of the sentence in pointwise fashion, resulting in a set of situation predicates which only differ from each other with respect to the chosen alternative. This set is called the "focus semantic value" (in addition to the "ordinary semantic value") of the sentence. Applying set union to the focus semantic value of a sentence returns a situation predicate that can then function as the restrictor of the respective Q-adverb. The formula in (2) shows the result of applying this mechanism to (1b):

(2) $\{s: \exists x\, [\text{person}(x) \wedge C(x) \wedge \text{go-to-beach-with}(x)(\text{John})(s)]\} \subseteq$
    $\{s: \text{go-to-beach-with}(\text{Mary})(\text{John})(s)\}$,
    where C is a free variable over predicates that is resolved on the basis of contextual information.

As already mentioned above, there is a controversy concerning the quantificational domain of A-quantifiers: in light of the fact that sentences like (3a) can be interpreted as paraphrased in (3b)—which exemplifies the Quantificational Variability Effects (QVEs) mentioned in the introduction—it has been suggested (cf. Lewis 1975; Kamp 1981; Heim 1982; and Kratzer 1995) that A-quantifiers do not exclusively quantify over situations, but are also able to bind individual variables, where those individual variables are taken to be introduced by indefinite DPs.

(3) a. A dolphin is usually SMART.
 b. Most dolphins are SMART.

It has been shown by Berman (1987), von Fintel (1994), and Herburger (2000), however, that QVEs can also be explained in a way that neither necessitates altering the semantics of A-quantifiers nor the semantics of the indefinite determiner. In addition to that, Endriss and Hinterwimmer (2008) and Hinterwimmer (2008) have shown that there are strong empirical arguments for taking Q-adverbs to be exclusive binders of situation (or event) variables. Berman (1987) and von Fintel (1994) assume that A-quantifiers quantify over *minimal* situations, that is, situations that contain only what is necessary to satisfy the respective predicate. Consequently, determining the restrictor of a sentence like (3a) on the basis of the focus semantic value, by default results in the interpretation given schematically in (4), which is equivalent to the paraphrase in (3b): since the situations quantified over are solely individuated on the basis of their containing dolphins (and nothing else), the situations necessarily vary with the dolphins.[2]

(4) Most s $[\min(s, \lambda s'.\exists x[\text{dolphin}(x)(s') \wedge \exists P[P(x)(s')]])]$
$[\exists s''[s \leq s'' \wedge \min(s'', \lambda s'''.\exists x[\text{dolphin}(x)(s''') \wedge \text{smart}(x, s''')])]]$,
where $\min(s, P)$ iff $P(s) \wedge \neg\exists s'[s' < s \wedge P(s')]$.
'Most minimal situations where a dolphin has some property (i.e. minimal situations containing a dolphin) can be extended to a minimal situation where a dolphin is smart.'

Let us now turn to the main topic of this paper: the interpretation of indefinite DPs which are grammatically marked as topics.

5.3 Indefinites as aboutness topics: The facts

5.3.1 *Grammatical Topic Marking in German: Left Dislocation*

Consider the examples in (5):

(5) a. Maria, die ist eine sehr talentierte Sängerin.
 Maria, RP-FEM.NOM.SG is a very talented singer
 'Maria is a very talented singer.'

[2] In formula (4) (and subsequent formulas), $s \leq s'$ means that situation s is extendable to a situation s'.

b. Peter, den hab ich lange nicht mehr gesehen.
Peter, RP-MASC.ACC.SG have I long not more seen
'I haven't seen Peter for a long time.'

c. Peter, ich hab ihn lange nicht mehr gesehen.
Peter I have him long no more seen
'As for Peter, I haven't seen him for a long time.'

The sentences in (5a, b) both exemplify so-called left-dislocation (LD) where an XP in fronted position is associated with a resumptive pronoun in the specifier position of CP. Sentence (5c), in contrast, is an example of Hanging Topic Left-Dislocation (HTLD), where an XP in fronted position is optionally (i.e. it does not have to be resumed at all) associated with an ordinary personal pronoun that can either remain in the middle field (as in (5c)) or occupy [Spec,CP]. Furthermore, in HTLD there is an intonational break between the fronted XP and the rest of the sentence, which is not the case in LD. We will not discuss HTLD in this chapter (see Frey 2004 for discussion and references).

We follow Frey (2004) in the assumption that only LD serves as an aboutness topic marking construction,[3] that is to say that we assume that German left-dislocated phrases which are not understood contrastively are necessarily interpreted as topics. We thus use left-dislocation as a topic test, comparable to Japanese *wa*-marking (see Portner and Yabushita 1998, Tomioka, this volume). Let us assume that left-dislocated phrases occupy the specifier of a functional projection above CP, which is exclusively reserved for topical phrases (cf. Rizzi 1997). Note that for reasons discussed in Section 5.3.3, we assume left-dislocated phrases to be base generated in this position. Accordingly, we do not take Spec,CP to be a topic position; rather, we follow Frey (2001) in assuming that (non-*wh*-)C^0 hosts a purely formal feature that can in principle be checked by any XP, be it topical or not (see below).

5.3.2 *The aboutness concept of topicality*

Intuitively, both sentences in (5a, b) are felt mainly to convey information about Maria and Peter, respectively: they are both fine as answers to questions like *What about Maria/Peter?* or commands like *Tell me something about*

[3] An important difference between LD and HTLD is that only the former allows an anaphor or pronoun contained within the fronted XP to be bound by a clause-internal DP, as shown by the contrast between (i) and (ii):

(i) Den Artikel über sich in der *Zeit*, den hat Hans gelesen.
the article about himself in the *Zeit* RP.-MASC.ACC.SG has Hans read
'Hans has read the article about himself in the *Zeit*.'

(ii) ??Den Artikel über sich in der *Zeit*, Hans hat ihn gelesen.
the article about himself in the *Zeit* Hans has it read

Maria/Peter, while they are odd as answers to questions such as *Who is a very talented singer?* and *Who haven't you seen for a long time?*, at least if they are read without an intonational break after the fronted DP and a strong accent on the RP (which might turn the sentences into instances of HTLD).[4] Note furthermore that the left-dislocated DPs in (5a, b) are both necessarily at least weakly familiar: being proper names, they can be used felicitously only if both speaker and hearer know what individuals they refer to.

Because of the prevalence of examples with proper names, definite descriptions and pronouns in the literature on topics, many linguists subscribe to the view that (weak) familiarity is a necessary property of topics (cf. Hockett 1958; Kuno 1972; Gundel 1988; Portner and Yabushita 1998). We will, however, follow Reinhart (1981; see also Molnar 1993 and Frey 2000, 2004) in assuming that familiarity is not a defining property of topics. This claim is based on the observation that not only individual denoting DPs can be sentence topics, but also singular and unmodified numeral indefinite DPs (while other quantificational DPs are excluded from topic positions; more on this below). This is shown by the examples in (6):

(6) a. EInen Song von Bob Dylan, den kennt
a/one[5] song by Bob Dylan RP-MASC.ACC.SG knows
JEder (nämlich *Blowing in the Wind*).
everyone namely *Blowing in the Wind*
'There is one/a certain song by Bob Dylan that everyone knows (namely *Blowing in the Wind*)'.

b. Eine neue Platte von Bob DYlan, die kommt
a new record by Bob Dylan RP-FEM.NOM.SG comes
meistens in die CHARTS.
usually in the charts
'A new record by Bob Dylan usually gets into the charts'.

Note furthermore that also propositions—that is, sets of situations—make good topics, as shown by (7):

(7) Dass die Polizei in Halberstadt nichts gemacht hat,
that the police in Halberstadt nothing done has
das kann ich kaum glauben.
RP-NEUT.SG can I hardly believe
'I can hardly believe that the police didn't do anything in Halberstadt'.

[4] But even in these cases, the respective individuals need to have already been established as discourse topics in the preceding context.

[5] German *ein* is ambiguous between the meaning of the indefinite article and the meaning of the cardinality predicate *one*.

In Section 5.4.2 we discuss other examples showing that sets of situations can function as topics and offer an account of how this can be brought in line with our basic understanding of topics. For the moment, (7) only serves to show that not only individuals, but also sets (of situations) can be topics.

As to indefinites, we will turn to the interpretation of topic-marked indefinites in Section 5.3.3. For the moment, suffice it to note that the acceptability of indefinites in topic position shows that any definition of topicality that is based on the notion of familiarity is doomed to fail, since it is well known that indefinite DPs have to be novel: they are not allowed to take up already existing discourse referents (cf. Heim 1982).[6] We therefore understand the term topic in the aboutness sense of Reinhart (1981), whose basic understanding of topichood is based on Strawson (1964). According to her, the topic of a sentence is simply understood as the centre of interest, the item the sentence is about. The topic, then, is the subject of the predication that the sentence expresses. The predication corresponds to the comment part of the assertion. Reinhart assumes that the topic of a sentence is the address where the rest of the information conveyed by the respective assertion is stored during the context update. Topicality thereby has an information structuring function in the literal sense of the term.

In cases where the respective DP is of type *e*, marking it as the topic of the assertion has only this pragmatic effect, but no truth conditional one: an address corresponding to the individual denoted by the respective DP is created, and the information conveyed by the rest of the sentence is stored there. Following Jacobs (1984) and Endriss (2009) (cf. also Searle 1969), we assume that sentences containing topics involve two speech acts: the first speech act establishes the (denotation of the) respective constituent as the topic, while the second speech act asserts that the (denotation of the) rest of the sentence, the comment, is true of the topic (see also Tomioka, this volume, for the view that topics always take scope outside of the assertion).

We propose to formalize this pre-theoretic aboutness understanding of topicality in such a way as to interpret topic–comment structures as generalized subject–predicate structures, where the topical DP (irrespective of case-marking, agreement relations, and thematic role) is the logical subject, and the comment is the predicate applying to this subject. A preliminary definition of the binary topic operator is given schematically in (8). We will refer to this interpretation scheme as the "topic principle". The topic operator is assumed to be located in the head of the above-mentioned functional projection.

[6] Note that this holds for specific indefinites, too, which may (but need to be) known to the speaker, but crucially not to the hearer.

It takes the denotation of the CP c-commanded by it as its first argument, and the denotation of the left-dislocated constituent in its specifier as its second argument. Note that existential quantification over α in combination with identifying α with the (denotation of) the topic-marked constituent X corresponds to the act of establishing X as the topic (i.e. address creation), while asserting that the comment P holds of α corresponds to the act of storing information under α. The topic operator defined in (8) thus conjoins two separate speech acts (cf. Krifka 2001 for the assumption that speech acts can be conjoined):

(8) $\lambda P \lambda X.\ \exists \alpha[\alpha = X\ \&\ \text{ASSERT}\ [P(\alpha)]]$

In Section 5.4, we will see that it is the requirement to create a generalized subject–predicate structure which is responsible for the truth-conditional effects of topic-marking with indefinites that we mentioned in the introduction, namely, the wide-scope interpretation and the generic as well as the QV-interpretation of topical indefinites. These effects will be looked at in detail in Sections 5.3.3 and 5.3.4. In a case like (9a), however, it is easy to see that the final interpretation given in (13b) is truth-conditionally equivalent to the case where *Peter* has not been topicalized:

(9) a. Den Peter, den mag ich.
the-ACC Peter RP-MASC.ACC.SG like I
'Peter, I like.'

b. $\lambda P \lambda X.\ \exists \alpha[\alpha = X\ \&\ \text{ASSERT}\ [P(\alpha)]]\ (\lambda x \lambda s.\text{like}'(x)(I)(s))\ (\text{Peter}) =$
$\exists \alpha[\alpha = \text{Peter}\ \&\ \text{ASSERT}\ [\lambda s.\text{like}'(\alpha)(I)(s)]]$.

Turning to the details of how the predicate functioning as the first argument of the topic operator in (9b) is generated, we simply assume that the resumptive pronoun in Spec,CP triggers lambda-abstraction (Heim and Kratzer 1998): the pronoun moved to Spec,CP leaves behind a trace (or copy) bearing the same index, as shown in (10). This trace is interpreted as a variable. At the same time, the presence of the resumptive pronoun in SpecCP triggers the insertion of a lambda-operator bearing the index of this pronoun. This lambda-operator binds any co-indexed variable in its c-command domain, thus turning the CP in (10) into the predicate shown in (9b).

(10) $[_{CP}\ \text{den}_i\ [_{C'}\ \text{mag}_j\ [_{TP}\ \text{ich}\ [_{T'}\ [_{vP}\ t_i\ t_j]\ t_j]]]]$

The claim that resumptive pronouns in left-dislocation structures behave like relative pronouns in triggering lambda-abstraction is supported by the fact that these pronouns are morphologically identical to the standard relative pronouns employed in German, as shown in (11):[7]

(11) a. Peter mag alle Filme, [$_{CP}$ die RP-PL Martin Scorsese gemacht hat].
Peter likes all movies RP-PL Martin Scorsese made has
'Peter likes all movies made by Martin Scorsese'.

b. Die besten Filme der letzten zwanzig Jahre, [$_{CP}$ die hat Martin Scorsese gemacht].
the best movies of-the last twenty years RP-PL has Martin Scorsese made
'The best movies of the last twenty years have been made by Martin Scorsese.'

Note that the CPs in (11a) and (11b) only differ with respect to the position of the finite verb: in (11b), it occupies C°, as required in assertive matrix clauses in German (den Besten 1977), while in (11a) it occupies the position at the right edge of the embedded clause (which we take to be T°, though nothing hinges on that), which is the typical situation in embedded clauses.

We take this near formal identity to be no coincidence, but rather as an indication that the two clauses are interpreted in full parallel, namely as the predicate *λx. has-made(x)(Martin Scorsese)*. In case of (11b), where the matrix verb directly follows the subject, the predicate is applied to its sister in order to create an assertion (cf. Lohnstein 2000 and the references cited therein). In case of (11a), in which the finite verb of the embedded clause is in final position, the predicate under discussion is combined with the predicate denotation of the modified head by way of predicate modification (cf. Heim and Kratzer 1998). Let us therefore state the following principle:

[7] Note, however, that this analogy is not perfect. There are cases where the resumptive pronoun in left-dislocation constructions and the relative pronoun of relative clauses are realized differently (cf. Gärtner 2001 for the observation that German employs different pronouns in ordinary German relative clauses as opposed to so-called integrated (relative-like) verb-second clauses).

(i) Peter zieht in eine Stadt, wo/*da es kein Kino gibt.
Peter moves to a city RelP-DAT.SG it no cinema gives.
'Peter will move to a city where there are no cinemas.'

(ii) In EINER Stadt in Deutschland, da/*wo gibt es kein Kino.
in a city in Germany, RelP-DAT.SG gives it no cinema
'There is one city in Germany where there are no cinemas.'

(12) A D-pronoun occupying Spec,CP at LF triggers lambda-abstraction.[8] Where *D-pronoun* is a neutral cover term encompassing all occurrences of pronouns of the type under discussion, that is, ones starting with a *d* that can also be used as relative pronouns.[9]

Having established a basic understanding of the concept "(aboutness) topicality" and an application of this concept to sentences containing referring expressions as topics, let us now turn to sentences with topical indefinites.

5.3.3 *The wide-scope reading of topical indefinites*

Consider again (6a), repeated here as (13a): note that the determiner *einen* (*a/one*) needs to be heavily accented to make the sentence sound appropriate.

(13) a. EInen Song von Bob Dylan, den kennt
a/one song by Bob Dylan RP-MASC.ACC.SG knows
JEder.
everyone
'There is one song by Bob Dylan that everybody knows.'

b. Nämlich *Blowing in the Wind.*
namely *Blowing in the Wind*

c. #Maria kennt *Visions of Joanna*, Peter kennt
Maria knows *Visions of Joanna* Peter knows
Everybody Must Get Stoned und Paula kennt
Everybody Must Get Stoned and Paula knows
Blowing in the Wind.
Blowing in the Wind

The oddity of the continuation in (13c) clearly shows that the topical indefinite in (13a) has to be interpreted as having scope over the universal quantifier in the matrix clause (see Cresti 1995, Jäger 1996, and Portner and Yabushita

[8] Note that there are cases where the D-pronoun does not occupy Spec,CP at the surface. Frey (2004) argues that cases like (i), in which the resumptive pronoun remains in the middle field, have to be analyzed as instances of left dislocation, because the anaphor *sich* can be bound by *er* (and this is only possible in left dislocation and not in hanging topic constructions, see n. 3):

(i) Einen Bericht über sich, er will mir den heute zeigen.
a-ACC report about himself he wants me-DAT RP.MASC.ACC.SG today show

To us, however, and to some other native speakers we consulted, such examples feel slightly degraded, though not completely out. For those cases, we assume that it is marginally possible for C^0 to come with a second specifier (i.e. a specifier above the one that is overtly filled by *er* in (i)) where the resumptive pronoun can be inserted at LF, triggering lambda-abstraction.

[9] Note that this excludes demonstratives like *dieser* (this one), which are not acceptable in left-dislocation constructions.

1998 for the claim that topical indefinites have to be interpreted with scope over other quantifiers): the songs are not allowed to vary with the people. This contrasts with the minimally different sentence in (14a), which involves no left-dislocation and where the indefinite occupies Spec,CP: in this case, both a wide-scope and a narrow-scope reading are available, showing that the indefinite can be reconstructed into its base position, where it is c-commanded by the universal quantifier in subject position.

(14) a. EInen Song von Bob Dylan kennt JEder.
a/one song by Bob Dylan knows everyone
'There is a certain song by Bob Dylan that everybody knows/everybody knows a/at least one song by Bob Dylan.'

b. Nämlich *Blowing in the Wind.*
namely *Blowing in the Wind*

c. Maria kennt *Visions of Joanna,* Peter kennt
Maria knows *Visions of Joanna* Peter knows
Everybody Must Get Stoned und Paula kennt
Everybody Must Get Stoned and Paula knows
Blowing in the Wind.
Blowing in the Wind

It is worth mentioning that it is often claimed (see Frey 2004 and the references cited therein) that left-dislocated DPs can be reconstructed into the base position of the resumptive pronoun in Spec,CP. Evidence for this claim comes from the fact that left-dislocated constituents may contain pronouns that are interpreted as bound by a quantifier contained within the matrix clause, as long as the quantifier c-commands the base position of the resumptive pronoun. Reconstruction would obviously result in a narrow-scope reading of the indefinite. If the indefinite could be reconstructed in (14a), this would result in a reading where *einen Song von Bob Dylan* "a song by Bob Dylan" takes narrow scope with respect to *jeder* "everyone". We assume, however, that even if the value of a DP that contains a bound pronoun varies with the quantificational value of a binding operator, this is not necessarily an instance of a genuine narrow-scope reading. We follow Endriss (2009) and Ebert and Endriss (2007) in the assumption that left dislocated DPs with dependent pronouns invoke so-called functional wide-scope readings. For reasons of space, we can go neither into the empirical facts supporting this conclusion nor into the details of how functional wide-scope readings are derived here, but rather have to refer the interested reader to Ebert and Endriss (2007) and Endriss (2009).

Following Frey (2004), we assume that left-dislocated XPs are base-generated in their surface position (see Wiltschko 1997, Grohmann 2000, and Grewendorf 2002 for different views). Concerning the question of how they receive their case-features, which are identical with the case features of the D-pronoun in Spec,CP, we tentatively make the following assumptions (as the focus of this chapter is on semantic issues, we remain rather sketchy here): the case features of the D-pronoun in Spec,CP are transmitted to C° under Spec–head agreement and to the head of the functional projection hosting the left-dislocated DP under strict locality with C°.[10] This enables the head of the higher functional projection to assign to the DP in its specifier the same case as the one carried by the D-pronoun in Spec,CP.

A further noteworthy fact concerning indefinites in left-dislocated position is that singular indefinites and unmodified numeral indefinites are the only quantificational DPs that can occur in this position, as shown in (15):

(15) a. (*Mehr als/ *weniger als/ *genau) zwei Songs von
more than/ less than/ exactly two songs by
Bob Dylan, die mag ich.
Bob Dylan RP-PL like I

b. *Keine/*wenige Songs von Bob Dylan, die mag ich.
no/ few songs by Bob Dylan RP-PL like I

In Section 5.4 we will see how this restriction as well as the obligatory wide-scope reading of topical indefinites (in sentences without adverbial quantifiers) can both be derived from the aboutness topicality concept. But let us first have a closer look at adverbially quantified sentences containing topical indefinites.

5.3.4 *Topical indefinites in adverbially quantified sentences*

Consider again example (6b), repeated as (16):[11]

[10] See n. 8 for a brief discussion of cases where the D-pronoun does not occupy Spec,CP at the surface.

[11] Note that adverbially quantified sentences with left-dislocated indefinites are also acceptable if the indefinite contains a reflexive pronoun that is bound by a CP-internal DP. This shows that we must be dealing with left dislocation constructions and not with hanging topic constructions. This is shown in (i):

(i) Einen Artikel über sich, den liest Hans meistens mit größter
an article about himself RP-MASC.ACC.SG reads Hans usually with greatest
Genugtuung.
satisfaction
'Hans usually reads an article about himself with greatest satisfaction.'

(16) Eine neue Platte von Bob DYlan, die kommt
a new record by Bob Dylan RP-FEM.NOM.SG comes
meistens in die CHARTS.
usually in the charts
'A new record by Bob Dylan usually gets into the charts.'

The sentence can be paraphrased only as *Most new records by Bob Dylan get into the charts*, in other words, the topical indefinite obligatorily induces the Quantificational Variability Effect already mentioned in sections 5.1 and 5.2: the quantificational force of the indefinite seems to depend on the quantificational force of the adverbial quantifier. Note, however, that such a reading is only available under the following condition: there is no heavy accent on the determiner *eine* as opposed to (13a), which was discussed in the last section, but the accent rather falls on some element within the NP-complement of the determiner (the most natural choice being *Dylan*).[12] If it falls on *eine*, the sentence becomes very odd for pragmatic reasons, since it can be interpreted only as saying that there is a particular new record by Bob Dylan which gets into the charts most of the time:

(17) ??EIne neue Platte von Bob Dylan, die kommt
a/one new record by Bob Dylan RP-FEM.NOM.SG comes
meistens in die CHARTS.
usually in the charts
'??There is a new record by Bob Dylan that usually gets into the charts.'

Sentence (18b), however, shows that there is nothing wrong with interpreting a left-dislocated indefinite with scope over a Q-adverb, as long as the resulting reading is not extremely implausible:

(18) a. Ein HUND, der ist meistens HUNGrig.
a dog RP-MASC.NOM.PL is usually hungry
'A dog is usually hungry.'

b. EIN Hund, der ist meistens HUNGrig.
a/one dog, RP-MASC.NOM.PL is usually hungry
'There is a certain dog that is usually hungry.'

Example (18) is instructive, since it exemplifies the two strategies that are available to interpret left-dislocated indefinites in sentences with adverbial

[12] Note that this accent is added to the main accent, which falls on the most deeply embedded constituent within the matrix VP by default, thereby signaling that the whole matrix-CP is to be interpreted as focal (cf. Selkirk 1995).

quantifiers: they can either be interpreted in the restrictor of the Q-adverb, which results in a QV-reading, or they can be interpreted with scope over the Q-adverb. The choice between the two readings depends on the accentuation pattern within the DP.

Note that the predicate *be hungry* in (18a) cannot be interpreted episodically, but only generically, that is to say that the sentence gets a reading that can be paraphrased as *most dogs are generally hungry*, and not as *most dogs are hungry at least once during their lifetime*. Furthermore, the sentence also gets a second, slightly different reading, which can be paraphrased as *in general, dogs are hungry most of the time*. This can easily be explained under the by now standard assumption that in addition to overt Q-adverbs, there also exists a covert generic operator, which has (roughly) the same quantificational force as the universal A-quantifier *always*, but differs from the latter in allowing exceptions (cf. Krifka, Pelletier, Carlson, ter Meulen, Chierchia, and Link 1995). With this assumption in place, (18a) can be analyzed as containing two instead of one A-quantifier: an overt one (*usually*), and the covert generic operator. The choice between the two readings paraphrased above then depends on the question which quantifier is interpreted with scope over the other: if *usually* is given higher scope, we get the first reading, and if the generic operator is given higher scope, we get the second one. In addition, assuming that there is a covert generic operator also enables us to analyze generic sentences like (19a, b), which are interpreted as making general assertions about elephants and lions, respectively, in analogy to sentences with overt Q-adverbs where the topical indefinite is interpreted in the restrictor of the Q-adverb:

(19) a. Ein EleFANT, der hat einen RÜSsel.
an elephant RP-MASC.NOM.SG has a trunk
'An elephant has a trunk.'

b. Ein LÖwe, der hat eine MÄHne.
a lion RP-MASC.NOM.SG has a mane
'A lion has a mane.'

Note that in these cases, too, a specific reading can be forced by placing an accent on the determiner instead of the noun:

(20) a. EIN Elefant, der hat einen RÜSsel.
a/one elephant RP-MASC.NOM.SG has a trunk
'There is one specific elephant that has a trunk.'

b. EIN Löwe, der hat eine MÄHne.
a/one lion RP-MASC.NOM.SG has a mane
'There is one specific lion that has a mane.'

We have seen that topical indefinites either receive wide scope (over other quantifiers or, in the presence of a Q-adverb, over the Q-adverb) or that they are interpreted in the restrictor of a (covert or overt) Q-adverb (cf. Partee 1995, Chierchia 1995, and Krifka 2001). The choice of the interpretation is dependent on where the nuclear accent falls. We will see in the next section that our concept of "aboutness topicality" naturally accounts for these two interpretative possibilities.

5.4 Indefinites as aboutness topics: The explanation

5.4.1 *Topical indefinites as direct aboutness topics*

Consider again example (6a), repeated as (21):

(21) EInen Song von Bob Dylan, den kennt JEder
A/One song by Bob Dylan RP-MASC.ACC.SG knows everyone
(nämlich *Blowing in the Wind*).
namely *Blowing in the Wind*
'There is one/a certain song by Bob Dylan that everyone knows (namely *Blowing in the Wind*)'.

As already mentioned in Section 5.3.1 (and as indicated by the paraphrase), sentences like (21), in which a left-dislocated DP is combined with a CP-internal quantificational DP, receive only an interpretation according to which the left-dislocated indefinite takes wide scope. At first sight, this seems to follow rather straightforwardly from the assumption that left-dislocated XPs are base-generated in their surface positions: as the indefinite cannot be reconstructed into the base position of the D-pronoun, the only way for the universal quantifier to have scope over it would be Quantifier Raising (QR). The standard assumption with respect to QR is, however, that it can target only IP/TP (cf. May 1985). Therefore, there is no way for a CP-internal quantifier to have scope over a left dislocated one. However, the account just sketched would be in conflict with the assumption introduced in Section 5.3.2, according to which the topic operator present in left-dislocation sentences has the denotation repeated here as (22):

(22) $\lambda P \lambda X. \exists \alpha [\alpha = X \,\&\, \text{ASSERT}\, [P(\alpha)]]$.

First, the left-dislocated DP cannot function as the second argument of the topic operator, and second, the denotation of the CP cannot apply to the denotation of this DP, as required by (22). In (23a, b), the denotations of the left-dislocated indefinite and the CP are given, and in (23c) we see the result of combining the two constituents directly, that is, without the intervening topic operator:[13]

(23) a. $\lambda P_{\langle e,\langle s,t\rangle\rangle}\lambda s.\ \exists x[\text{song-by-Bob-Dylan}(x)(s) \wedge P(x)(s)].$

b. $\lambda z.\lambda s.\ \forall y[\text{person}(y)(s) \rightarrow \text{know}(z)(y)(s)].$

c. $[\lambda P_{\langle e,\langle s,t\rangle\rangle}\lambda s.\ \exists x[\text{song-by-Bob-Dylan}(x)(s) \wedge P(x)(s)]].$
$(\lambda z.\lambda s.\ \forall y[\text{person}(y)(s) \rightarrow \text{know}(z)(y)(s)]) =$
$\lambda s.\ \exists x[\text{song-by-Bob-Dylan}(x)(s) \wedge \forall y[\text{person}(y)(s) \rightarrow$
$\text{know}(x)(y)(s)]].$

While combining the denotation of the left-dislocated indefinite with the denotation of the CP directly via functional application gives us the right result—namely a wide-scope interpretation for the indefinite—it does not only come at the price of ignoring the meaning contribution of the topic operator in (22), but it is in conflict with the intuition that aboutness topics are the logical subjects of a predication contributed by the comment: as the indefinite denotes (the characteristic function of) a set of sets of relations between individuals and situations, while the CP denotes (the characteristic function of) a set of relations between individuals and situations, it is the denotation of the indefinite which applies to the denotation of the CP, that is, the object denoted by the indefinite functions as the predicate, while the object denoted by the CP functions as the argument. This is counter to the aboutness concept of topicality and to the proposed topic principle, where the topic serves as the subject of predication.

Rather, we assume (following Ebert and Endriss 2004 and Endriss 2009) that only individuals and sets can legitimately serve as addresses for storing information, in other words, as aboutness topics. This is evidenced by the fact that there are discourse referents and corresponding anaphoric expressions for individuals and sets, but not for sets of sets. Therefore, the type of a topic-marked generalized quantifier has to be lowered to the type of sets at least, in order for it to serve as an address where the information conveyed by the rest of the sentence can be stored. Recall from Section 5.3.2 as we have seen in (7), that a set, namely a set of situations (i.e. a proposition), can function as the aboutness topic of a sentence. Therefore, the topic operator has to be flexible enough to accommodate the type of sets anyway.

One way to achieve the goal of turning the generalized quantifier denoted by an indefinite DP into a set (another one will be discussed in Section 5.4.2) is by creating a representative of the quantifier in the form of a minimal witness set

[13] Note that we have to complicate the semantics of generalized quantifiers slightly by adding an additional situation argument in order to be consistent with what will be proposed in Section 5.4.2.

(in the sense of Barwise and Cooper 1981; see Endriss 2009; cf. also Szabolcsi 1997). A minimal witness set of a quantifier is an element of this quantifier that does not contain any "unwanted" elements. The formal definition is given in (24):

(24) Definition of a minimal (witness) set X of a generalized quantifier G
$\min(X)(G) = \lambda s\,[G(X)(s) \wedge \forall Y\,[G(Y)(s) \rightarrow \neg(\lambda x.Y(x)(s) \subset \lambda x.X(x)(s))]]$.

In the case of a quantifier like *three dogs* (in a situation *s*), for instance, a minimal witness set of this quantifier is a set that contains three dogs (in *s*) and nothing else. Such a minimal witness set can then function as the address where the information conveyed by the comment is stored. In order for this to be possible, however, the denotation of the topic—which now is an object of type $\langle e\langle s,t\rangle\rangle$—has to be combined with the denotation of the comment, which is of the same type. Furthermore, as noted above, the intuition formalized in (22) has to be respected, in other words, the comment has to be applied to the topic. This can be achieved in the following way: an operator which turns the elements of the minimal witness set into a (sum) individual (cf. Link 1983) is applied to the respective minimal witness set.

Taking all this together, we assume that in (non-generic, non-adverbially quantified) sentences that contain a topical indefinite, the topic operator in (22) is shifted as shown in (25), which gives us an object that can apply to the quantifier and the comment property, respectively:

(25) $\lambda P_{\langle e,\langle s,t\rangle\rangle}\lambda X_e.\ \exists\alpha[\alpha = X\ \&\ \text{ASSERT}\ [P(\alpha)] \Rightarrow$
$\lambda P_{\langle e,\langle s,t\rangle\rangle}\lambda X_{\langle\langle e,\langle s,t\rangle\rangle,\langle s,t\rangle\rangle}\exists\alpha[\exists s[\min(\alpha)(X)(s)\ \&\ \text{ASSERT}\ [P(\sqcap\{x : \alpha(x)(s)\})]]]$.

In the case of example (21), above, repeated here as (26a), this yields the result in (26b):

(26) a. EInen Song von Bob Dylan, den kennt JEder
a/one song by Bob Dylan RP MASC.ACC.SG knows everyone
(nämlich *Blowing in the Wind*).

b. $\lambda P_{\langle e,\langle s,t\rangle\rangle}\lambda X_{\langle\langle e,\langle s,t\rangle\rangle,\langle s,t\rangle\rangle}\exists\alpha[\exists s[\min(\alpha)(X)(s)\ \&\ \text{ASSERT}[P(\sqcap\{x : \alpha(x)(s)\})]]]$.
$(\lambda z\lambda s.\ \forall y[\text{person}(y)(s) \rightarrow \text{know}(z)(y)(s)])$
$(\lambda P\lambda s.\ \exists x[\text{song\_by\_Bob\_Dylan}(x)(s) \wedge P(x)(s)]) = \exists\alpha[\exists s[\min(\alpha)$ $(\lambda P\lambda s'.\ \exists x[\text{song\_by\_Bob\_Dylan}(x)(s') \wedge P(x)(s')])(s)\ \&\ \text{ASSERT}$ $[\lambda s'.\forall y[\text{person}(y)(s') \rightarrow \text{know}(\sqcap\{x : \alpha(x)(s)\})(y)(s')]]]$.

This is the correct result. It reflects the wide-scope reading for the indefinite and at the same time respects the principle underlying our formalization of the topic operator: it allows the creation of an address corresponding to the minimal witness set where the information conveyed by the comment can be stored.

In cases like (26), sum formation is trivial, as any minimal witness set contains just one element, anyway. In examples like (27), however, we get a true *sum* individual, as a minimal witness set of the quantifier under discussion has more than just one element. This has the consequence that in order to obtain the correct result, a distributivity operator (cf. Link 1983) has to be inserted in cases where the respective predicate is non-collective in order to distribute the atomic elements of the sum individual (i.e. the members of the minimal witness set) over the elements of the set denoted by the predicate.

(27) ZWEI Mitglieder der Beatles, die kennt JEder.
two members of-the Beatles RP-PL knows everyone
'There are two members of the Beatles that everyone knows.'

We follow Ebert and Endriss (2004) and Endriss (2009) in their assumptions about when a topical interpretation of a quantifier is possible and when it is prohibited. We can only briefly sketch the main ideas here and refer the reader to the cited papers. The authors assume that a quantifier can be interpreted as a sentence topic only if two constraints are fulfilled: (1) topic interpretation may not change the actual content of the sentence under discussion, and (2) topic interpretation may not eliminate anaphoric potential—both compared to the case when the quantifier under discussion is not interpreted as topical, but with similar scope relations as in the topic interpretation case. Formally, this can be spelled out as a condition on the lexical semantics of quantifiers, which checks whether the quantifier under discussion can serve as topic or not. A quantifier is called "topic-able" if and only if it passes this so-called "Topic Condition" that is divided into the two constraints pointed out above (see Endriss 2009 for further formal details). A generalized quantifier Q is topic-able if applying Q to a non-complex predicate P (i.e. one that does not contain any scope operators) leads to the same result as applying P to the representative of Q (i.e. the sum individual corresponding to a minimal witness set of Q). For our concerns here this means that the topic operator in (25) applied to a non-complex property P and a quantifier Q must yield the same outcome as would result from applying Q to P. The reasoning behind this is the following: the first constraint ensures that topic-marking is first and foremost a means of structuring information that is to be conveyed by the speaker, and not an operation that is intended to alter the truth conditions. The second constraint makes reference to dynamic binding possibilities and demands that in addition to the truth conditions, also the binding possibilities may not be affected by the topic interpretation of quantifiers. This is in line with the assumption made in dynamic semantics (e.g. Groenendijk and Stokhof 1990) that the meaning of an expression is not exhausted by its truth conditional content, but rather by the combination of truth conditional content and binding potential.

Note that these assumptions concerning the interpretation of topical indefinites explain immediately why downward entailing generalized quantifiers (like *weniger als n NP* "less than n NP") cannot be topicalized (see Section 5.3.3): in these cases, the corresponding minimal witness set would be the empty set, obviously not a sensible representative for the quantifier. The truth conditions of the non-topic case, where the quantifier "less than n NP" is applied to a predicate P, and the topic case, where predicate P would be applied to the sum individual corresponding to the empty set, would differ massively. Hence the GQ "less than n NP" is not topic-able. If such a DP was overtly topic-marked (e.g. by a left-dislocation construction in German), this would result in a deviant sentence because the quantifier does not pass the Topic Condition and hence the speech act corresponding to the creation of a storage address would fail.

With modified quantifiers of the form *genau n NP* "exactly n NP" or *mehr als n NP* "more than n NP", the problem is again that these GQs do not meet the Topic Condition, because the truth conditions of the topic and the non-topic case would not be equivalent for non-complex predicates P. As expected, (28a) is odd:

(28) a. ??Genau ZWEI Mitglieder der Beatles, die kennt JEder.
exactly two members of-the Beatles RP-PL knows everyone

b. ??Mehr als ZWEI Mitglieder der Beatles, die kennt JEder.
more than two members of-the Beatles RP-PL knows everyone

Assume that there are in fact three members of the Beatles that have property P, say, "sleep" (namely John Lennon, Paul McCartney, and George Harrison). In this case, the proposition *exactly two members of the Beatles sleep* would be false. Under a topic interpretation of *exactly two members of the Beatles*, however, the sentence would be true, since it is only required that there is a set containing two members of the Beatles and each member of this set sleeps. This leaves open the possibility that actually more than two members of the Beatles sleep. The oddity of such examples can thus be accounted for under the assumption that topic interpretation is excluded here, as it would not lead to a truth conditionally equivalent interpretation.

In the case of quantifiers of the form *mehr als n NP* "more than n NP" in (28b), the problem is that anaphoric possibilities would be destroyed that were available if the indefinite did receive its ordinary interpretation, in other words, the second constraint on topic interpretation is violated. It is well known that if the quantifier *mehr als zwei Gäste* "more than two guests" in a sentence such as *Mehr als zwei Gäste schlafen* "more than two guests sleep" binds a pronoun in a subsequent clause via dynamic binding (cf. Groenendijk and Stokhof 1990), this pronoun is understood to refer back to the maximal number of individuals that are guests and sleep (cf. Kadmon 1985). Creating a minimal witness set of the quantifier *mehr als zwei Gäste* "more than two guests", however, would have the consequence that a pronoun in a subsequent clause could only denote *three* individuals that satisfy the predicate *guests* (and that sleep), but not four or more. The oddity of (28b) can thus naturally be accounted for by reference to the second above-mentioned constraint on the topical interpretation of quantifiers: in addition to the truth conditions, also the binding possibilities may not be affected by topicalizing a quantifier.

Note finally that universally quantified DPs seem to pose a problem for this account only at first glance. The minimal witness set of a universally quantified DP like *jeder Hund von Paulas Onkels* "every dog of Paula's uncle" is the set containing all dogs of Paula's uncle and nothing else. This is intuitively a decent representative for the quantifier and the quantifier passes the Topic Condition. The quantifier should thus be one that can be interpreted as sentence topic and hence be able to appear in left-dislocated position. However, (29) shows that this is not the case.

(29)	*Jeden	Hund	von	Paulas	Onkel,	den	mag	Peter.
	every	dog	of	Paula's	uncle	RP-MASC.ACC.SG	likes	Peter

Now note that there is a mismatch between the semantic plurality of the object created from the generalized quantifier in (29) and the pronoun in Spec,CP. That this mismatch is (partially) responsible for the ungrammaticality of (29) becomes evident by comparing it to the minimally contrasting example (30), where the left-dislocated universally quantified DP is morphologically plural and thus comes with a corresponding plural D-pronoun in Spec,CP.

(30) ?Alle Hunde von Paulas Onkel, die mag Peter
all dogs of Paula's uncle RP-PL likes Peter

While still odd, (30) is considerably better than (29), thus showing that the (mis)match between semantic and morphological plurality plays a role here. That (30) is still far from perfect might be due to the fact that after applying the sum operator to the minimal witness set derived from the generalized quantifier, the resulting object has exactly the same denotation as the plural definite *The dogs of Paula's uncle*, namely, the sum individual consisting of all the dogs Paula's uncle has. In view of the fact that *The dogs of Paula's uncle* makes a perfect topic, it would be extremely uneconomical to convey the same information by using the topicalized *all dogs of Paula's uncle*. Following Chierchia (1998), it can be assumed that covert typeshifting operations like the ones required to turn a universally quantified DP like the one in (30) into an object of type *e* are blocked whenever the same result could have been achieved via a lexical item—namely, the definite determiner. This explains why not only morphologically singular, but also morphologically plural universally quantified DPs cannot felicitously be left-dislocated.

Taken together, this leaves indefinites and unmodified numerals as the only quantifiers that may be topicalized. Note finally that left-dislocated indefinites shifted to minimal witness sets can be seen as *direct* aboutness topics insofar as the comment is directly predicated over the element(s) of the minimal witness set. In this respect, they behave like topicalized DPs that denote individuals (i.e. proper names and definite descriptions). In the next section, we will see that topical indefinites can also function as indirect aboutness topics.

5.4.2 *Topical indefinites as indirect aboutness topics*

In Section 5.3.4, we have seen that left-dislocated indefinites in adverbially quantified sentences can be interpreted in two ways: either with scope over the Q-adverb, or in the restrictor of the Q-adverb, depending on the accentuation pattern within the left-dislocated DP. The wide-scope interpretation can be accounted for by assuming that the mechanism described in the last section applies to the topical indefinite and the predicate denoted by the CP containing the Q-adverb. An example like (18b), which is repeated here as (31a), thus receives the interpretation in (31b), which is paraphrased in (31c):

(31) a. EIN Hund, der ist meistens HUNGrig.
a/one dog RP-MASC.NOM.PL is usually hungry
'There is a certain dog that is usually hungry.'

b. $\lambda P_{\langle e,\langle s,t\rangle\rangle}\lambda X_{\langle\langle e,\langle s,t\rangle\rangle,\langle s,t\rangle\rangle}\exists\alpha[\exists s[\min(\alpha)(X)(s)$
$\& \text{ ASSERT}[P(\sqcap\{x : \alpha(x)(s)\})]]]$
$(\lambda x\lambda s.\text{ Most } s'[s' \leq s \wedge \min(s', C)]\exists s''[s' \leq s'' \wedge \min(s'',$
$\text{hungry}(x))]])$
$(\lambda P\lambda s.\ \exists x[\text{dog}(x)(s) \wedge\ P(x)(s)])=\exists\alpha[\exists s[\min(\alpha)(\lambda P\lambda s'.\exists x[\text{dog}(x)$
$(s') \wedge P(x)(s')])(s)\ \&\ \text{ASSERT}[\lambda s'''.\text{ Most } s'\ [s' \leq s''' \wedge$
$\min(s', C)][\exists s''[s' \leq s'' \wedge \min(s'', \text{hungry}'(\sqcap\{x:\ \alpha(x)(s)\}))]]]]]$

c. There is a situation *s* such that there is a minimal witness set α of the quantifier *a dog* in *s*, and it is asserted that most contextually restricted minimal situations can be extended to a minimal situation where the element in α is hungry.

Under the assumption that the C-variable in the restrictor is resolved to (the characteristic function of) the set of situations containing the dog in the minimal witness set α that are situations where this dog might possibly be hungry (i.e. situations where it is not asleep, etc.), this is intuitively the correct result. Note that in such cases the topicalized indefinite functions as a direct aboutness topic in the sense discussed above: the predicate denoted by the CP is applied to the element contained in the minimal witness set of the quantifier, the only difference to the cases discussed in the last section being that this predicate now contains a quantifier over situations, not over individuals.

Now consider example (16), which is repeated here as (32):

(32) Eine neue Platte von Bob DYlan, die kommt
a new record by Bob Dylan RP-FEM.NOM.SG comes
meistens in die CHARTS.
usually in the charts
'A new record by Bob Dylan usually gets into the charts.'

With the accentuation pattern indicated, the sentence gets a QV-reading that can be paraphrased as "Most new records by Bob Dylan get into the charts". We have already seen that QV-readings can be accounted for by assuming that the respective indefinite is interpreted in the restrictor of the Q-adverb.

But now the problem is that such an interpretation seems to be in conflict with our assumption that left-dislocated indefinites always need to be interpreted as aboutness topics: if the indefinite is interpreted in the restrictor of the Q-adverb, the CP containing this Q-adverb cannot be interpreted as a predicate applying to (the denotation of) this indefinite.

It is, however, possible to reconcile our view of left-dislocated indefinites as aboutness topics with the fact that such indefinites receive QV-readings in the presence of Q-adverbs if we view quantification as a higher-order predication process. Seen this way, the restrictor set—that is, the set quantified over—is the logical subject of a higher-order predication, where this higher-order predication consists in specifying the degree to which the restrictor set is contained within the set denoted by the respective matrix predicate (see Löbner 2000 for a similar view). A sentence like *Most children sleep* could thus be viewed as a higher-order quantification construction where the subject of the predication are the children, the predication is the property of sleeping, and *most* gives the degree to which this *sleep*-property holds of the subject *children*. In the case of such quantificational DPs as *most children* this relation is masked, however, by the fact that quantificational determiners form constituents with NPs, which function as their restrictors (see Section 5.2). Accordingly, the restrictor in these cases cannot be marked as an aboutness-topic via separating it from the rest of the clause, which could then function as the comment. In the case of Q-adverbs, on the other hand, this is possible, as Q-adverbs do not form constituents with their restrictors, but rather—occupying vP-adjoined (base) positions—with their nuclear scopes.

Let us therefore assume that Q-adverbs always come in two, closely related, varieties: the first variety is the one already discussed. Q-adverbs of this variety take the denotation of the clause they are contained in as their nuclear scope, while the restrictor is given in the form of a free variable ranging over situation predicates. This is the variety that is employed in cases where (a) no overt material c-commanding the Q-adverb is given; (b) this material is to be interpreted with scope over the Q-adverb (as in (31a)); or (c) this material is reconstructed into a position where it is c-commanded by the Q-adverb at LF (see Hinterwimmer 2006, 2008).[14]

Q-adverbs of the second variety take two arguments, like quantificational determiners, but they take their arguments in reverse order (seen from the perspective of determiner-quantification; cf. Chierchia 1995): they combine with the set of situations denoted by the vP-segment they c-command at LF first, forming a predicate that can be applied to the respective topical set (see Hinterwimmer 2008 for details).

Consider the example in (33a), where a *when*-clause has been left-dislocated. We assume that in such cases a topical set of situations is given directly in the form of the left-dislocated *when*-clause. In order to account for such examples, the types of the arguments that the topic operator in (22) takes need to be adjusted, as shown in (33b), and the sentence can be

[14] Clause (c) is relevant for examples where a focus-marked constituent that c-commands the Q-adverb overtly is interpreted in the nuclear scope. Hinterwimmer (2006, 2008) argues that in these cases the focus-marked material is reconstructed at LF.

interpreted as shown in (33c) (under the assumption that the subject *Maria* is reconstructed into its vP-internal base position at LF). Note that we assume the D-pronoun *dann* to be a resumptive pronoun (cf. Iatridou 1993 and Bhatt and Pancheva 2001) and thus to trigger lambda abstraction over sets of situations:

(33) a. Wenn Paul in seinem Büro ist, dann ist Maria meistens
when Paul in his office is then is Maria usually
glücklich.
happy
'When Paul is in his office, Maria is usually happy.'

b. $\lambda P_{\langle s,t\rangle,\langle s,t\rangle} \lambda X_{\langle s,t\rangle}.\ \exists \alpha[\alpha = X\ \&\ \text{ASSERT}\ [P(\alpha)]].$

c. $\lambda P_{\langle s,t\rangle,\langle s,t\rangle} \lambda X_{\langle s,t\rangle}.\exists \alpha[\alpha = X\ \&\ \text{ASSERT}\ [P(\alpha)]].$
$(\lambda Q_{\langle s,t\rangle}.\ \lambda s.\ \text{Most } s'\ [s' \leq s \wedge \min(s', Q)]$
$[\exists s''[s' \leq s'' \wedge \min(s'', \text{happy(Maria)})]]]).$
$(\lambda s.\ \text{in\_his\_office(Paul)}(s)) =$
$\exists \alpha[\alpha = \lambda s.\ \text{in\_his\_office(Paul)}(s)$
$\&\ \text{ASSERT}\ [\lambda s.\ \text{Most } s'\ [s' \leq s \wedge \min(s', \text{in\_his\_office(Paul)})]$
$[\exists s''[s' \leq s'' \wedge \min(s'', \text{happy(Maria)})]]]]]$

In an example like (33a), the left-dislocated *when*-clause denoting a set of situations is thus the *direct* aboutness topic, being the "subject" of the higher-order predication expressed by the comment. In a case like (32), on the other hand, no such *direct* aboutness topic is given, as the left-dislocated indefinite denotes a set of sets of relations between individuals and situations, not a set of situations, as shown in (34). We assume that in order to fix this mismatch, there is a second possibility available (in addition to the one discussed in Section 5.4.1) to turn an indefinite into a set that can serve as an address for storing information: it can be turned into a set of situations via a simple type-shift, namely by applying the predicate $\lambda x \lambda s.in(x)(s)$ to it (see Hinterwimmer 2008 for details). This gives us a set of situations each of which contains an individual of the respective kind, as shown in (34) for the left-dislocated indefinite of example (32):[15]

(34) $\lambda P_{\langle e,\langle s,t\rangle\rangle} \lambda s.\ \exists x[\text{new-record-by-Bob-Dylan}(x)(s) \wedge P(x)(s)]$
$(\lambda y \lambda s.\text{in}(y)(s)) = \lambda s.\ \exists x[\text{new-record-by-Bob-Dylan}(x)(s) \wedge \text{in}(x)(s)]$
$= \lambda s.\ \exists x[\text{new-record-by-Bob-Dylan}(x)(s)].$

[15] We assume that in the case of left-dislocated bare plurals basically the same mechanism applies, modulo the fact that bare plurals denote kinds which have to be turned into plural indefinites in cases where they are to be combined with non-kind-level predicates (see Hinterwimmer 2008 and the references cited therein for further discussion).

This set of situations can then function as the aboutness topic in a case like (32), and the left-dislocated indefinite can be seen as the *indirect* aboutness topic of such sentences, as the *direct* aboutness topic is a set of situations that has been derived from the denotation of the respective indefinite.

One additional complication has to be taken into account in order to derive the reading we are after in cases like (32): the D-pronoun cannot be interpreted in Spec,CP, where it triggers lambda-abstraction, as this would result in a reading where the indefinite has scope over the Q-adverb (as in (31)). In order to overcome this problem, we assume that the D-pronoun in the specifier position of CP can optionally be reconstructed into its vP-internal base position, where it is interpreted as a free variable, in other words, just like an ordinary pronoun.[16] As such, it can be dynamically bound by the indefinite in the restrictor of the Q-adverb. This gives us a higher-order predicate that can be applied to the topical set, as shown in (35), resulting in an interpretation that can be paraphrased as "Most (minimal) situations that contain a new record by Bob Dylan can be extended to a minimal situation where this record gets into the charts":

(35) a. $\lambda P_{\langle\langle s,t\rangle,\langle s,t\rangle\rangle}\lambda X_{\langle s,t\rangle}.\exists\alpha[\alpha = X \;\&\; \text{ASSERT}[P(\alpha)]]$
$(\lambda Q_{\langle s,t\rangle}\ \lambda s.\ \text{Most } s'[s' \leq s \wedge \min(s', Q)]$
$[\exists s''[s' \leq s'' \wedge \min(s'', \text{gets\_into\_charts}(x))]])$
$(\lambda s.\ \exists x[\text{new\_record\_by\_Bob Dylan}(x)(s)]) =$
$\exists\alpha[\alpha = \lambda s.\ \exists x[\text{new\_record\_by\_Bob Dylan}(x)(s)]\ \&$
$\text{ASSERT}[\lambda s.\ \text{Most } s'[s' \leq s \wedge \min(s', \alpha)]$
$[\exists s''[s' \leq s'' \wedge \min(s'', \text{gets\_into\_charts}(x))]]]]$

In this section, we have seen how our concept of aboutness topicality can account for the QV-readings of adverbially quantified sentences with topical indefinites.

5.5 Conclusion

In this chapter, we have shown how the two readings that topical indefinites receive (the wide-scope reading and the QV-reading) can be derived from a concept of aboutness topicality, according to which the topic of a sentence has to be interpreted as the logical subject of a predication. In the case of

[16] It is well known that D-pronouns in German can function like ordinary pronouns insofar as they can either be dynamically bound, or be assigned a value on the basis of contextual information (what is excluded is binding by c-commanding quantifiers).

the wide-scope reading, we are dealing with a first-order predication, as the predicate denoted by the comment applies to an individual that is derived from the original denotation of the indefinite as a generalized quantifier. In the case of the QV-reading, we are dealing with a second-order predication, as the comment denotes a second-order predicate that applies to a set of situations which is derived from the original denotation of the indefinite.

6

Contrastive topics operate on speech acts

SATOSHI TOMIOKA

6.1 Overview

The semantic notion of contrastiveness is frequently mentioned in studies of focus, particularly in connection with such empirical issues as exhaustive answers in question–answer pairs, overtly contrasting statements ("not A but B"), correcting statements, clefts/pseudo-clefts, and association with focus with adverbs such as *only* and *always*. The main empirical phenomenon to be examined in this chapter is, however, contrastiveness associated with topicality, which is typically placed as a polar opposite of focus in the pragmatics literature. The combination of the two notions is known to create a particular semantic/pragmatic effect that is often characterized as a sense of incompleteness, non-finality and/or uncertainty. The aim of this paper is to discover how this effect comes about.

Linguistic expressions that generate a sense of incompleteness were noted and discussed fairly early in the generative tradition. Perhaps, A- vs. B-accents in English (cf. Jackendoff 1972) are the first empirical phenomenon that falls

Much of the material in this paper was developed when I taught a seminar as a guest lecturer at Universtät Potsdam in Spring 2006. I am grateful to Caroline Féry and Gisbert Fanslow for inviting me there. Earlier versions of the paper were presented at various venues; Universtät Potsdam, University of British Columbia, University of Stony Brook-SUNY, Kyoto University, Nanzan University, and Kanda University of Foreign Studies. I would like to thank the audiences at those occasions for comments and suggestions. The following individuals deserve special mention for their contribution to the chapter though none of them should be held responsible for the content; Nick Asher, Beata Gyuris, Nobuko Hasegawa, Shin Ishihara, Kyle Johnson, Katalin É. Kiss, Yoshihisa Kitagawa, Sachie Kotani, Manfred Krifka, Chung-min Lee, Mats Rooth, Yuki Takubo, Reiko Vermulen, Katsuhiko Yabushita, Masahiro Yamada, and, most importantly, Yurie Hara, whose University of Delaware dissertation got me interested in the topic. Working with her was both stimulating and rewarding. Two anonymous reviewers and the editors (Caroline Féry and Malte Zimmermann) gave me valuable comments. This research was in part supported by National Science Foundation Grant #BCS-0650385 awarded to the author.

into this category. The connection between this accent pattern and its semantic–pragmatic effect has been the topic of many subsequent works (e.g. Carlson 1983; von Fintel 1994; Kadmon 2001; Büring 2003). The Rise–Fall contour in German elicits effects similar to the English A-/B-accent cases, as discussed extensively in Büring (1997) and Krifka (1998). While English and German can produce the "incompleteness" effects via prosodic cues, there are languages that make use of morphological means, in addition to prosody, to get comparable results. Japanese and Korean are two such languages. As a matter of fact, the connection between contrastiveness and topicality is more transparently seen in these languages precisely because of this morphology. The morpheme (*wa* in Japanese and (*n*)*un* in Korean) is used to mark what is often called a sentence topic or a thematic topic. The distinction between the two uses of the topic morphology was noted as early as in Kuno (1973), who popularized the term Contrastive Topic (henceforth abbreviated as CT).[1]

This chapter is not meant to be a comprehensive study of CTs across languages. Rather, I will draw certain conclusions primarily from Japanese data, which are to a large extent comparable to Korean facts, and leave as an open question how possible cross-linguistic variations can be accommodated. In what follows, I review the key characteristics of Japanese CTs. During this process, I will make reference to some of the analyses of CTs that have been proposed in the past and examine whether they can meet the challenges that Japanese presents. Section 6.2 begins with the general semantic/pragmatic effects that Japanese CTs bring about. It will become clear that Japanese CTs induce the same kind of incompleteness effect that the English and German constructions do. Prosodic properties of Japanese contrastive topics will be reviewed next. The most significant in the discussion is the fact that a CT in Japanese can be the sole focalized element in a sentence, which makes it harder to employ for this language the kind of analysis that was proposed based on the presence of two distinct accents (e.g. A-/B-accents in English and Rise–Fall contour in German). While most of the discussions of CTs (of any language) have focused primarily on declarative sentences or assertions, Japanese CTs can be abundantly found with such non-assertive speech-act sentences as questions, imperatives, performatives, and exhortatives. This fact

[1] Although discussions of CTs in Japanese and Korean are not infrequent in the literature, they are often not connected to the idea of incompleteness and uncertainty. The description I find occasionally is that the use of a CT means that there are contrasted entities in the context, and if the CT is replaced by any of those entities, the sentence becomes false. This description is not precise, and we will see that there are many examples that do not match it. There are some predecessors whose characterizations are more accurate; see C.-M. Lee (1999, 2006) and Hara (2006).

is particularly challenging for an analysis based on knowledge states of a speaker (e.g. Hara 2006). In addition, a systematic account for the morphology of a CT remains elusive. Why should the same particle appear both in the contrastive and the thematic topic environments? No existing analyses present an insightful answer to this query. Section 6.3 will present a brief introduction of the main idea that I will pursue. I argue that a CT necessarily involves a set of alternative speech acts. A focal accent on a CT evokes a non-singleton focus value, and the topic marker *wa* functions as the guarantor of the maximal scope of this focus value: the focus alternatives of a CT cannot be used up until the speech act level. The effect of incompleteness/uncertainty is generated out of a set of alternative speech acts with help of the typical Gricean reasoning and inferences. In order to account for subtle differences among the various uses of CTs, however, it is necessary to augment the analysis with an additional ingredient, namely an economy-like principle that makes focusing a more preferred option than CT-marking. Section 6.4 discusses a few cases where this principle is at work. While my proposal solves many of the puzzles and challenges that Japanese CTs present, it also leads to a new set of questions and problems. In Section 6.5, I will address some of them. Although most of the discussions will remain inconclusive and no concrete answers will be given, I will lay out general directions or paths that one can pursue.

6.2 Japanese contrastive topics

6.2.1 *Basics*

Let me first establish the background understanding that Japanese CTs have effects of incompleteness or uncertainty similar to those found in English and German. Consider the English example (1).

(1) FREd ate BEAns.
    B-ACC A-ACC

The sentence (1) can be uttered as a partial answer to the question *What did the students eat?* provided that Fred is one of the relevant students. In such a situation, (1) leaves the impression that the speaker does not know what the other students ate. Hence, a sense of incompleteness/uncertainty on the speaker's part. The Japanese counterpart of (1) takes the following form.

(2) ERIka-wa MAME-o tabe-ta (kedo).
    Erika-TOP beans-ACC eat-PAST (but)
    'Erika ate beans (but...).'

This sentence can be used as a partial answer in the same situation described for (1). Importantly, if the topic marker *wa* is replaced by the canonical subject marker (i.e. the nominative particle *ga*), the sentence is infelicitous. Such a sentence can, of course, be used as a complete answer to the multiple *wh*-question *Who ate what?*, but no matter how one attempts to manipulate its prosody, the nominative counterpart of (2) can never induce the same effect that (2) has.

CT sentences also affect scope interpretations in the same way that the Rise–Fall contour in German does. In the example (3a), for instance, the universal quantifier, *alle Politiker* "all politicians", can take either wide or narrow scope with respect to negation. The use of the Rise–Fall contour in this sentence, however, disambiguates the sentence in such a way that it only has the negation-wide-scope interpretation.

(3) a. Alle Politiker sind nicht korrupt. (= Büring 1997, (1a))
all politicians are not corrupt
'All politicians are such that they are not corrupt' OR 'Not all politicians are corrupt.'

b. /ALLE Politiker sind NICHT\ korrupt. (= Büring 1997, (2a))
'Not all politicians are corrupt.'

In Büring (1997), this disambiguation phenomenon is tied to the incompleteness or partiality that CTs bring about. The surviving interpretation is weaker than the eliminated one in the sense that it leaves more open questions. In terms of the proportion of corrupt politicians, the eliminated interpretation (i.e. the universal-wide-scope reading) would give a final, definitive answer to it whereas the surviving one is still a partial answer and possibly invites further questioning (e.g. *if not all, are MOST politicians corrupt?*). Japanese displays patterns similar but perhaps not identical to German (cf. Hara 2006). Unlike German, almost all native speakers judge a sentence like (4a) unambiguous with the universal-wide-scope being the only available reading.[2] However, the effect of a contrastive topic is identical: it reverses the scope relation, and as a result, (4b) has the negation-wide-scope interpretation only.

(4) a. Minna-ga ko-nak-atta.
all-NOM come-NEG-PAST
'All people were such that they didn't come.'

[2] This gap between German and Japanese is based on the judgment reported in Büring (1997). However, this view does not seem to be universally endorsed. Krifka (1998, s. 3.3) points out that it is commonly assumed that the sentence is unambiguous, just as is the case with the Japanese counterpart.

b. MINNA-wa/Minna-WA ko-nak-atta.
all-TOP come-NEG-PAST
'Not all people came.'

From the facts above, we can safely conclude that Japanese has a way to express the incompleteness or partiality that the English A-/B-accents or the German Rise–Fall contour produces, and that the strategy necessarily involves the topic morphology.

6.2.2 *Prosody*

While Japanese CTs may give the impression that they should be treated on a par with the A-/B-accents in English and/or the Rise–Fall contour in German, they have one property that is not shared with the English and the German counterparts. As the example (2) shows, a Japanese CT can co-occur with a focused element, but the presence of an additional focus is not required. In (4b) in the previous subsection, for instance, the VP *ko-nak-atta* "did not come" is prosodically reduced, which indicates the absence of focus in the VP. As far as its prosodic characteristics are concerned, a Japanese CT behaves just like a proto-typical focus. In focusing in Japanese, a high pitch accent is placed on the focused element, and the pitch accent of the material on its right is radically lowered. This lowering, called post-focus reduction in Ishihara (2003), is more drastic than the typical down-step phenomenon that takes place within a phonological unit (i.e. a major phrase or an intermediate phrase). A CT does not seem any different from focus in this regard.[3]

It is my impression, although not statistically proved, that CTs without any other focus are more common or more frequently found than CTs with additional foci, just as may be the case that single focus sentences are more common than multiple foci sentences. Here are some more examples where the CTs are the sole focalized phrases.

(5) A: Who passed?
B: KEN-wa/Ken-WA ukat-ta.
KEN-TOP/Ken-TOP pass-PAST
'(At least) Ken passed.'

[3] Many people have pointed out to me that they prefer placing pitch peak on the particle itself to placing it on the focused phrase itself. Some claim that both patterns are possible but with different semantic/pragmatic effects. At this point, it is too premature to have a conclusive answer to this debate. It is worth noting, however, that other focus particles, such as *-dake* "only", *-sae* "even", or *-mo*, show the same pattern (cf. Tomioka 2007), which suggests that *-wa* is not exceptional.

(6) A: How many people will come to the party?

B: SAN-NIn-wa/San-nin-WA kuru-desyoo.
THREE-CL-TOP/three-CL-TOP come-EVID
'(At least) Three people will come, (as far as I can tell).'

(7) A: How much does a new hybrid car cost?

B: NIMAN-GOSEN-DORU-wa/Niman-gosen-doru-WA suru.
25,000 dollars-TOP/25,000 dollars-TOP costs
'It costs at least $25,000.'

Unfortunately, Büring's (1997, 2003) analyses, based on the English and German facts, require two different types of accents to generate what his analyses need.[4] Couched within the Alternative Semantics for Focus (Rooth 1985, 1992), Büring's accounts have two distinct levels of alternative generation, as briefly summarized below.

(8) a. A focus accent elicits a non-singleton set of propositions (= a focus value).

b. A topic accent operates on a focus value and elicits a set of sets of propositions (= a topic value), which is identified as a set of questions (cf. Hamblin 1973, Karttunen 1977).

c. In Büring (1997), the Disputability Condition requires that the ordinary value of the sentence not answer all the questions in the topic value. In Büring (2003), the utterance context of the sentence must furnish one of the questions as a discourse topic.

The Japanese prosodic patterns do not go along well with this type of analysis, as already pointed out by Hara (2006).

Although not introduced as a counter-argument against Büring's analysis, Hara (2006) further notes that a Japanese CT can be embedded, and furthermore, when it is embedded, it shows an ambiguity in terms of whose point of view is relevant for incompleteness. Consider the following example.

[4] Büring (2003: 532) anticipated a challenge of this kind, citing an English example like (i), where there does not seem to be any focused material other than the CT.

(i) a. Can Jack and Bill come to tea? $[\text{Bill}]_{CT}$ can. (= Büring 2003, (38), p. 532)

b. $\{\{x \text{ can come to tea}\} / x \in D_e\}$ (= Büring 2003, (39c), p. 532)

The suggestion made in connection to (ia) is to treat its topic value as a set of polar questions, which Büring identifies as a set of singleton sets of propositions, as in (ib). This may work for an example like (i), but perhaps not for (6) or (7) above, where the CT is a measure expression.

(9) a. [$_{CP}$ MARI-wa/Mari-WA kita-to] Erika-ga sinzite-iru.
MARI-TOP/Mari-TOP came-C Erika-NOM believe-PRES

b. Erika believes that Mari came, but Erika is not certain whether those other than Mari came.

c. Erika believes that Mari came, but the speaker is not certain whether Erika believes anyone other than Mari came.

I take this finding by Hara as another piece of evidence against the kind of approach Büring proposes. The reading (9b) requires the partiality/incompleteness from the point of view of Mari, the matrix subject. Büring's analysis, however, relies on the presence of a topic (a question) at the discourse level. This difficulty becomes more acute in (10).

(10) A: What did Fred say?

B: [$_{CP}$ MARI-wa/Mari-WA kuru-to]-**dake** (pro) it-ta.
MARI-TOP/Mari-WA come-C-only say-PAST
'(He) only said that [Mari]$_{CT}$ will come.' Or paraphrasable as '(He) only said that at least Mari will come.'

Notice that B's sentence gives a complete answer to A's question, as indicated by the exclusive expression "only". The use of a CT is nonetheless licensed.

The Japanese facts, therefore, encourage us to pursue an account that derives the effect of incompleteness without making appeal to two distinct accents.

6.2.3 *Contrastive topics and speech acts*

The second point that I would like to make concerns the appearances of Japanese CTs in a variety of speech-act sentences. As far as I can tell, almost all studies of CTs, of Japanese or otherwise, have focused almost exclusively on declarative sentences or sentences that correspond to assertion acts. We need not look hard, however, to find CTs in other types of speech-act sentences in Japanese.[5]

(11) Interrogative
... Zyaa Erika-WA/ERIka-wa doko-e itta-no?
... then Erika-TOP/ERIKA-TOP where went-Q
'..., well then, where did ERIka go?'

[5] Curiously, exclamative sentences cannot host a CT.

(i) #Kyooto-WA/KYOOto-wa nan-te kirei-nan-daroo
Kyoto-TOP/KYOto-TOP what-GER pretty-be-EVID
'#At least, how beautiful KYOto is!'

I confess that I do not understand why.

(12) Imperative
Eego-WA/EEGO-wa tyanto yatte-ok-e.
English-TOP/ENGLISH-TOP without-fail do-prepare-IMP
'At least, prepare yourself for ENGLISH.'

(13) Exhortative
Kyooto-NI-WA/KYOOto-ni-wa iko-o.
Kyoto-LOC-TOP/KYOTO-LOC-TOP go-exh
'At least, let's go to KYOto.'

(14) Perfomative
Sutoraiki-no-tame, KYOO-wa/kyoo-WA yasumi-to suru.
labor strike-GEN-due TODAY-TOP/today-TOP off day-COMP do
'Due to the labor strike, we make it that there be no work TODAY.'

The significance of CTs across speech acts is that they cannot be easily accounted for within the analyses of CTs based on knowledge states of speakers/attitude holders, such as Hara (2006), van Rooij and Schulz (2004), Schulz and van Rooij (2006), and Hara and van Rooij (2007). Hara (2006), for instance, makes a proposal that is summarized in (15).

(15) a. Pitch accent of a CT generates a set of scalar alternatives, which are ordered in terms of their semantic strength (cf. Sauerland 2004).
b. The appearance of a CT in a sentence α presupposes that, among the set of scalar alternatives to $[[\alpha]]$, there must be a proposition that is stronger than $[[\alpha]]$.
c. The appearance of a CT in a sentence α conventionally implicates that the speaker of/attitude holder of $[[\alpha]]$ believes that the stronger proposition is possibly false.

Hara's analysis is one step closer to the minimal theory of contrastive topics in the sense that, unlike Büring's account, it requires no extra semantic objects like topic values. As Hara herself acknowledges, however, her analysis primarily targets CTs in assertions where the notion of the speaker's knowledge state is most directly relevant. The extension of her analysis to other speech-act sentences with CTs is not impossible perhaps, but it is certainly not straightforward. For instance, take the imperative sentence (12), repeated here:

(12) Eego-WA/EEGO-wa tyanto yatte-ok-e.
English-TOP/ENGLISH-TOP without-fail do-prepare-IMP
'At least, prepare yourself for ENGLISH.'

Suppose that this sentence is given as a reply to someone's query of what he should do to increase his chance of passing the entrance exam to some university. As a speaker, I know that English, math, and natural sciences are equally important for the final outcome. Even then, I can still say (12). What would I mean in such a context? Possibly I think that preparing all three subjects would be too demanding for the poor examinee, and that focusing on English would be a better option than spreading his time and energy thin on all three. Or I may even be convinced that the examinee has no chance of passing and believe that studying English will have more practical advantage in the examinee's future than the other two subjects. These are some possible ways of my reasoning in saying (12). It is not clear how my epistemic state alone can derive these effects.

Even with assertions, on which epistemic knowledge is most directly reflected, a CT does not necessarily induce the kind of effect that a theory like Hara's predicts. Consider (5) again.

(5) A: Who passed?

B: KEN-wa/Ken-WA ukat-ta.
KEN-TOP/Ken-TOP pass-PAST
'(At least) Ken passed.'

Imagine the following scenario. B is one of the examiners, and A assumes that B has full, complete, knowledge of the outcome of the exam. Hara's implicature guarantees that B thinks it possible that the people other than Ken, let's say Mari and Erika, have failed. In this scenario, however, A would draw a conclusion that is much stronger than that. The assumption that B knows the outcome of the exam, coupled with the general Gricean principle that requires B to be as informative as possible, would lead to the conclusion that Mari and Erika did not pass. This is something that B can easily foresee that A would conclude under the circumstances. Then, why did B say what he said? Why did he not say (16), where *Ken* has the nominative *ga*.

(16) A: Who passed?

B: KEN-ga ukat-ta.
KEN-NOM pass-PAST
'KEN passed.'

Utterance (16B) would induce the same result as the CT marking in the scenario described above; the answer renders itself as the complete answer to A's question, and it therefore implicates that no one other than Ken passed. Thus, as far as the hearer's understanding of the speaker's epistemic state and its impact on the discourse context are concerned, the two versions should be no different from each other. Why, then, are both strategies allowed? The fact is, the two sentences are not the same. The CT version (5) may get the hearer to suspect that the speaker, even though he has full knowledge, wishes not to communicate the outcomes of the others to the hearer. I am not certain whether this kind of effect is derivable within the knowledge-based accounts of CTs.

6.2.4 *Topic morphology*

The final piece of the CT puzzle is perhaps the most obvious: Japanese uses the same morphology for a contrastive topic and a thematic topic. There has not been a serious attempt to explain why this morphological fact exists. In his (1997) account, Büring's use of the term *contrastive topics* has a theory-internal justification. He takes the position that a sentence topic is identified with a question in the discourse (Question-under-Discussion; QUD). This approach has been popular among formal semanticists and seen a lot of development in recent years (cf. von Fintel 1994, C. Roberts 1996, among many others). A topic value generated by a topic accent is a set of questions, and those questions are connected to a sentence topic (=QUD) in some ways. In this sense, we can see some connection between a contrastive topic and a sentence topic. The topic marking in languages like Japanese and Korean has encouraged a rival theory to this approach to emerge. Portner and Yabushita (1998) call it the "Topic as an Entity" approach, which has its root in a more pragmatics-oriented framework, such as the Prague School pragmatics. It was made popular by Kuno (1973), and Portner and Yabushita (1998) themselves endorse a version of it in their analysis of Japanese topics. Ironically, this approach makes it harder to see a correlation between CTs and sentence topics or thematic topics (henceforth, TTs).

Apart from the use of the same particle, CTs and TTs do not seem to share a lot of characteristics. Indeed, their differences are more noticeable than their similarities.

(17) a. A TT does not receive focal accents whereas a CT must.
  b. A TT is most typically found in the sentence-initial position whereas a CT can stay in situ.
  c. A TT must be nominal or quasi-nominal (i.e. NP, CP, or PP) whereas a CT can be of any category, including V(P), Adj(P), and Adv(P).
  d. A TT refers to a contextually familiar or recoverable entity whereas a CT can be familiar or novel.

If one pursues a uniform semantic theory of *wa*, these facts are certainly discouraging. (17d), in particular, defies the usual understanding of what a sentence topic is. It definitely refutes a seemingly sensible way of analyzing a CT as contrasted old information. To see how unlikely this idea is, we only have to look at one of our old examples.

(7) A: How much does a new hybrid car cost?
B: NIMAN-GOSEN-DORU-wa/Niman-gosen-doru-WA suru.
25,000 dollars-TOP/25,000 dollars-TOP costs
'It costs (at least) $25,000.'

Under no reasonable criterion of familiarity could "$25,000" be construed as old information in this context.

6.2.5 *Summary*

To sum up this section, we have seen the following characteristics of Japanese CTs.

(18) a. CTs induce the sense of incompleteness or non-finality in a way similar to the English A-/B-accents and the German Rise–Fall contour.
b. In Japanese, a CT can be the only focalized element in a sentence.
c. Japanese CTs can appear in speech act sentences other than assertions.
d. The same particle is used for a CT and a TT in Japanese.

None of the currently available accounts of CTs can accommodate all the properties without major modifications or additions to their machinery. What would the minimal and hence ideal analysis of Japanese CTs look like? It would make use of the semantic contribution of focal accent on a CT and combine it with the function of the topic particle. Coupled with some independently needed principles, these two ingredients should be sufficient to derive the effect of incompleteness across speech acts. Such an account may sound too idealistic, but I believe that it is an obtainable goal.

6.3 Contrastive speech acts

In this section, I will lay out the main proposal. It is based on a very simple idea: CTs operate at the level of speech act, and the effect of incompleteness/non-finality is a result of a general principle of conversation in the Gricean sense, augmented by the notion of competition between a CT and an ordinary focus in terms of informativity.

First of all, let me spell out one of the background assumptions needed for my proposal. I take a position, following Krifka (2001, 2002, 2004b), that speech acts are within the boundary of sentence grammar. Thus, they are integrated into syntactic representations (i.e. the presence of Speech Act Phrases (SAPs)) and compositional interpretations in the semantics component, can be quantified into when certain conditions are met (cf. Krifka 2001), and can even be embedded (cf. Krifka 2002, 2004b). More specifically;

(19) a. A speech act is one of the basic types (type *a*).

b. A speech act operator is a function from the type of the sentence radical it selects to type *a*. For instance, the Assertion Operator assert is type ⟨*st*, *a*⟩.

The second ingredient is a straightforward adaptation of the Alternative Semantics for Focus. A focal accent on a CT elicits a set of alternatives, or more generally, such an accent makes the focus value of a constituent containing a CT a non-singleton set.[6] The combination of the two background assumptions creates a new possibility, however. We can now have a set of alternative speech acts, and this is the idea I would like to exploit for CTs. Specifically, I propose (20).

(20) A focus on a CT is not closed off until the speech act level.

(20) means that the appearance of a CT necessarily leads to the existence of a set of alternative speech acts. Let me illustrate how the proposal works with a specific example.

(5) A: Who passed?

B: KEN-wa/Ken-WA ukat-ta.
KEN-TOP/Ken-TOP pass-PAST
'(At least) Ken passed.'

B's sentence in (5) has the LF structure shown in (21a). Its ordinary value, (21b), is straightforward. As for the focus value, I adopt Kratzer's (1991) idea of using designated variables and distinguished assignments although the way I implement it was inspired by Beck's (2006) re-interpretation of Kratzer's idea.[7]

[6] This set of alternatives can be ordered in terms of scale of informativity or semantic strength, as Hara (2006) assumes, but for the current proposal, it does not have to. See Section 6.4 for more discussion.

[7] Alternatively, a CT-marked constituent moves and adjoins to a SAP, resulting in a kind of structure that is used in the Structured Meaning Approach to Focus (cf. Krifka 1995, von Stechow 1989, among

(21) a.

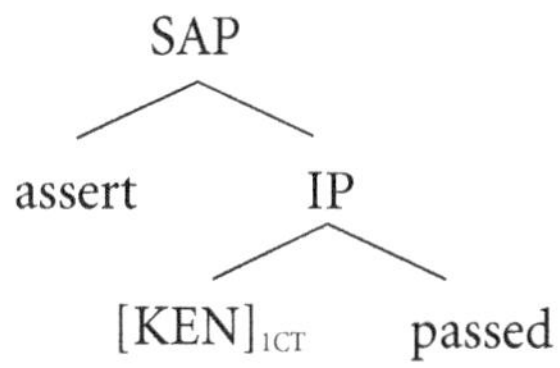

b. The ordinary values [[ ]]: For any ordinary assignment function g;
$[\![[\textbf{KEN}]_{1\,CT}\ \textbf{passed}]\!]^{(g)} = \lambda w.$ Ken passed in w
$[\![\textbf{assert}\ [\textbf{KEN}]_{1\,CT}\ \textbf{passed}]\!]^{(g)} = \text{assert}(\lambda w.$ Ken passed in w)

c. The focus values $[\![\ ]\!]_f$: For any ordinary assignment function g,
$[\![[\textbf{KEN}]_{1\,CT}\ \textbf{passed}]\!]_f^{(g)} = \{p: \exists h$, h is a distinguished assignment, $p = [\![[\textbf{KEN}]_{1\,CT}\ \textbf{passed}]\!]_f^{(g),\ h}\}$
$= \{p: \exists h$, h is a distinguished assignment, $p = \lambda w.\ h(1)$ passed in w}
$= \{p: \exists x \in D_e.\ p = \lambda w.$ x passed in w)}
$[\![\textbf{assert}\ [\textbf{KEN}]_{1\,CT}\ \textbf{passed}]\!]_f^{(g)} = \{a: \exists x \in D_e.\ a =$
assert(λw. x passed in w)}

Let us assume a context in which Ken, Mari, and Erika are under consideration. The set of alternatives would in effect be {assert(λw. Ken passed in w), assert(λw. Mari passed in w), assert(λw. Erika passed in w)}. From this point on, we make use of a typical rule of conversation:

(22) a. The speaker asserted that Ken passed.

b. There are three possible assertions that she could have made, but she only asserted one of them.

c. There must be a reason for her not asserting the remaining two.

The difference between the current proposal and the knowledge-based approach is that the speaker's knowledge state is only one of the possible reasons for not making the alternative assertions. Recall the scenario I described in the previous section. B is one of the examiners and is expected to have full knowledge of the outcome of the exam. Then, A, the hearer, deduces that the reason for B's using CT in his answer is not that he does not know whether Mari and Erika passed. Perhaps, B thinks that it is impolite to advertise Mari's

others). This is certainly a possibility, but there are a few reasons why I prefer the "in-situ" approach. First, when a CT is on a quantifier, it actually does not get wide scope (e.g. a CT-marked universal quantifier with negation) although this problem can be circumvented by assuming that the trace left by the movement is necessarily of the same type as the moved constituent. Second, a CT can be just on a verb or an adjective stem, as briefly summarized in (15c). This seems to necessitate movement of an X-category to an XP adjunction position. This problem, like the first one, could be solved. For instance, some kind of pied-piping operation would make the movement more syntactically acceptable. Unless we have good reason for postulating movement to SAP, however, I follow the in-situ approach.

and Erika's failure in the exam. Or else, B is just being coy or suspenseful, waiting for A to ask "How about Erika and Mari?" These exemplify possible ways of reasoning that A would make, and the use of a CT in this kind of situation invites such speculations. The use of the canonical nominative *ga* does not warrant such thoughts since it is supposed to be a final, complete answer to the question *Who passed?*

The other speech act sentences work in a similar fashion. Take the imperative example again.

(12) Eego-WA/EEGO-wa tyanto yatte-ok-e.
English-TOP/ENGLISH-TOP without-fail do-prepare-IMP
'At least, prepare yourself for ENGLISH.'

The use of a CT in (12) leads to the following reasoning: The speaker engaged in the imperative act "Prepare yourself well for English", and did not engage in the other possible imperative acts, such as "Prepare for Math", or "Prepare for Natural Sciences". We can easily imagine that the possible motivations for the speaker's choice include those I described in Section 6.2.3. All in all, the presence of alternative speech acts allows conversation participants to make the kinds of conjectures that would bring about the sense of incompleteness, uncertainty or non-finality.

Another distinct advantage of the current proposal over its competitors is that we now have a clue for the morphological marking. First of all, if there is any linguistic expression that can be outside the scope of speech act, it is one that corresponds to a topic. Krifka (2001) suggests, following Jacobs (1984), that topics can, or even must, be outside the scope of speech acts.

> Going one step further, one could argue that topics even *have to* scope out of speech acts. Topic selection is a speech act itself, an initiating speech act that requires a subsequent speech act, like an assertion, question, command, or curse about the entity that was selected. This was suggested, for example, in Jacobs (1984), where topics are assigned illocutionary operators of their own. (Krifka 2001: 25)

More recent advocates of "outside-of-speech-act" topics include Endriss (2009) and Ebert and Hinterwimmer (this volume). With this understanding of what topics are, we have a new perspective on the morphology of CTs. The requirement (20) provides that a set of alternatives should survive until it reaches outside the scope of a speech act. It is no accident that this task is carried out by the presence of the morpheme that typically signals topicality. Of course, the correspondence between the two types of topics is completely parallel if a CT moves and adjoins a SAP, as discussed in note 7. But even with the in-situ approach that I am taking in this chapter, we can consider *wa* as an indicator of "outside the scope of a speech act". Thus, one of the outstanding puzzle of CTs is explained.

The general idea I put forward is quite simple. With speech acts being fully represented in syntax, we can make use of speech acts as semantic objects within the sentence grammar. In particular, the mechanics used for focusing effects is extended to speech acts, and the notion of a set of alternative speech acts becomes available for pragmatic reasoning. This way of thinking allows us to see the morphological parallelism between CTs and TTs. However, this is not the end of my story. To get certain subtle differences among various uses of CTs, the general idea I presented in this section needs further refinement.

6.4 Fine-tuning the analysis

Recall how the pragmatic reasoning works for a set of alternative speech acts. A CT in a speech act α elicits a set of speech acts that includes α and its alternatives. Since the speaker engaged only in the speech act α, the hearer is invited to speculate possible reasons for the speaker's not engaging in the other speech acts. In a typical assertion context, not carrying out the other assertion acts can indicate that the speaker does not know whether the propositions in those assertions are true or false, or else that she knows that they are false. In other words, the pragmatic reasoning with a set of alternative speech acts does not preclude the possibility that the speaker is fully knowledgeable of the relevant facts. However, this possibility is often unavailable, particularly with CTs on measure phrases. Consider (6) and (7) again.

(6) A: How many people will come to the party?

B: SAN-NIn-wa/San-nin-WA kuru-desyoo.
THREE-CL-TOP/three-CL-TOP come-EVID
'(At least) Three people will come, (as far as I can tell).'

(7) A: How much does a new hybrid car cost?

B: NIMAN-GOSEN-DORU-wa/Niman-gosen-doru-WA suru.
25,000 dollars-TOP/25,000 dollars-TOP costs
'It costs (at least) $25,000.'

The following are the steps of pragmatic reasoning for (6).

(23) a. There is a set of alternative speech acts generated by the CT-marking on *san-nin* "three people". Let it be {assert(λw. one person will come in w), assert(λw. two people will come in w), assert(λw. three people will come in w), assert(λw. four people will come in w), assert(λw. five people will come in w), assert(λw. six people will come in w), ...}

b. The speaker asserted that three people will come.
c. There is a reason for the speaker's not engaging in the other acts.

The first two acts (i.e. assert(λw. one person will come) and assert(λw. two people will come)) are easily eliminated: the propositions in those speech acts are entailed by the proposition that three people will come. So, these assertions would be less informative than what the speaker's knowledge allows. How about the rest? Could it be that the remaining acts were not acted because the speaker knows that no more than three people will come? It should be possible, but then, such reasoning would lead to the "exactly three" interpretation that is not present in the example. With the utterance of (6B) comes the strong sense that the speaker believes it possible that more than three people will come. Thus, the most suitable translation of the sentence is "At least three people will come." In other words, the strengthening of its meaning to "exactly three" has to be blocked.

To explain this effect, I introduce the idea of competition between a CT and a focus. In the same context as in (6), B could have said (24).

(24) B: SAN-NIn kuru-desyoo.
THREE-CL come-EVID
'Three people will come.'

Without *wa*, the measure phrase *san-nin* is a typical focus, and it generates the usual "exactly three" implicature associated with a numeral expression. (24) was a possible response for B to make, but in reality, B used a CT in (6) instead. This leads to the addition of an extra step to (23):

(23) d. The speaker could have avoided using a CT, which would allow the implicature that three but no more than three people will come. There must be a reason for the speaker's choosing a CT.

This would successfully eliminate from the hearer's reasoning the possibility that the speaker has full knowledge of the facts concerning the question of how many people will come. Among the alternatives listed in (22a), the first two acts (i.e. assert(λw. one person will come) and assert(λw. two people will come)) are eliminated for the same reason as before. However, the remaining alternatives would not get the same reasoning as before because of the weakening effect that comes out of (23). Hence, the result is the obligatory "at least three" interpretation for (6B).

The competition between a CT and a usual focus can also deal with the kind of case for which Hara (2006) used a scalar presupposition. Recall that she had three conditions, given in (15), for using a CT.

(15) a. Pitch accent of a CT generates a set of scalar alternatives, which are ordered in terms of their semantic strength (cf. Sauerland 2004).

b. The appearance of a CT in a sentence α presupposes that, among the set of scalar alternatives to $[\![\alpha]\!]$, there must be a proposition that is stronger than $[\![\alpha]\!]$.

c. The appearance of a CT in a sentence α conventionally implicates that the speaker of/attitude holder of $[\![\alpha]\!]$ believes that the stronger proposition is possibly false.

The necessity of scalar alternatives and the presupposition (15b) comes from infelicitous examples like (25).

(25) #MINNA-wa/Minna-WA kita.
ALL-TOP/all-TOP came
'[Allpeople]$_{CT}$ came.'

The CT-marking on *minna* "all (people)" would generate the following set of scalar alternatives.

(26) {All people came, most people came, some people came, no one came}.

The presupposition (15b) requires that there be an alternative stronger than the ordinary value of the sentence *all people came*. However, there is none in (26). The propositions *most people came* and *some people came* are entailed by the ordinary value, and the last one contradicts it. So, the condition is not met, and that is why (25) is infelicitous.

In the previous section, we have seen that the conventional implicature based on the speaker's knowledge (= (15c)) can be dispensable once we make use of a set of alternative speech acts. However, the presupposition (= (15b)) seems still necessary because the following reasoning is valid.

(27) a. The CT-marking on *minna* generates a set of speech acts; {assert(λw. all people came in w), assert(λw. most people came in w), assert(λw. some people came in w), assert(λw. no people came in w)}

b. The speaker asserted that all people came.

c. There must be a reason for the speaker's not engaging the other acts.

d. The reason for not asserting no one came is that it is false. The reason for not asserting the rest is that they are entailed by the asserted proposition and are therefore not as informative as the asserted proposition.

With the addition of the competition between a CT and a focus, however, the presupposition (15b) is no longer necessary. The speaker could have avoided the use of a CT by simply focusing on *minna* "all (people)", and there should be a reason for this avoidance. Notice, however, that the nominative counterpart, shown below, gives exactly the same interpretation.

(28) MINNA-ga kita.
all-NOM came
'All people came.'

Therefore, the use of a CT becomes infelicitous. The idea behind this way of analyzing (25) is that a CT is a more marked option than an ordinary focus. The use of a CT is justified only when the result is distinct from that of a focus.[8] In general, Hara's presupposition says that a CT cannot be used in a context where the sentence with the CT is the strongest among the scalar alternatives. Instead of using comparison among semantic objects (= alternatives), I am suggesting to use comparison between two different focal strategies; a CT vs. a focus. The latter should be stronger than the former, and when there is no difference, the CT option becomes infelicitous. This strategy comparison is needed to ensure weaker interpretations in sentences with measure phrases, and replacing Hara's presupposition with it would make our analysis one step closer to the minimal theory of CTs.[9]

6.5 More speculations and further issues

Let me summarize what I have so far argued for. I introduced the notion of alternative speech acts as a possible focus semantic value in natural language. This new semantic object is derived from the two essential facets of CTs; focal accent and the topic particle *wa*. A set of alternative speech acts interacts with the general Gricean reasoning and inferences, one of which is the "economy-like" principle that prefers the focus strategy to the CT when the results are indistinguishable. The current proposal makes many of the construction-specific properties and/or principles of CTs totally dispensable;

[8] It is interesting to note that Büring (2003) arrives at exactly the opposite conclusion (Büring 2003, (36), p. 531) although the empirical phenomena he derives this conclusion from are rather different from our current concerns.

[9] In Tomioka (2009), I discuss two additional cases where the competition between focus and CT is relevant. First, when the focus option is independently blocked, the CT option does not necessarily undergo pragmatic weakening and is compatible with the exhaustive interpretation. Second, when a disjoined DP is marked for a CT, the focus-related exhaustivity (i.e. A or B but not the others) disappears, but the exhaustivity associated with *or* itself (i.e. A or B but not both) is retained. These facts are correctly predicted under the current proposal.

topic values and the disputability requirement of Büring (1997) and the scalar presupposition and the conventional implicature of Hara (2006) have no roles to play. Having said this much, I am the first to admit that there are many unresolved issues in connection with the new idea, and I would like to use this penultimate section for brief discussions of them.

6.5.1 *Focus vs. CT*

First of all, the competition between a CT and a focus needs further theorizing. I take it for granted that focusing induces a kind of exhaustivity implicature that affects possible interpretations of a CT. I did not even attempt, however, to show how this exhaustivity should be derived. One of the most intriguing trends in the formal semantic and pragmatic literature is to question the long-established boundaries between sentence grammar and pragmatics. Particularly interesting for the purpose of this paper is the treatment of certain types of pragmatic strengthening mechanisms. Fox (2006), for instance, advocates the view that the exhaustivity associated with disjunction and other scalar items is derived via the exhaustivity operator (Exh).[10]

(29) $\text{Exh}\ (A_{\langle st,t\rangle})(p_{st})(w) = p(w) \& \forall q \in NW(p, A) : \neg q(w)$ (= Fox 2006, (15))

Notes: $A_{\langle st,t\rangle}$ = a set of (scalar) alternatives
NW (p,A) = a set of alternatives that are not weaker than p

Although Fox does not specifically discuss the exhaustivity associated with contrastive focus, its potential to be extended to contrastive focus is quite obvious.[11] What this exhaustivity operator does is quite simple. It is a lot like *only*, except that the denotation of its argument (the proposition denoted by the sentence Exh selects) is part of its assertive content whereas it is presupposed with *only*. What I find remarkable is that this operator performs a job that is almost completely parallel to the pragmatic reasoning that we use on a set of alternative speech acts. Of course, the parallelism is not complete because speech acts are not negated in the same way as propositions are and we did not appeal to the scalarity of alternatives for speech acts. Nonetheless, the similarity is there. Our pragmatic reasoning says that, out of a given set of alternative speech acts, the ordinary value is the only one that was carried out, and the speech act agent did not engage in the others in the set. Therefore, it is not inconceivable that CTs use the same operator, Exh, as focus does, and that the only difference is the matter of scope: Exh for focus takes scope over

[10] (29) is not the final version that Fox endorses, but it is simpler and serves our purpose perfectly well.

[11] Fox himself is quite receptive to this possibility (Fox 2006, n. 13).

propositions whereas Exh for a CT sits above a Speech Act Phrase. This is an interesting possibility that deserves close examination. Whether Exh is the right operator for CTs or not, the presence of some operator is clearly needed. Although I have concentrated on the cases where CTs are the only focalized phrases, a CT can co-occur with a focus. Recall that we had the following sentence at the beginning of the paper.

(2) ERIka-wa MAME-o tabe-ta (kedo).
Erika-TOP beans-ACC eat-PAST (but)
'Erika ate beans (but...).'

Imagine that (2) was uttered as an answer to the question *What did the students eat?* Then, *MAME-o* "beans-ACC" is the focus of the sentence whereas *ERIka-wa* "Erika-TOP" is a CT that induces the non-finality effect. Under the mechanism that I suggested in Section 6.3, both phrases bear focus indices. However, the two indices should not be bound in the same way. In particular, the index on a CT cannot be bound below the speech act level, and if one is committed to not using movement of a CT at LF, the best option seems to use two operators that selectively bind the indices, as in (30).

(30) $[_{SAP}$ OP_1 $[_{SAP}$ Assert $[_{IP}$ OP_2 $[_{IP}$ $[\text{ERIka-wa}]_1$ $[\text{MAME-o}]_2$ tabe-ta]]]]

Attractive though it may look, this idea is too simplistic. The complexity of the problem goes beyond the scope of this speculative section, but here is the bottom line: With (30), the speaker asserted Erika ate beans (with the implicature that she ate nothing else), and the speaker did not engage any assertion about anybody other than Erika that they ate beans (and nothing else). This would lead to the speculation that the speaker does not know whether the others ate beans (and nothing else) or not. This is obviously wrong. What we should get is, the speaker does not know *what* the others ate, but unfortunately (30) cannot deliver it. In Tomioka (2009) I offer an account based on Wold's (1996) idea of double-indexing of a focus, as illustrated in (31).

(31) $[_{SAP}$ $OP_{1,2}$ $[_{SAP}$ Assert $[_{IP}$ OP_3 $[_{IP}$ $[\text{ERIka-wa}]_1$ $[[\text{MAME-o}]_3]_2$
tabe-ta]]]].

In this representation, the focus is interpreted twice; at the propositional and the speech act levels. Since the focus, as well as the CT, generates alternatives at the speech act level, we can arrive at the correct uncertainty meaning. The implementation of the indexing in (31) requires too much background information and technical details for this chapter to accommodate. The reader is referred to Tomioka (2009) for a full analysis.

The idea of competition between focus and CT also predicts that there should be no cases where both the focus and the CT options are available under the same pragmatic condition. This prediction seems to be falsified by an example like (32).

(32) Q: Who brought what to the party?

MAri-ga/-wa	SARAda-o,	Ken-ga/-wa	PIZA-o,	sosite
Mari-NOM/CT	salad-ACC	Ken-NOM/CT	pizza-ACC	and then

ERIKA-ga/-wa	KEEki-o	motteki-ta.
Erika-NOM/CT	cake-ACC	bring-PAST

'Mari brought salad, Ken a pizza, and Erika a cake.'

Both answers seem perfectly appropriate, and presently I cannot give a coherent story about the difference(s) between the two answers.

6.5.2 *CTs in complex sentences*

As some readers might have noticed, sentences like (32) pose a different, and perhaps more serious, challenge to my proposal. Since an instance of a CT signals the presence of a speech act and there is one CT in each of the three conjuncts, I am forced to say that there are three speech acts conjoined in (32). It is a part of a larger challenge, namely appearances of CTs in complex sentences. Not only do CTs in Japanese show up in conjunctions, but they can also appear in embedded clauses, as we have seen in Section 6.2.2. Recall that Hara (2006) noted that an embedded CT creates an ambiguity with respect to whose point of view the notion of uncertainty is related to. I repeat the example below.

(9) a.

[$_{CP}$ MARI-wa/Mari-WA	kita-to]	Erika-ga	sinzite-iru.
MARI-TOP/Mari-TOP (CT)	came-C	Erika-NOM	believe-PRES

b. Erika believes that Mari came, but Erika is not certain whether those other than Mari came.

c. Erika believes that Mari came, but the speaker is not certain whether Erika believes anyone other than Mari came.

Under the proposal presented in this chapter, the interpretation (9b), where Erika, the agent of believing, is not certain about the others, has to be derived via pragmatic reasoning on embedded speech acts, rather than matrix ones. Thus, it must be concluded that speech acts can be embedded, as Krifka (2002, 2004b) argues based on phenomena other than the ones discussed here. As far as embedded CTs are concerned, there are a few encouraging signs for such a direction. First, as Hara (2006) notes, not all embedded sentences can host CTs. In particular, some adjunct sentences like *when*- and *if*-clauses cannot have CTs and relative clauses. Those are kinds of embedded sentences that do not seem to license speech acts within themselves.[12] Second, non-assertive speech-act sentences in Japanese are most often marked with particular sentence-final particles; *-ka* for a question sentence, *-(y)oo* for an exhortative one, and *-e*, *-ro* for an imperative one. These particles also show up in embedded sentences. Here is an example of an embedded imperative.

(33) Syachoo-wa Erika-ni kinoo-made-ni kare-no-hisho-ni
president-TOP Erika-DAT yesterday-by him-secretary-DAT
hookoku-si **ro**-to meeree-sita-noni...,
report-do **imp**-COMP order-did-but...
'Although the president ordered Erika to report to his secretary by yesterday, ...'

Since the pronoun *kare* in (33) is meant to refer to the president and the indexical *kinoo* "yesterday" points to the day prior to the time of utterance of the entire sentence (not the time of the president ordering, which would make his order impossible to carry out), the embedded sentence should not be considered as a direct quote. The appearance of the imperative marker *-ro* is nonetheless necessary in this sentence. I admit that appearances of CTs in conjunctions are perhaps harder to justify. However, I can point out a few facts that suggest that the idea of conjoined speech acts is not as far-fetched as it may sound. First, unlike conjunctions, disjunctions do not seem to support CTs. Or at least, I have not managed to find any examples of disjunctions with CTs that are acceptable. This fact is in accordance with Krifka's (2001) claim that speech acts cannot be disjoined. Second, the sentence-final particles that signal conjunctions, such as *-si* (and) and *-ga/-kedo* (but), can conjoin heterogeneous speech-act sentences, as witnessed in (34).

(34) a. *Assertion-but-Question*
Tanaka-san-no kotoba, chotto kawatte-iru-**kedo**,
Tanaka-san-GEN speech a little strange-**but**
doko-no syussin desu-ka?
where-GEN birth-place be-Q
'Tanaka-san speaks a bit strange, but (???) where is he from?'

[12] Of course, those embedded clauses can host CTs when they contain attitude predicates like *think*, *believe*, *say*, etc. One interesting fact, also noted in Hara (2006), is that, unlike *when*- or *if*-clauses, *because*-clauses can sometimes contain CTs. This remains as a mystery to our current analysis. See Hara (2006, ch. 3) for more discussions.

b. *Assertion-and-Assertion-and-Question*
Uti-ni-mo i-nai-**si**, keetai-ni denwa-site-mo
home-also be-NEG-**and** mobile-to phone-do-also
de-nai-**si** ittai doko-ni i-ta-no?
answer-NEG-**and** on-earth where be-PAST-Q
'You weren't home, and you didn't answer when I called your mobile, and (???) where the hell were you?!'

What we see above indicates that Japanese conjunction is significantly more liberal, compared to the English counterpart, in allowing mismatched speech-act sentences to be conjoined.

6.5.3 *Variation across languages*

Another sticky issue is cross-linguistic variability. One of the main motivations for the current proposal is the fact from Japanese prosody: a CT can be the only focalized element in a sentence and does not require the presence of another focus. This fact is, in my opinion, quite solid, and I do not think it is wise to look for some kind of hidden focus in the prosodically reduced portion that follows a CT. It seems genuinely true, on the other hand, that the German CT must be accompanied by a focus. Thus, the natural question is whether the two seemingly contradictory facts can be reconciled. Putting it differently, is it possible to come up with a uniform analysis of both types? Considering that the two types of CTs yield practically the same kind of semantic/pragmatic effects, one might feel uncomfortable at not having a unifying account for both. Such an account may be ideal, but I do not think that it is necessarily so. For instance, my proposal relies crucially on the morphological cue for the "outside the scope of a speech act" property. There are many languages that are not equipped with such a strategy. Then, we can speculate that the system I propose is not available for those languages that cannot reliably indicate the ultimate wide scope. They may use two distinct accentual/prosodic patterns to elicit the same effects, which may be better analyzed in the way that Büring (1997, 2003) or his competitors (e.g. Krifka 1998) propose.

6.6 Closing remarks

As I said at the beginning, this chapter is not meant to put an end to the discussion of CTs. On the contrary, it is more likely to stir further debates and discussions on the topic, particularly from cross-linguistic perspectives. I nonetheless believe that the analysis has answered all the questions and challenges that we started out with, and I hope to have shown that, as far as those answers are concerned, my proposal does a decent job. It also makes use of the minimal amount of machinery; the alternative generation by a focal accent and the preservation of the alternatives to the speech act level by the topic morphology, coupled with the Gricean reasoning, which we can all agree is needed independently. The comparison between the focus and the CT strategies also plays an important role in determining possible pragmatic effects for CTs. As is the case for any new proposal, my analysis generates a new set of questions and challenges, some of which I managed to address in the previous section.[13]

It is indeed ironic that this chapter itself is a lot like a CT: it makes a few main points but leaves other issues unsaid or unresolved. Hence, it epitomizes the sense of incompleteness/non-finality. I do believe, however, that it presents a coherent portrayal of CTs that can at least serve as a viable alternative to the existing analyses.

[13] Towards the completion of this chapter, several new analyses of CTs came to my attention. Although I cannot do justice to them in this chapter, some deserve mention for their relevance to the discussion here. Yabushita (2008) proposes an account of Japanese CTs based on Partition Semantics of interrogatives (cf. Groenendijk and Stokhof 1984). One of the key observations by Yabushita is the multiple occurrences of CTs within one sentence, and I believe that this fact can be easily accommodated into the current proposal. **?** (2009) observes a variety of interesting ordering asymmetries and constraints with CTs and foci. When both a CT and a focus appear in a sentence, for instance, the CT must precede the focus (see also Wagner 2008 for similar observations in German). As one can probably guess, my analysis in the chapter has very little to offer for such syntactic effects. It remains to be seen whether and how the current proposal can be enriched so as to accommodate the facts presented by Vermeulen.

7

Biased questions, intonation, and discourse

BRIAN REESE AND NICHOLAS ASHER

7.1 Introduction

This chapter explores how intonation contributes to what is said through an examination of a particular sentence type, in other words, a coincidence of grammatical structure and discourse function (Sadock and Zwicky 1985). Specifically, we discuss a range of interrogative constructions, each characterized, at least in part, by subject–auxiliary-verb inversion, that appear to have properties of both questions and assertions. We refer to these sentences as biased questions, as they all convey an expectation, or bias, on the part of the speaker toward a specific answer to the question. We show that biased questions exhibit characteristics typical of assertions. Furthermore, we argue that prosody, that is, intonational phrasing, intonation, and stress, plays an important role in the interpretation of these questions.

The biased questions that the chapter addresses are exemplified by the examples in (1)–(3). Tag questions, as in (1), provide a natural starting point, as they wear their illocutionary force(s) on their sleeve, so to speak.

(1) a. Jane isn't coming, is she?
    b. Jane is coming, isn't she?

As a matter of clausal syntax, they possess both declarative and interrogative components. It is not unexpected, then, that they have properties of both assertions and questions. However, a number of more nuanced issues arise regarding their interpretation. First, tag questions are not always biased. Second, the exact discourse function of the interrogative component is influenced by the final pitch movement over the tag.

The authors wish to thank the two anonymous reviewers as well as the editors, Caroline Féry and Malte Zimmerman, for helpful comments.

Negative polar questions as in (2) are also biased toward a specific answer (Ladd 1981; van Rooy and Šafářová 2003; Romero and Han 2004; Reese 2006a).

(2) a. Isn't Jane coming *too*?
 b. Isn't Jane coming *either*?

We will argue that on what is called the outside-negation reading of (2a) (cf. Ladd 1981) negative polar questions convey an assertion. We link the biased reading of negative polar questions to the neutral reading of tag questions and discuss the weaker form of bias present on the so-called inside-negation reading, (2b).

The examples in (3) convey a bias toward a negative answer. Borkin (1971) links negative bias to the presence of a strong negative polarity item (NPI) (cf. Ladusaw 1979, Krifka 1995, van Rooy 2003, Guerzoni 2004).

(3) a. Did John *lift a finger* to help Mary?
 b. Is John EVER going to help Mary?
 c. Did I TELL you writing a dissertation would be easy?

But it also occurs when a weak NPI like *any* or *ever* is pronounced with emphatic stress, as in (3b), and in certain examples of narrow focus, as in (3c) from Sadock (1971). As far as we know, the examples in (3) have not received a unified account (see Asher and Reese 2005 for a recent attempt). It seems to us that such an account is desirable and we attempt to provide one here.

Although the issue of bias does not necessarily intersect with most mainstream research on information structure, we feel that consideration of lesser-studied sentence types like biased questions reveals a need for more fine-grained notions of information structure which include rich information about the beliefs, intentions, and goals of conversational agents beyond the given–new partitioning that most work in information structure addresses. Much current research on intonation addresses these issues (cf. Steedman 2000, Gunlogson 2003, Nilsenová 2006, among others). In addition, the theory of discourse interpretation adopted here provides information beyond that offered by theories that structure discourse in terms of questions and sub-questions under discussion (cf. C. Roberts 1996, Büring 2003).

The research presented here falls within the tradition of information structure research focusing on what Krifka (2007) refers to as information management, that is, those aspects of discourse understanding that address the informational needs of discourse participants. Questions, of course, are the quintessential means of indicating these informational needs, typically requesting information about what the world is like. But a lot of research stops short, failing to note that there are additional kinds of informational needs that a conversational agent might have. The discussion below addresses precisely this issue, as biased questions are a conventionalized means of conveying specific types of informational needs. In addition to information about the world, a conversational agent might need information about the status of a proposition with respect to the common ground, or information about the cognitive state of his or her conversational partners. It is precisely these discourse functions that biased questions conventionally serve and which we discuss in detail below.

7.2 Types of biased question

The present section provides more detail on the constructions introduced in Section 7.1.

7.2.1 *Tag questions*

Although tag questions have received a fair amount of attention in descriptive grammars of English (Quirk, Greenbaum, Leech, and Svartvik 1985; Huddleston and Pullum 2002) and from syntacticians (Culicover 1992; Kay 2002), there is relatively little formal semantic and pragmatic work, and certainly little or no recent work.[1] Nevertheless, tag questions provide an interesting case study of how intonation contributes to what is said.

Syntactically, tag questions consist of a declarative clause, or anchor, paratactically related to a reduced interrogative clause, or tag, as in (1) from Section 7.1. While these surface syntactic features contribute to the presence of both an assertion and a question at the level of discourse logical form, (i) they do not guarantee it, and (ii) they do not provide information about the rhetorical contribution of the tag. The most one could conclude from these surface syntactic cues is that the assertion blocks the default communicative goal associated with the question, viz. to know an answer. Additional lexical and phonological cues provide information for the computation of more fine-grained discourse functions.

We assume the description of intonational tunes provided by the Tones and Break Indices labeling conventions (ToBI, Beckman and Elam 1997), which are based on Pierrehumbert (1980). Within the ToBI framework, intonational tunes are constructed according to a simple generative grammar. An intonational phrase—the largest prosodic unit in ToBI—consists of one or more

[1] Older treatments of the semantics and pragmatics of tag questions include Sadock (1974), Millar and Brown (1979), Rando (1980), Ladd (1981), Oehrle (1987).

intermediate phrases followed by a high or low boundary tone, L% or H%. An intermediate phrase consists of one or more pitch accents followed by a phrase accent, L− or H−. The ToBI annotation scheme as described in Beckman and Elam (1997) assumes five pitch accents: L*, H*, L + H*, L* + H, H*+!H. Pitch accents are tonal targets aligned with stressed syllables. The last pitch accent in an intermediate phrase is the nuclear pitch accent, which, at least in English, marks focus.

There are two phonological distinctions relevant to understanding the discourse contribution of tag questions. First, the sequences of phrase accent and boundary tone—that is, final falling vs. final rising intonation—on the tag itself is claimed to have interpretive effects (cf. Rando 1980, Quirk *et al.* 1985, Huddleston and Pullum 2002, among others.).[2] Most, if not all, descriptions of tag questions note this fact and associate some interpretation with the fall–rise distinction. Descriptions of these interpretations are remarkably consistent, and we have no reason to dispute them here. Tag questions with falling intonation ask for acknowledgment from the addressee that the communicative goal of the anchor has been achieved (cf. Huddleston and Pullum 2002, for example). In the case of declarative anchors the speech-act related goal, or SARG, is simply that the addressee comes to believe the truth of the anchor.

The constructed dialogue in (4) provides an illustration. Imagine that *A* and *B* are trying to complete a task at which neither is proficient, but at which Julie is known to be.[3]

(4) a. A: Julie wouldn't do it that way.

b. B: Well, Julie isn't here, / is she.

Intuitively, *B*'s turn in (4b) does not convey any doubt regarding the truth of the anchor, but rather is used to get *A* to acknowledge that Julie is not present (and therefore that how Julie would accomplish the task is irrelevant to the present situation).

Tag questions with final rising intonation are still biased toward an answer confirming the content of the anchor, but express some uncertainty on the part of the speaker. The dialogue in (5) illustrates.

(5) a. A: Can Julie do it for us?

b. B: Julie isn't here, / is she?

[2] It is an empirical question how best to characterize the fall–rise distinction. For example, Gunlogson (2003) distinguishes between falling and non-falling. As a result, she counts final plateaus, i.e. H-L% sequences, as rises. We will not address these issues in any detail here and follow Gunlogson in drawing the line between falling and non-falling tunes.

[3] We follow a common orthographic convention of marking final falling intonation with a period, and a final rising intonation with a question mark.

B's turn in (5b) conveys a weak belief that Julie is not present (and thus answers *A*'s question). The tag itself, however, expresses doubt or uncertainty, in other words, the speaker is open to the possibility that he is wrong. On this use, then, the tag acts as a request for confirmation of the anchor. If the addressee has evidence to the contrary, he should provide it; if not, then he should acknowledge the truth of the anchor.

Both of these interpretations are biased as we defined bias above, in that the anchor is asserted (see the discussion in Section 7.4), blocking the default intention associated with the interrogative component of the utterance, viz. to know an answer. However, tag questions can function as neutral requests for information, as shown by (6).

(6) a. A: We need to find someone who has done this before.

b. B: Julie isn't here = is she?

Several aspects of linguistic form appear to be necessary for a neutral reading to be possible. First, it's possible only with a negative anchor and when there is little or no rhythmic break between the anchor and the tag (Ladd 1981; McCawley 1988; Huddleston and Pullum 2002). Moreover, the anchor of a neutral tag question like (6b) is more likely to contain a H– phrase accent than either (4b) or (5b).

Ladd (1981) refers to tag questions like (4b) and (5b) as nuclear tag questions, indicated with a slash between the anchor and the tag, and to the neutral reading in (6) as a postnuclear tag question, indicated with an equals sign. Ladd's description of postnuclear tag questions corresponds in the ToBI scheme to an utterance with just a single intonational phrase, which itself consists of a single intermediate phrase (and a boundary tone). The nuclear pitch accent, the last pitch accent in the intermediate phrase, occurs somewhere in the anchor. On this view, there is no pitch accent on any element of the tag itself. Nuclear tag questions, on the other hand, consist of at least two complete intonational phrases, or one, which itself contains at least two intermediate phrases. The crucial feature is a rhythmic break between the anchor and the tag. We find this description of postnuclear tags dubious. It is difficult in our experience not to hear a pitch accent on the auxiliary verb in the tag on either a biased or neutral reading. Of course, one could posit the existence of postnuclear pitch accents (which is what Ladd appears to have in mind), but this is a controversial claim. We do, however, agree with Ladd and others, notably McCawley (1988) and Huddleston and Pullum (2002), that neutral readings of tag questions contain a *weaker* boundary between the anchor and tag than nuclear tag questions. For this reason, we prefer to recast the nuclear/postnuclear distinction in terms of intonational phrasing as follows: nuclear tag questions consist of at least two complete intonational phrases, at least one for the anchor and one encompassing the tag. Postnuclear tag questions consist of one intonational phrase that is constructed from at least two intermediate phrases for the anchor and tag. It is our contention (as discussed in more detail below) that these prosodic facts conspire with syntax and semantics to yield two speech acts or one. In either case, the computation of the discourse function of the tag relative to the anchor proceeds in a similar fashion. However, postnuclear prosody allows a neutral interpretation, which nuclear prosody does not.

7.2.2 *Negative polar questions*

Many analyses of the semantics of interrogatives predict that positive and negative closed interrogatives like those in (7a) and (8a) are equivalent. On these approaches a question is a partition of the space of logical possibilities, each element of the partition being a proposition expressing a direct answer to the question (cf. Hamblin 1958 and more recently Groenendijk and Stokhof 1997). At first glance, the prediction appears correct; the same propositions count as direct answers to both types of sentence, as shown by intuitions about the simple *yes* and *no* answers to (7a) and (8a) below.

(7) a. A: Is Jane coming?
    b. B: Yes, she is. (= Jane is coming.)
    c. B: No, she isn't. (= Jane is not coming.)

(8) a. A: Isn't Jane coming?
    b. B: Yes, (of course) she is. (= Jane is coming.)
    c. B: No, she isn't. (= Jane is not coming.)

However, negative polar questions differ from positive polar questions in at least two important respects. First, negative questions convey a backgrounded attitude on the part of the speaker toward the proposition expressed by a positive answer (Ladd 1981; Büring and Gunlogson 2000; Han 2002; van Rooy and Šafářová 2003; Romero and Han 2004; Reese 2006a). Sentence (9b), for example, is a felicitous continuation of (9a), which conveys a stance of epistemic-neutrality by the speaker regarding the issue raised by the question.

(9) a. I have no beliefs on the matter. I just want to know...
    b. Did the President read the August 6 PDB?
    c. #Didn't the President read the August 6 PDB?

Example (9c) is infelicitous in the same context, as it conveys a prior belief toward the issue raised by the question that conflicts with the neutrality expressed in (9a), namely, that the president read (or ought to have read) the briefing.

The second way in which the two types of polar question diverge is that negative questions are ambiguous in a way that positive questions are not (Ladd 1981; Büring and Gunlogson 2000; Huddleston and Pullum 2002; Romero and Han 2004). The two readings can be disambiguated by including a polarity item. Negative questions that contain a positive polarity item (PPI), as in (10), have a strong bias toward a positive answer. Ladd (1981) dubs this interpretation the outside-negation reading. Negative questions that contain a negative polarity item, as in (11), on the other hand, are biased toward a negative answer, Ladd's inside-negation reading.

(10) a. Didn't Kim read the report *too*?
b. Aren't there *some* vegetarian restaurants around here?

(11) a. Didn't Kim read the report *either*?
b. Aren't there *any* vegetarian restaurants around here?

Section 7.4 shows that outside-negation negative questions like those in (10) involve an assertion and are consequently complex speech acts, whereas inside-negation negative questions as in (11) are not. We argue that outside-negation, like negation in the anchor of a neutral tag question, is metalinguistic.

7.2.3 *Emphatic focus questions*

Questions that contain a strong, or idiomatic, negative polarity item, such as those in (12), convey a bias toward a negative answer. Of the kinds of biased question that we consider here, these have received the most attention from formal semanticists (cf. Borkin 1971, Ladusaw 1979, Krifka 1995, Abels 2003, van Rooy 2003, Guerzoni 2004, Asher and Reese 2005). These previous analyses, for the most part, center around the interpretation of polarity items, that is, their lexical semantic properties, in combination with certain well-attested pragmatic principles.

(12) a. Did Fred contribute *a red cent* to the campaign?
b. Did John *lift a finger* to help Mary?
c. Does Fred do *a damn thing* at the office?
d. Did Mary *bat an eye* when you threatened her?

These questions have an intonational component, as well, in at least an indirect sense: their interpretation appeals to focus, which in English is marked primarily by the placement of the nuclear pitch accent. Krifka (1995) is explicit on this point, noting that since NPIs introduce alternatives over the denotation of the NPI, they resemble items in focus. Krifka also notes that strong NPIs necessarily require emphatic focus, which he associates with an emphatic assertion operator mirroring the presupposition of the adverb *even*, and whose meaning others, notably van Rooy (2003) and Guerzoni (2004), assume is shared (in part) by NPIs. This raises the question of whether it is the semantics of strong NPIs which is directly responsible for bias, or whether certain intonational facts are primarily responsible.

The examples in (12) all contain strong NPIs. Consequently, the presence of such lexical items appears to be a sufficient condition for bias to exist. But it is not a necessary condition. Questions with domain wideners like *any* and *ever* (Kadmon and Landman 1993; Krifka 1995) are neutral, unless read with the same emphatic stress as the minimizers in (12), as demonstrated by the minimal pair in (13a) and (13b) and the similar pair in (14). The existence of minimal pairs like (13a) and (13b) suggests that intonation plays some role in the derivation of bias.

(13) a. Did Fred contribute anything to the campaign?
 b. Did Fred contribute ANYthing to the campaign?

(14) a. Has John ever voted for a democrat?
 b. Has John EVER voted for a democrat?

More importantly, marked intonation *alone* can produce negative bias, as shown by the examples in (15b) and (16) (from Sadock 1971). Both of these examples have narrow focus, intuitively marked with a L* or L* + H nuclear pitch accent, on the matrix verb.[4]

(15) a. Do you need that pork chop?
 b. Do you NEED that pork chop?

(16) Did I TELL you that writing a dissertation was going to be easy?

Examples (15) and (16) establish that the presence of a strong NPI is not a necessary condition for negative bias. The foregoing discussion, we believe,

[4] The location of the nuclear pitch accent need not fall on the matrix verb. Take the example in (i) in which the nuclear stress falls on *writing*.
(i) Did I tell you that WRITING a dissertation was going to be easy?
(i) expects a negative answer and might be followed up by an utterance like *No, I told you that* defending *it would be easy.*

establishes that intonation is the prime mover in deriving the negative bias of (12)–(16). While it is tempting to adopt Krifka's analysis in terms of emphatic assertion, we note that it is insufficient, as it does not establish the existence of an *assertion*, which we argue is necessary given the evidence introduced in Section 7.4.

7.3 Previous approaches: bias as implicature

Previous accounts treat bias as a kind of generalized conversational implicature. Krifka (1995), van Rooy (2003), and Guerzoni (2004) address the negative bias of questions with strong negative polarity items, and van Rooy and Šafářová (2003) and Romero and Han (2004) provide accounts of the positive bias inherent to negative polar questions. Each of these approaches share the assumption that bias is a calculable inference based on the semantic content of an interrogative sentence and pragmatic principles about rational and cooperative behavior. While some authors, notably Guerzoni (2004) and Romero and Han (2004), posit hidden lexical content or covert syntactic ambiguities to jump start such calculations, these approaches are in broad details no different from the other accounts mentioned above.[5]

We begin our review of the literature with the account of questions containing strong negative polarity items in Krifka (1995) and the refinements in van Rooy (2003) and Krifka (2003). We suggest possible extensions of Krifka's analysis to cover what we have termed emphatic focus questions in general. We then address Romero and Han's 2004 analysis of negative polar questions and its account of Ladd's ambiguity.

7.3.1 *Emphatic focus questions*

Borkin (1971) notes that interrogative sentences that contain a strong negative polarity item are necessarily biased toward a proposition expressing a negative answer. For the most part, previous research has not been concerned with bias *per se*, but with why NPIs are licensed in questions in the first place. Krifka (2003) and Guerzoni (2004) are notable exceptions. Accounts of bias, however,

[5] There is relatively little formal work on the interpretation of tag questions. For this reason, we do not discuss them here. There is reason to suspect, however, that whatever analysis one has of negative polar interrogatives applies equally to tag questions (or vice versa). This is especially true of analyses that treat interrogative tags as illocutionary force indicating devices, i.e. as functions that take an assertion and return a question (Culicover 1992). For current purposes, since Romero and Han (2004) provide the only detailed analysis of the interpretation of negative polar questions, any critique of it serves as a critique of analyses that view tags as illocutionary force indicating devices in the sense described above.

are intimately tied to accounts of NPI licensing, so some discussion of the broad outline of these approaches is necessary here.

Krifka's (1995) account of NPI licensing assumes that strong NPIs are associated with a set of alternative utterances—assertions or questions—that must be pragmatically motivated. Krifka (2003) invokes the principle below, motivated introduction of alternatives (MIA), to effect the required pragmatic licensing.

Motivated Introduction of Alternatives

If U (... α ...) is uttered, where α introduces an alternative set A, then for alternative utterances U (... α' ...), where $\alpha' \in A$, the speaker must have reason (i) to introduce the alternatives and (ii) reason not to utter an alternative instead of the actual utterance (Krifka 2003).

MIA not only applies to strong negative polarity items, but to any expression that is associated with a set of alternatives, for instance, to any constituent that is "in focus" (Rooth 1985, 1992). Consider the infelicitous example in (17), which contains the strong NPI *a drop*. According to Krifka (1995) (following Fauconnier 1975), strong NPIs denote a minimal element on a scale and introduce a set of ordered alternatives. *A drop*, for example, refers to minimal quantities of liquid and introduces a set of alternative quantities.

(17) #Larry drank *a drop* of whiskey.

Given MIA, the interpreter assumes the speaker has some reason to introduce the alternative assertions associated with (17). Generally, Krifka supposes, these reasons involve (i) the congruence of the alternatives with the current question under discussion (C. Roberts 1996) or (ii) the speaker's awareness of being able to make stronger or weaker claims. More central to the current argument, MIA states that the speaker has some reason *not* to assert any of the alternative assertions. If the alternatives are not ordered by entailment, the speaker might believe the alternatives to be false, or at the very least does not believe them to be true and thus obeys Grice's (1975) Maxim of Quality. If the alternative propositions are all logically weaker than the asserted proposition, the interpreter can assume that they are not the strongest defensible claims consistent with the speaker's knowledge, and therefore avoids violating the Maxim of Quantity. Finally, if the alternatives are logically stronger, the interpreter infers by the familiar scalar reasoning that the speaker believes that the alternative assertions are all false.

Returning to (17), each alternative assertion is at least as strong as the actual assertion, if not stronger. This follows from the additional premises that a person who consumes a quantity of liquid consumes all of its parts and that every quantity of liquid has a minimal part, that is, a part that has no proper sub-parts (see Krifka 1995). The constraints imposed by MIA can be satisfied in this situation by assuming that all of the alternatives are false. But this still does not explain why (17) is infelicitous.

In order to rule out assertions like (17), Krifka and others assume that strong NPIs have an "emphatic" component to their interpretation. The intuition is that assertions like (17) are weak and, consequently, not very informative (cf. Israel 2001). Some authors model this interpretive component by incorporating some or all of the presuppositions of the scalar additive adverb *even* into the denotation of strong NPIs (Heim 1984; van Rooy 2003; Guerzoni 2004; among others). Krifka himself assumes a pragmatic operation of emphatic assertion. The approaches have similar affects in so far as they require, at a minimum, the actual assertion to be stronger than any of the alternatives.[6] The fact that the actual assertion is at most as strong as any of the alternatives, if not weaker, contradicts the constraints imposed by the emphatic component of interpretation. The infelicity of the strong NPI in (17) follows, then, from the assumption that strong NPIs denote minimal elements on a scale, for example, minimal quantities of liquid, and are generally involved in emphatic, or strong, assertions. Negation reverses the entailment relations between the actual assertion and the alternatives. In a negative assertion, then, the actual assertion is logically stronger than any of the alternatives. The alternative assertions are not asserted, in this scenario, because they are not the strongest claim consistent with the speaker's beliefs. Furthermore, the actual assertion qualifies as emphatic, since it is stronger than any alternative.

The negative bias of questions like (18), according to Krifka (1995, 2003), results from a negative implicature based on MIA. MIA requires that the speaker have reasons (i) to introduce the alternative questions and (ii) not to *ask* any of these alternatives. As for the first requirement, Krifka argues that the speaker is aware that the alternative questions are potentially relevant, for example as questions under discussion or sub-questions to some superordinate question under discussion (C. Roberts 1996; Büring 2003), or—in cases where the alternatives form a scale—that he is aware of being able to ask more or less inquisitive questions.

(18) Did Larry drink *a drop* of whiskey?

According to Krifka, there are two reasons a speaker might choose not to ask any of the invoked alternative questions. Following Borkin (1971), the speaker

[6] Krifka (1995), in addition, requires the actual assertion to be stronger than the conjunction of the alternatives.

believes that the answers to the alternative questions are already known. If the actual question is not settled, then the presumed answers to any alternative question must be negative, as a positive answer to any alternative entails a positive answer to the actual question, contradicting the assumption that the actual question is not settled. Example (18) therefore implicates, or presupposes on some accounts, that Larry did not drink any substantial amount of whiskey, and asks whether he drank even the minimal amount.

More generally, Krifka (2003) suggests that the speaker might assign very low, non-zero, probabilities to the propositions *Larry drank amount* α *of whiskey*, for alternatives $\alpha \in A$. In this case, the speaker is so certain that the answer to her question is negative, that she deliberately asks a "risky" question, that is, one that has a high a priori likelihood of receiving a positive answer. Following van Rooy (2003), Krifka provides an information-theoretic formulation of this idea. Van Rooy argues that, as a general conversational strategy, speakers strive for a kind of equilibrium in the questions they ask. Specifically, they ask questions whose answers on average yield roughly the same amount of information.[7] This strategy ensures that speakers receive the most "bang for their buck". A "risky question" might have a single answer that yields a large amount of information but whose likelihood is quite low. Thus answers to this question on average yield very little information. Imagine that someone has chosen a card at random from a standard deck of playing cards and your task is to learn by asking as few polar questions as possible what card they are holding. It makes more sense in this situation to begin by asking whether it is a black card than it does to ask whether it is the six of clubs. For even though a positive answer to the latter question has a high potential payoff, you will receive that payoff on average only once in every fifty-two attempts.

The measure needed to formalize this idea is the notion of "entropy" familiar from information theory. The information value of a proposition q is defined according to Bar-Hillel and Carnap (1953) as:

$$\inf(q) = -\log_2 P(q)$$

This definition ensures that as the probability of a proposition q approaches 0 its information value increases. Conversely, when the probability of q is 1, its information value is 0, capturing the intuition that learning q in this situation provides no new information. The entropy E of a question Q is the weighted

[7] Van Rooy (2004) shows that this is a special case of the more general decision-theoretic strategy of maximizing the expected utility value $EUV(Q)$ of a question.

average of the informativity of its answers:

$$E(Q) = \sum_{q \in Q} P(q) \times \inf(q)$$

$E(Q)$ is maximized when each answer in Q is equally likely.[8]

The upshot of this approach is that interrogators are assumed to be "entropy maximizers" in that they choose questions from the pool of potential questions whose answers on average yield the most information. Returning to (18), if the interpreter assumes the speaker is an entropy maximizer, then they will assume that the speaker asks the actual question rather than any of the alternative questions because the entropy of the actual question is greater than the entropy of any the alternatives:

$$E(Q(\ldots \alpha \ldots)) > E(Q(\ldots \alpha' \ldots)), \text{ for } \alpha' \in A$$

This inequality holds just in case the likelihood of a positive answer to any of the invoked alternatives is less than the likelihood of the negative answer. That is, for any substantial amount of whiskey α, the speaker considers it less likely that Larry drank quantity α of whiskey, than that he did not drink that amount. Asking the "extreme" question with the minimizer increases the entropy of the question, because the probability that Larry drank the minimal amount of whiskey is greater than the probability that he drank any substantial amount.[9] Asking the question with the minimizer reduces bias by adding probability to the positive answer, but at the same time shows that the speaker considers it unlikely that Larry drank any substantial amount.

This analysis might apply to other kinds of emphatic focus question, though Krifka (1995, 2003) does not address these cases directly. Consider, for example, (15b), repeated in (19). This question, like questions with strong NPIs, is biased toward a negative answer.

(19) Do you NEED that pork chop?

However, it is hard to see at first glance how this type of question is amenable to the analysis described above. In particular, note that the alternative set

[8] Note that these definitions presuppose that the denotation of an interrogative sentence is a partition over the space of logical possibilities, e.g. as proposed by Groenendijk and Stokhof (1984).

[9] This follows from the fact that entailment between propositions reduces to set-theoretic inclusion,

$$\phi \models \psi \text{ iff } [\![\phi]\!] \subseteq [\![\psi]\!],$$

and from the manner in which the probabilities of propositions are computed:

$$P(\phi) = \sum_{\omega \in \phi} P(\omega).$$

associated with the focused verb, for instance, *want*, *yearn for*, does not form a scale with the item in focus. That *x needs y* does not entail that *x wants y* and vice versa. So the speaker may simply choose not to ask any of the alternative questions because she believes them to be settled. However there is no way to decide based on this information which way the alternatives are settled: in favour of the positive or of the negative answer.

A possible solution to this puzzle hinges on the observation that the pitch accent on *need* in (19) is not a simple focus marker, e.g. an H* tone. To our ear, (19) is most naturally read with an L* + H accent. Krifka (1995) argues that stressed versions of *any* and *ever*, which also yield a negative bias when they appear in questions, are emphatically focused. He goes on to define a pragmatic notion of emphatic assertion, whose effect mirrors that of the focus sensitive adverb *even*. That is, if an assertion has an emphatic focus, then any alternative assertion is considered to be more likely in the discourse context than the actual assertion.[10] In (20), for example, the actual assertion—that John would distrust Mother Theresa—is less likely than the proposition that John would distrust any other person.

(20) John would distrust MOTHER THERESA.

The main effect of the emphatic assertion operator is to induce a partial order over the alternative assertions based on their relative likelihood. It also contributes the information that the actual assertion is more informative than any of the alternatives. So given the emphatic focus in (19), it follows that the entropy of the actual question is greater than the entropy of any of the alternatives, as in questions with a strong NPI. The speaker is in effect conceding that the person reaching for the pork chop may want it, yearn for it, etc., but (probably) does not need it.

The most important aspect of the analysis described above is that the bias is a conversational implicature. It follows from the meaning of the question and the pragmatic constraint formulated in MIA. Furthermore, it is a rather weak bias. In effect, the bias states that all of the invoked alternatives are settled in a particular way, leaving the actual question open.

7.3.2 *Negative polar questions*

Romero and Han (2004) provide an analysis of negative polar interrogatives that attempts to account for (i) the backgrounded speaker attitude associated with their use and (ii) the outside-negation–inside-negation ambiguity. Their

[10] Krifka's (1995) emphatic assertion operator is slightly stronger than this, since it also requires that the actual assertion is less likely than the conjunction of all of the alternative assertions.

analysis is based on an observation that negative polar questions and questions containing the epistemic adverb *really* exhibit similar biases. Based on this similarity, they suggest that *really* and preposed negated auxiliaries introduce an operator VERUM into the logical form of sentences. VERUM is defined in (21). $\text{Epi}_x(w)$ is the set of worlds accessible from w and compatible with x's knowledge in w. $\text{Conv}_x(w)$ is the set of worlds where the conversational goals of x in w' are met. Finally, $\text{CG}_{w''}$ is the set of propositions representing the shared beliefs of the discourse participants in w'', the common ground.

(21) $[\![\text{VERUM}]\!]^{g[x/i]} = \lambda p_{\langle s,t\rangle}\lambda w.\forall w' \in \text{Epi}_x(w)[\forall\, w'' \in \text{Conv}_x(w')$
$[p \in \text{CG}_{w''}]]$.

VERUM says of a proposition φ that some agent x is certain that in all of the worlds in which the conversational goals of x are met, φ is part of the common ground, which Romero and Han gloss as FOR-SURE-$\text{CG}_x\varphi$.

According to the definition of VERUM in (21), it is an epistemic modal operator that embeds a "meta-conversational" modal statement, that is, a modal statement that refers to the conversational goals of the discourse participants rather than to the state of the world. Romero and Han (2004) argue that the use of meta-conversational moves, or speech acts, is governed by the following maxim.

Principle of Economy

Do not use a meta-conversational move unless necessary (to resolve epistemic conflicts or to ensure Quality) (Romero and Han 2004: 629).

They derive the backgrounded speaker attitude associated with negative questions, which they treat as a conversational implicature, via (21). Negative questions are for them meta-conversational moves by definition, since VERUM embeds the modal operator $Conv_x$. Assuming that discourse participants are cooperative and obey (21), then, it follows either that the speaker holds some belief φ which is inconsistent with the addressee's beliefs or believes φ but lacks sufficient evidence to assert it, in other words, the speaker wishes to avoid violating the Maxim of Quality (Grice 1975).

The outside-negation–inside-negation ambiguity on the VERUM account amounts to a simple scope ambiguity, in keeping with Ladd's (1981) original intuition. The outside-negation reading results when negation outscopes VERUM and the inside-negation reading results when VERUM outscopes negation. The logical form and denotation of the outside- and inside-negation readings of the negative polar question in (8) from Section 7.2.2 are provided in (22) and (23) respectively.

(22) a. [[[$_{\text{CP}}$ Q not [VERUM [$_{\text{IP}}$ Jane is coming]]]]]=

b. {FOR-SURE-CG$_x$ Jane is coming,
¬FOR-SURE-CG$_x$ Jane is coming }.

(23) a. [[[$_{\text{CP}}$ Q VERUM [not [$_{\text{IP}}$ Jane is coming]]]]] =

b. {FOR-SURE-CG$_x$ ¬ Jane is coming,
¬FOR-SURE-CG$_x$ ¬Jane is coming }.

Outside-negation polar questions on Romero and Han's analysis ask the addressee whether or not he is certain that φ should be added to the common ground, where φ is the descriptive content of the question. Inside-negation polar questions make the same inquiry about the proposition $\neg\,\varphi$.

We see a number of problems with this analysis. Romero and Han correctly note that negative questions often challenge some aspect of a previous utterance, and build this discourse function into the meaning of the question via the semantic operator defined in (21), which explicitly mentions the conversational goals of the speaker/hearer. There are two concerns with this approach: (i) conversational goals do not appear to be the right level of description and (ii) one cannot rely on epistemic conflict or uncertainty to derive the bias as they do. Both points are illustrated by dialogue in (24).

(24) a. A: Some of the regents support a tuition increase.

b. B: Don't all of them support an increase?

Sentence (24a) implicates that not all of the regents support a tuition increase, which *B* intuitively challenges with (24b). But it seems incorrect to say that *A* has as a conversational goal that the implicature be added to the common ground. While implicatures may become common ground, they need not do so. Indeed, it is an identifying characteristic of implicatures that they can be explicitly canceled by the speaker, indicating that they do not issue in a public commitment.[11] More importantly, the asserted content of (24a) is consistent with either a positive or negative answer to (24b), short-circuiting any attempt to derive a conversational implicature corresponding to the bias of (24b) by appealing to an epistemic conflict.

Furthermore, the bias inherent to negative questions, or more properly the backgrounded speaker attitude, is not necessarily an epistemic one, as assumed in Romero and Han (2004). The examples in (25) are from Huddleston and Pullum (2002). (25a) and (25b) convey a *deontic* and a *desiderative* bias respectively.

[11] It may be this very fact about conversational implicatures that constitutes their usefulness (cf. Pinker 2007, ch. 8).

(25) a. Aren't you ashamed of yourselves?

b. Don't you like it?

There are also issues associated with the scope account of the outside-/inside-negation readings. It is a standard assumption in the literature at this point that questions denote the set of propositions that count as direct answers (Groenendijk and Stokhof 1984). The most direct answers to a polar question are expressed with the answer particles *yes* and *no*. Intuitively, a simple *yes* response to (8) on an outside-negation reading conveys the proposition that Jane is coming, not the weaker one predicted by Romero and Han (2004) that FOR-SURE-CG_x Jane is coming.[12] Likewise, *no* means that Jane is not coming, not the weaker statement that ¬FOR-SURE-CG_x Jane is coming.

The denotation of the inside-negation polar question given in (23b) is even more problematic. A simple positive answer to (8) on the inside-negation reading is most naturally understood as expressing the proposition that Jane is coming, not FOR-SURE-CG_x ¬Jane is coming, which would seem to express just the opposite. And again, a negative answer intuitively means that Jane is not coming, not ¬FOR-SURE-CG_x ¬Jane is coming, which is more congruent with the proposition that Jane *is* coming. As we already suggested, it would seem that the positive questions in (7) and (8) are equivalent and that the denotations in (22b) and (23b) do not accurately reflect the denotation of (8).

7.3.3 *A preview of the current proposal*

Both Krifka (1995) (and later van Rooy 2003) and Romero and Han (2004) analyze the bias inherent to these types of question as conversational implicatures. Specifically, bias follows from the meaning of the question in conjunction with the assumption that the speaker is obeying certain principles of rational and cooperative behavior. If bias is indeed an implicature, one would expect it to be cancelable. After all, cancelability is the hallmark of conversational implicatures. However, there is reason to doubt these approaches precisely because bias does not appear to be cancelable. Consider (26) (which is a variant of (9), adopted from Gunlogson 2003), which would be a very odd sequence of utterances in a context that presupposes neutrality, say, at a hearing.

(26) a. I have no beliefs on the matter. I just want to know...

b. #Aren't you a member of the Communist Party?

[12] As pointed out by an anonymous reviewer, simple yes answers to outside-negation questions are somewhat marked and normally require some kind of amplification, as in *Yes indeed* or *Yes, of course*. The point made here still stands, however. These amplified *yes* answers do not convey the proposition predicted by the account in Romero and Han (2004).

If the bias in (26b) were truly a conversational implicature, there is no reason that it should not simply be canceled in the context above.

The situation appears even worse when you consider examples like (27). The context favors a neutral question reading: *A* and *B* are interested to know which individuals in a given domain satisfy a certain property, viz. the property of being a teetotaler.

(27) [*A* and *B* are trying to decide who to invite to their teetotaler's party.]

a. A: #Doesn't Bob drink? / Does Bob not drink?

b. B: No. He doesn't.

c. A: #Doesn't Larry drink? / Does Larry not drink?

d. B: Yes. He does.

e. A: #Doesn't Constance drink? / Does Constance not drink?

Negative questions with a preposed, negated auxiliary verb cannot be used as neutral requests for information. Furthermore, the second member of each pair, which is also a negative question, *can* be used in this context. Examples (7) and (8) provide evidence that these two kinds of question are semantically equivalent.[13] But the inference leading to an implicature is supposed to be based on meaning, rather than form, that is, it is non-detachable (Grice 1975). It is mysterious, then, why one is necessarily biased and the other is not.

Admittedly, the constraints on the use of NPIs imposed by their meaning and MIA plausibly make their implicature more difficult to cancel. One might assume, even, that these questions presuppose that every alternative answer is settled in favor of the negative answer (van Rooy 2003). However, these questions are felicitous in contexts where it is clearly the case that this presupposition does not hold, and where the speaker can not simply assume that it will be accommodated. Consider (28) from Asher and Reese (2005). Clearly *A* does not presuppose that John does not do any substantial amount of housework given (28a).

(28) a. A: John is a pretty decent husband.

b. B: Does he do the dishes?

c. A: No.

d. B: Does he do the laundry?

e. A: Well... no.

f. B: Does he do a damn thing around the house?

[13] Romero and Han (2004) do not maintain this equivalence, as described above.

We argue in Section 7.4 that bias essentially involves a speaker commitment, which we characterize as an assertion. This explains the non-cancelability facts above. Romero (2005) moves toward this position when she suggests that the bias of negative questions is a conventional implicature (see Kratzer 1999 and Potts 2005), which also involves a commitment by the speaker.[14] We prefer to treat biased questions as a species of indirect speech act, and model the semantics of tag questions, negative polar questions, and emphatic focus questions using complex speech-act types. First, we provide more evidence for the dual nature of biased questions, showing that they have distributional properties of both questions and assertions.

7.4 Evidence for multiple speech acts

Sadock (1971, 1974) introduces several tests for illocutionary force that appeal to the co-occurrence restrictions between specific discourse markers and utterances used as assertions and questions. The sentence-initial parenthet-icals *after all* and *yet* co-occur with assertions, for example, but not neutral questions. *After all*, for example, can be prefixed to an assertion, but not to a neutral question, as in (29a) and (29b) respectively.

(29) It's fine if you don't finish the article today.

   a. After all, your adviser is out of the country.

   b. #After all, is your adviser out of the country?

Likewise, utterances prefixed with *yet* can follow an assertion, cf. (30b), but not a neutral question, as in (31b).

(30) a. John is always late for work.

   b. Yet, he continues to be promoted.

(31) a. Is John always late for work?

   b. #Yet, he continues to be promoted.

There are parallel tests for questions. Sadock (1971, 1974) notes that sentence-initial *tell me* and the expression *by any chance* occur with questions, but not assertions, as established in (32) and (33).

(32) a. #John, by any chance, owns a car.

   b. Does John, by any chance, own a car?

[14] Reese (2006b) discusses Romero (2005) in more detail.

(33) a. #Tell me, John owns a car.

b. Tell me, does John own a car?

Furthermore, we argue that *tell me* and *by any chance* also distinguish between distinct sub-types of question. *Tell me*—as a simple request for a response from the addressee—selects for any type of question. *By any chance*, on the other hand, as an expression of epistemic uncertainty, only selects neutral questions. As such, it does not appear with biased questions, which we believe convey a commitment by the speaker.

In the following section we apply these tests to the kinds of construction discussed in the previous section, showing that they are indeed complex speech acts consisting of both a question and an assertion.

7.4.1 *Tag questions*

7.4.1.1 *Nuclear tag questions* Unsurprisingly, nuclear tag questions involve both an assertion and a question according to Sadock's diagnostics. Examples (34) and (35) establish that nuclear tag questions do assert the anchor according to the *after all* test.

(34) a. A: The conference should be exceptional this year.

b. A: After all, Julie is coming / isn't she.

(35) a. A: The conference might be sub-par this year.

b. A: After all, Julie isn't coming / is she.

The examples in (36a) and (36b), on the other hand, show that nuclear tag questions are not neutral questions, as they do not pass the *by any chance* test, but that they are questions, as they *do* pass the *tell me* test.

(36) a. Tell me, Jane {is/isn't} coming / {isn't/is} she.

b. #Jane {is/isn't} coming, by any chance / {isn't/is} she.

7.4.1.2 *Postnuclear tag questions* Postnuclear tag questions exhibit more variation in discourse function than nuclear tag questions. The polarity of the anchor appears to affect the available interpretations. Tag questions with positive anchors behave much like nuclear tag questions: they are assertions, as shown by (37) and (38), and (*tell me*) questions, (39). They are not neutral questions, however, as demonstrated by (39).

(37) a. A: Why is Nicholas so sure the conference will be dull?

b. A: After all, Julie is coming=isn't she?

(38) a. A: Pascal's not coming, so why is Nicholas so sure the conference will be a success?

b. A: After all, Julie isn't coming {#too/either}=is she?

(39) Tell me, Jane {is/isn't} coming={isn't/is} she?

The inclusion of a positive polarity item in a postnuclear tag question with a negative anchor coerces a neutral-question reading for examples such as (40). The anchor is no longer asserted under these circumstances, as established by (38b).

(40) Jane isn't coming {too/#either}, by any chance=is she?

The disambiguating role of the PPI is an important clue to understanding how this neutral reading arises. We believe that the negation in these examples scopes over the speech act itself, that is to say that it is a metalinguistic operator. Metalinguistic negation, as has been noted by Horn (1989), neither licenses NPIs, nor anti-licenses PPIs.

7.4.2 *Negative polar questions*

According to Sadock's diagnostics, outside-negation questions are assertions, while inside-negation negative questions are not. Note that (41b) can be preceded by *after all* when it contains the PPI *too*, but not when it contains the NPI *either*. Either version of (41b) is felicitous in the discourse context established in (41) if *after all* is left off.

(41) a. A: Sue can't attend, so there'll be no syntacticians there.

b. B: What do you mean? After all, isn't Jane coming {too/#either}?

Similarly, (42b) can follow (42a) when it contains *too*, but not when it contains *either*.

(42) a. A: Isn't Jane coming {too/*either}?

b. A: Yet, Mary claims there will be no syntacticians there.

Again, if *yet* is left off of (42b), then either the outside- or inside-negation reading of (42a) is available. Because outside-negation questions pass the *after all* and *yet* tests, we maintain that they characteristically make assertions. This is not true of inside-negation questions.

Outside- and inside-negation questions, however, are still questions: they can be answered with *yes* or *no* and they co-occur with the discourse marker *tell me*.

(43) Tell me, isn't Jane coming {too/either}?

Outside-negation questions, consequently, are *both* questions and assertions, as demonstrated by the discourse in (44). The *tell me* prefixed to the utterance in (44a) requires it to be a question.

(44) a. A: Tell me, isn't Jane coming too?
b. A: Yet, Mary claims there will be no syntacticians there.

At the same time, prefixing *yet* to (44b) requires (44a) to be understood as an assertion. In order for the typing constraints of *tell me* and *yet* to be satisfied in (44), then, the negative interrogative in (44a) must be simultaneously typed as an *assertion* and *question*. In other words, the type associated with (44a) is a complex semantic type (Asher and Pustejovsky 2004).

7.4.3 *Emphatic focus questions*

Finally, applying the tests to the type of interrogative sentences exemplified in (3) shows that they too are complex speech acts, as shown in (45)–(48) from Asher and Reese (2005).

(45) a. After all, does John lift a finger to help around the house?
b. Does John lift a finger to help around the house? Yet you continue to reward him.
c. Does John, by any chance, lift a finger to help around the house?
d. Tell me, does John lift a finger to help around the house?

(46) I don't understand why you think that John is a liberal.
a. After all, has he EVER voted for a democrat?
b. #After all, has he ever voted for a democrat?

(47) a. Has John EVER voted for a democrat? Yet you still claim that he is a liberal.
b. #Has John ever voted for a democrat? Yet you still claim that he is a liberal.

(48) a. Has John, by any chance, EVER voted for a democrat?
b. Tell me, has John EVER voted for a democrat?

(49) [Nicholas is reaching for the last porkchop, after already having had three.]
a. You should have some fruit instead. After all, do you NEED that porkchop?
b. Tell me Nicholas, do you NEED that pork chop?

Again, it is not the case that the illocutionary force of these questions is ambiguous or underdetermined. Rather, it is overdetermined. Biased questions are simultaneously assertions and questions as shown by the discourse continuation in (50).

(50) a. A: After all, has John EVER voted for a democrat?
b. B: No. He hasn't.

If *after all* is restricted to assertions and *B*'s turn in (50b) functions as an *answer* to (50a), (50a) must be both an assertion and a question.

7.5 Toward an analysis of bias

The current section provides some formal background on the interpretation of utterances in a discourse and on complex speech acts. In particular, we introduce the formal framework of Segmented Discourse Representation Theory (SDRT: Asher and Lascarides 2003) and provide an outline of how this formal framework can be used to analyze biased questions.[15]

7.5.1 *The basics of SDRT*

SDRT is a dynamic semantic framework for the interpretation of discourse and dialogue that takes the rhetorical connections between utterances and their semantic consequences seriously. Formally, a segmented discourse representation structure, or SDRS, is a triple $\langle A, \mathcal{F}, LAST\rangle$, where:

- A is a set of labels,
- $LAST$ is a distinguished label in A (intuitively, the content of the last clause added to the discourse logical form), and
- $\mathcal{F}$ is a function that assigns each label in A a formula of the SDRS language, which includes formulas of some version of dynamic semantics (DRT, DPL, Update Semantics, Martin Löf Type Theory, among others).

To get a sense of the kind of structures posited by SDRT and their semantic consequences, consider the temporal interpretation of the discourse in (51). The temporal structure of a text is more elaborate than suggested by many existing semantic analyses of tense. Examples (51c)–(51d), for example, describe in more detail the event introduced in (51b), which itself elaborates on the event introduced in (51a). (51e) continues the elaboration of John's evening that (51b) begins, forming a narrative with it. The order that these

[15] We refer the reader to Reese and Asher (2006) for additional details.

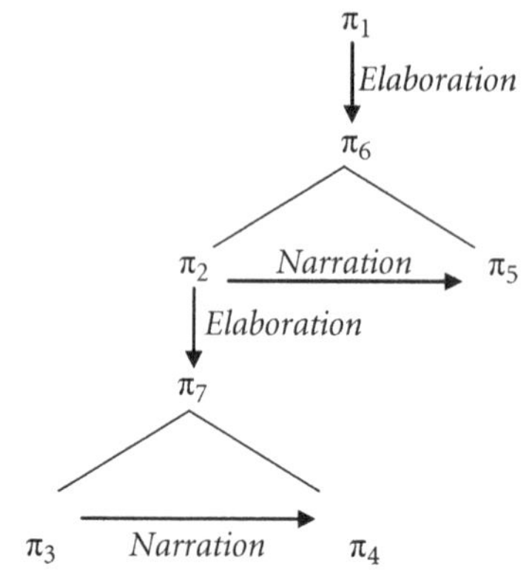

FIGURE 7.1 SDRS graph for the discourse in (51)

events occur in does not follow from the order in which they are presented, but obeys constraints imposed by the discourse structure. Thus eventualities that are understood as elaborating on others are temporally subordinate to them (temporal inclusion) and those events that represent narrative continuity are understood as following each other (temporal progression).

(51) a. (π_1) John had a great evening last night.
b. (π_2) He had a great meal.
c. (π_3) He ate salmon.
d. (π_4) He devoured lots of cheese.
e. (π_5) He then won a dancing competition.

The SDRS for (51) can be graphically represented as in Figure 7.1. On the basis of this discourse structure and the semantic consequences of the rhetorical relations *Elaboration* and *Narration* (see Asher 1993 and Asher and Lascarides 2003), one can infer that the dancing competition described in π_5 follows the meal in π_2, and that both of these events are temporally included within the evening described in π_1.

In addition to providing a formalism for describing the content of a discourse, SDRT provides a logical system for computing discourse structure on the basis of information available from syntax and compositional and lexical semantics. Crucially, intonation and prosody can be added as additional sources of information. There are two parts to this system. The first is a glue logic containing axioms for inferring rhetorical relations between discourse constituents. The second is a logic for reasoning about the belief, intentions, and goals of participants in a dialogue.

Each constituent in a discourse is assigned a unique label. The glue logic axioms exploit information provided by a description of the current SDRS $\top$, in addition to descriptions of the new discourse constituent β and the discourse constituent α to which it is to be related. These descriptions describe discourse structures by specifying which constituents are related to each other and by indicating in which constituent that information is to be found. Thus, a binary discourse relation like *Acknowledgement* is expressed in the description language as a three place predicate symbol *Acknowledgement* (α, β, λ), meaning that the constituent labeled by β serves as an acknowledgement to α and that this information is contained within the SDRS formula associated with the label λ.

The glue logic axioms exploit standard propositional logic connectives and a weak conditional operator $>$, which serves to represent defeasible rules about discourse structure. Glue logic axioms take the following form:

$$(?(\alpha, \beta, \lambda) \wedge Info(\alpha, \beta, \top)) > R(\alpha, \beta, \lambda)$$

These rule schemas state that if the discourse segment labeled by β is to be attached in some way to the segment labeled by α within the larger discourse segment λ, and certain information about α, β, and the whole discourse structure $\top$ is available, then normally β attaches to α with the rhetorical relation R. These normality conditionals support a defeasible version of modus ponens. Thus, when the left-hand-side formula holds, we can defeasibly infer $R(\alpha, \beta, \lambda)$. Asher and Lascarides (2003) provide a complete specification of the glue logic and of the defeasible consequence relation $\mid\!\sim$.

Finally, in order to compute relations in dialogue SDRT makes use of an extension of the glue logic for reasoning about the cognitive states of discourse participants known as the logic of cognitive modelling. This extension not only contains predicates relevant to computing discourse structure, propositional connectives, and the weak conditional operator $>$, but indexed belief and intention operators $\mathcal{B}_A, \mathcal{B}_B, \ldots, \mathcal{I}_A, \mathcal{I}_B, \ldots$ (for discourse participants *A*, *B*, etc.). The contribution of prosody to calculating the discourse function of the various types of biased question described above is expressed in both the glue logic and the logic of cognitive modelling.

7.5.2 *Complex speech acts*

In Section 7.4, we argued that tag questions, outside-negation polar questions, and emphatic focus questions involve not only a question, but an assertion as well. In the present section we argue that biased questions are assigned a complex speech act type as a matter of linguistic convention. Following Asher and Lascarides (2001), we model complex types using the notion of a "dot type" (Asher and Pustejovsky 2004). An utterance is a conventionalized complex speech act (CSA) if,

> (a) the grammar assigns it a complex speech act type of the form $s_1 \bullet s_2$, such that s_1 and s_2 are distinct (incompatible) types of semantic objects; and (b) Gricean-style principles of rationality and cooperativity link the constituent type s_1 to the type s_2
>
> (Asher and Lascarides 2003: 310).

The data in Section 7.4 provides evidence that biased questions are assigned a complex speech act. Moreover, the component types identified by that data are associated with distinct, incompatible semantic objects. On most compositional semantic accounts the content of assertions and questions are associated with incompatible model-theoretic objects (or context-change potentials in a dynamic semantic setting). Assertions, for example, are associated with propositions, or sets of possible worlds, whereas questions are associated with *sets* of propositions (Hamblin 1973) or propositional concepts (Groenendijk and Stokhof 1984). According to clause (a) in the above quotation, then, biased questions are conventionally assigned a complex speech act type *assertion* • *question*.

The grammar can exploit both constituent types of a complex type when computing the rhetorical contribution of an utterance by applying a rule of Dot Exploitation. If an utterance β attaches to an utterance α (with some undetermined rhetorical relation) in the discourse context λ and β is assigned a complex type $t_1 \bullet t_2$ by the grammar, then new speech act referents γ_1 and γ_2 of type t_1 and t_2 respectively are introduced. These new discourse labels are related to the original label β by a relation *O-Elab*, or "dot elaboration".

Clause (b) of the definition of conventionalized CSAs requires Gricean-style reasoning about rationality and cooperation to link the constituent types of the complex type. This requirement is formalized in Coherence Constraint on Complex Types (Asher and Lascarides 2001). $\mathcal{C}$ encodes the *linguistic competence* of the discourse participants and therefore contains information about the conventional mapping between linguistic form and compositional and lexical semantics, in addition to glue logic axioms for inferring rhetorical connections between utterances. $\mathcal{R}$ consists of axioms for reasoning about the cognitive states—the beliefs, intentions, and goals, of discourse participants—in addition to domain-specific and general world knowledge.

Coherence Constraint on Complex Speech Act Types:

Suppose that:

- $?(\alpha, \beta, \lambda)$
- $\beta : t_1 \bullet t_2$
- $O\text{-}Elab\ (\beta, \gamma_1) \wedge O\text{-}Elab\ (\beta, \gamma_2)$
- $\gamma_1 : t_1 \wedge \gamma_2 : t_2$

Then:

$$\mathcal{R}, \mathcal{C}, ?(\alpha, \gamma_1, \lambda), ?(\gamma_1, \gamma_2, \lambda'), \ Info(\gamma_1, \gamma_2) \mid\!\sim \mathcal{R}(\gamma_1, \gamma_2, \lambda'),$$

where λ' labels an SDRS that results from attaching γ_1 to α in the SDRS labeled by λ.

The coherence constraint thus ensures that the constituent types of a complex speech act are related by a discourse relation R, inferred on the basis of conventional linguistic knowledge and Gricean-style reasoning about rationality and cooperation as modeled by $\mathcal{R}$. In this regard, it closely resembles the analysis of indirect speech acts in Searle (1975).

Before addressing how the constituent types of a biased question are rhetorically linked, a few more words need to be said about the content of $\mathcal{C}$ and $\mathcal{R}$, and what it means for an utterance α to have the type *assertion* in SDRT. To reiterate what was said above, $\mathcal{C}$ encodes the linguistic competence of the discourse participants; it therefore provides information about syntax, prosody, and lexical and compositional semantics, in addition to information about the semantic contribution of specific rhetorical relations and SDRT's axioms for inferring them. As such, $\mathcal{C}$ includes the information that the negation in neutral tag questions and outside-negation polar questions is metalinguistic (however the notion "metalinguistic negation" is cashed out formally). It also contains information about the intonational tune of an utterance and how this tune is interpreted. The direction of the F0 trend at the end of an intonational phrase, for example, is often assumed to convey information about the speaker's relation to a proposition and its relation to the common ground (Gussenhoven 1984). The placement and choice of nuclear pitch accent provides similar information by marking content as new or backgrounded (Steedman 2000), and by introducing a (partially ordered) set of alternative propositions. Intonation thus provides the interpreter with a rich source of information for reasoning about the cognitive state of the speaker, or at least information about the speaker's "take" on the content and structure of the discourse context.

These observations segue naturally into a discussion of the content of $\mathcal{R}$, the axioms for reasoning about the cognitive states of discourse participants based on what has been said in the course of a discourse or dialogue and on who said it. $\mathcal{R}$ includes, for example, axioms that model Gricean-style reasoning about the rationality of discourse participants, in addition to axioms that associate certain default goals with utterances based on their linguistic form. These goals are known as speech act related goals, or SARGS. Question Related Goals (QRG), for example, states that the default speech-act related goal of an interrogative sentence is that the speaker believe an answer to it. Known Answers blocks this inference when the speaker already believes an answer.

QRG: $Sanswer\,(\alpha,\ p) > \text{SARG}(\alpha,\ \mathcal{B}_{S(\alpha)}p)$

Known Answers: $(Sanswer\,(\alpha,\ p) \wedge \mathcal{B}_{S(\alpha)}p) > \neg\text{SARG}(\alpha,\ \mathcal{B}_{S(\alpha)}p)$

Similar axioms associate SARGS with indicative and imperative sentences.

Finally, we formalize the notion of *assertion* in SDRT. Our characterization is uncontroversial, but is captured in a very specific way in a discourse-based framework like SDRT.

Searle and Vanderveken (1985: 37) note that with an assertion "the speaker presents a proposition as representing an actual state of affairs in the world of utterance", that is, a word-to-world fit. In other words, the proposition conveyed by an assertion should be true. Based on this understanding, we provide the definition of assertion in (52).

(52) β : *assertion* just in case (i) $R(\alpha, \beta, \lambda)$ and *right-veridical* (R) or (ii) β is the first constituent entered in the discourse structure and $\top \models_\iota \mathrm{K}_\beta$.

A right-veridical rhetorical relation is one that entails the content of its right argument:

$$R(\alpha, \beta) \rightarrow \mathrm{K}_\beta, \text{ where } \mathrm{K}_\beta \text{ is the SDRS formula assigned to } \beta \text{ by } \mathcal{F}$$

The first clause of (52), then, simply requires that the speaker's utterance attach to the discourse context with a relation that entails its content. The second clause covers cases where the discourse segment is the first utterance in a discourse. It says that if the label for the discourse as a whole $\top$ immediately outscopes β and $\top$ entails the content K_β of β, then β is an assertion.

Rhetorical relations like Narration, Explanation, and Correction are right-veridical relations, and so on our definition are all types of assertion; relations like Q-Elab or Narration$_q$, on the other hand, are not right-veridical. These and similar relations require their right-argument to be a question. We discuss below how the complex speech act types assigned to tag questions, outside-negation polar questions and emphatic focus questions satisfy the coherence constraint on complex types.

7.6 Tag questions

Tag questions may or may not instantiate a complex speech act type depending on their specific prosodic and intonational properties. In the case of nuclear tag questions, we believe, clausal syntax and semantics, intonational phrasing, and the alignment rules of SDRT suggest the presence of two illocutionary acts: an assertion (based on the declarative anchor) and a question (derived from the tag). In the case of postnuclear tag questions, it is plausible to assume a complex type *assertion • question*. Recall that the discourse functions available to postnuclear tag questions are a super-set of those available to nuclear tag questions. Whereas both types of tag question function as requests for acknowledgment or confirmation, postnuclear tag questions can also function as neutral questions. The neutral use of postnuclear tag questions, however, has peculiar lexical semantic properties, viz. the presence of a metalinguistic negation operator.

The reasoning that links the anchor and tag of a nuclear tag question mirrors exactly that which links the constituent types of a postnuclear tag question. We therefore focus on the latter below, since we are interested for the most part in the analysis of biased questions as complex speech act types. As an illustration, we focus on the interpretation of tag questions as requests for acknowledgement. This interpretation, recall, is associated with falling intonation over the tag, a phonological feature that we assume provides no essential semantic information (cf. Reese 2007 and Reese and Asher 2006 for more discussion).

The axiom schema in (53) provides the semantic content of acknowledgement questions. $Acknowledgement_q$ links α to β just in case the answer to β entails that the SARG of α has been accepted or achieved. This semantic information, we assume, is sufficient to infer that $Acknowledgement_q$ links an utterance to the prior discourse context. The axiom is part of SDRT's glue logic, so it is included in the set of conventional linguistic information $\mathcal{C}$.

(53) *Axiom on Acknowledgement Questions*
$(?(\alpha, \beta, \lambda) \wedge \textsc{sarg}(\alpha, \varphi) \wedge Sanswer(\beta, p) \wedge (\mathcal{B}_{H(\alpha)}(p) > \mathcal{B}_{H(\alpha)}\varphi)) >$
$Acknowledgement_q(\alpha, \beta, \lambda)$

The rhetorical link between the constituent types of a postnuclear tag question follows from compositional semantics of the anchor and tag and default information provided by the logic of cognitive modelling. Let β be a postnuclear tag question. Assume that $?(\alpha, \beta, \lambda)$, where β is assigned the complex speech act type *assertion* • *question*.[16] Since $?(\alpha, \beta, \lambda)$ assumes that β has a simple type, Dot Exploitation is called upon, yielding:

$$O\text{-}Elab(\beta, \gamma_a) \wedge O\text{-}Elab(\beta, \gamma_t),$$

where γ_a: *assertion* provides the semantic contribution of the anchor and γ_t: *question* provides the contribution of the tag. The coherence constraint on complex types, then, requires a rhetorical link between γ_a and γ_t. As we argued above, the requisite link is one of two relations: *Acknowledgement*$_q$ or *Confirmation*$_q$.

Because γ_a has the type *assertion*, it must attach within λ with a right-veridical relation as required by (52). Given this constraint, the SARG of γ_a is that the addressee believe its propositional content. This means that in the schema in (53), ϕ is instantiated with $\mathcal{B}_{H(\gamma_a)}(p_{\gamma_a})$. It also follows from certain axioms of cognitive modeling that $\mathcal{B}_{S(\gamma_a)}(p_{\gamma_a})$[17] and from the compositional semantics of questions and answers that *Sanswer* $(\gamma_t, \; p_{\gamma_a})$. Finally, it is a theorem of the logic of cognitive modeling that $\mathcal{B}_{H(\gamma_a)}(p_{\gamma_a}) > \mathcal{B}_{H(\gamma_a)}\mathcal{B}_{H(\gamma_a)}(p_{\gamma_a})$, as belief is a K45 modality. As a result, in the absence of conflicting information the addressee infers that *Acknowledgement*$_q$ $(\gamma_a, \gamma_t, \lambda')$.

We do not go into the details of the derivation of confirmation questions here, except to note that we assume that final rises make a semantic contribution through either a modal expression of uncertainty (Šafářová 2005), or by expressing "ownership" in some sense of the underlying proposition expressed by the utterance (Steedman 2000; Gunlogson 2003). Reese and Asher (2007) and Reese (2007) provide proofs that this information blocks the default inference to *Acknowledgement*$_q$ described above. The reason is that the final rise commits the speaker to inconsistent intentions (or, equivalently, SARGS), which we assume is ruled out by principles of rational action (see for example Cohen and Levesque 1990).

Neutral readings of postnuclear tag questions, as already stated, have a peculiar lexical feature, namely, a metalinguistic negation operator in the anchor. As such, the computation of their discourse function is a separate matter from that of the postnuclear tag questions described above. We adopt the analysis of metalinguistic negation common to multi-valued logics (see for example Bochvar 1981 as discussed by Beaver and Krahmer 2001) in which $\sim K_\pi$ is equivalent to $\neg(\pi : \textit{assertion})$, at least with respect to declarative sentences. Given our characterization of assertion, this means that it is not the

[16] The argumentation that follows holds for nuclear tag questions, as well, except that there is no need in the case of nuclear tag questions to employ Dot Exploitation.

[17] For tag questions it holds that $S(\gamma_a) = S(\gamma_t)$.

case that π attaches to the discourse context with a right-veridical relation. If π does not attach with a right-veridical relation, then it is allowed to attach with a rhetorical relation pertinent to a neutral question. Note that in the cases discussed above, the association of the anchor with an assertion blocks the default SARG of a question: if the speaker (of a tag question) believes the content of the anchor—which follows from cognitive modeling and the fact that it is asserted—then Known Answer will fire with respect to the tag's SARG. However, if the anchor contains a metalinguistic negation operator, the interpreter can no longer infer that the speaker believes the content of the anchor and there is nothing blocking QRG.

A remaining issue involves the relationship between postnuclear intonational phrasing and neutral interpretations: why can't nuclear tag questions have a neutral interpretation? The answer, we maintain, lies in our assumption that nuclear phrasing forces two speech act discourse referents, one for the anchor and one for the tag. Postnuclear phrasing, on the other hand, assigns the tag question a dot type and Dot Exploitation fires only when there is a type clash. This is normally the case, but metalinguistic negation—which forms part of the linguistic form of all neutral interpretations—cancels the assertion as described above.

7.7 Outside-negation polar questions

Outside-negation polar questions, unlike their inside-negation counterparts, are also assigned a complex type *assertion* • *question* by the grammar. The connection between the constituent types varies according to the use to which the utterance is put. Outside-negation polar questions are felicitous in two types of situation, what Romero and Han (2004) call "contradiction" and "suggestion" scenarios. In the former situation, outside-negation polar questions are often prosodically marked in the same way as corrections, in which one finds some combination of higher mean pitch, greater pitch range, higher mean intensity and increased duration on the nuclear pitch accent (Swerts and Krahmer 2008). The sentences in (54) provide an example of the contradiction use. *A*'s turn in (54a) biases the context against the proposition that John is in Hawaii.

(54) a. A: John is coming to the party tonight.
   b. B: Isn't John still in Hawaii?

Reese (2006b) provides a number of examples which show that the discourse function of outside-negation negative polar interrogatives often patterns with the use of positive assertions as denials. This is to be expected on our analysis, since we maintain that outside-negation polar questions involve a positive assertion. In these cases, it is natural to assume that the assertion obtained through Dot Exploitation attaches to the preceding discourse context with a divergent rhetorical relation like *Correction* or *Counterevidence*. The presence of this assertion requires, on pragmatic grounds, a reinterpretation of what question is being asked. A number of possibilities exist for attaching the question to the assertion. For example, the constituent speech acts may be related via $Acknowledgement_q$ or $Confirmation_q$, as with tag questions. Another possibility is that a stronger relation like $Counterevidence_q$ holds, in which case the question functions as a challenge to the addressee to back up a previous commitment by supplying counterevidence to the speaker's assertion (see Reese 2006a).

Outside-negation polar questions also occur in neutral contexts, in which case they function as polite suggestions. Example (55), where (55b) serves as an answer to the question in (55a), illustrates this use.

(55) a. A: Who wrote Gravity's Rainbow?
b. B: Wasn't it Thomas Pynchon?

In this and similar cases, the component assertion, viz. that Thomas Pynchon wrote *Gravity's Rainbow*, attaches to the speech act discourse referent introduced by (55a) with *QAP* (Question–Answer Pair), a right-veridical relation. The question component of the complex speech act type again attaches to the answer with $Acknowledgement_q$ or $Confirmation_q$ depending on the certainty conveyed by B.

7.8 Emphatic focus questions

Our treatment of emphatic focus questions is similar to the treatment of outside-negation polar questions above. One difference, however, is that emphatic focus questions involve a negative assertion instead of a positive one. As with the use of outside-negation questions in contradiction scenarios, the assertoric component of the complex type assigned to emphatic focus questions attaches to the prior discourse context with a divergent rhetorical relation, that is, *Correction* or *Counterevidence*. With respect to the dialogue in (56)—repeated from (28)—*B*'s utterance in (56f) disputes *A*'s assertion in (56a) to the extent that it suggests that John does at least a minimal amount of housework.

(56) a. A: John is a pretty decent husband.
b. B: Does he do the dishes?
c. A: No.

d. B: Does he do the laundry?

e. A: Well... no.

f. B: Does he do a damn thing around the house?

Question (56f) challenges *A* to either provide counterevidence to *B*'s (negative) assertion (indirectly providing evidence for the original claim in (56a)) or to explain why she said it in the first place. These discourse functions are captured in SDRT with the relations $Counterevidence_q$ and $Explanation^*_q$ respectively.

The intonational properties of emphatic focus questions align with the characterization of their discourse function above. Emphatic focus, to our ears, is marked with an L*+H or L+H* nuclear pitch accent, often followed by a lowrising boundary—L*+H L–H% or L+H* L–H% in the ToBI annotation scheme. Liberman and Sag (1974) refer to this tune as the "contradiction contour" and Ward and Hirschberg (1985) argue that it conveys speaker incredulity, especially when occurring with marked spectral features. Along similar lines, Steedman (2000, 2003) maintains that L*+H marks contested thematic constituents, making it a natural marker of corrections. Consider the emphatic focus question (57)—repeated from (19)—and its declarative cousin, which is a standard correction move, in which the speaker corrects the addressee's choice of words in the previous utterance.

(57) Do you NEED that pork chop?

(58) a. A: Oh, I need that pork chop.

b. B: You WANT that pork chop.

The same tune is operative in both (58b) and (57). In fact, we also believe that it plays a role in the interpretation of outside-negation polar questions. Asher and Reese (2005), following Ward and Hirschberg (1985) and Pierrehumbert and Hirschberg (1990), argue that this tune contributes the following meaning:

(59) $\lambda C \exists p.[p \in C \wedge {}^{\vee}\neg p]$

C is a set of alternative propositions which include the core proposition associated with the utterance and one other proposition just like the core proposition except that some other element of the appropriate type is substituted for the element in focus. We suppose this second proposition to be supplied by the non-linguistic context or by discourse attachment.[18] For

[18] Our treatment of *C* differs and, we think, improves upon the treatment in Asher and Reese (2005).

simple corrections, C consists of the discourse constituent to which the correcting constituent is attached—namely, in (58) it consists of the singleton $\{||(58a)||\}$ and we derive the meaning of a corrective discourse relation (see Asher and Lascarides 2003) thus from the intonation alone.[19] In the case of the emphatic focus interrogative, the set C could consist of a proposition made salient by the non-linguistic situation that has a different element substituted for the accented element in the core proposition of the interrogative. The effect of the placement of the pitch accent singles out the element to which the speaker objects—that is, the proposition here would be something like: addressee needs the pork chop he just grabbed. In other cases like (56), C picks up the topic of the exchanges in (56b)–(56e). Once again the focused lexical element enables us to determine in part what that is, viz. it's the proposition that John does something around the house, and it is this topic that is corrected. As with other types of biased questions, this correction based on intonation forces one to reinterpret the question, which by our tests is still conveyed by emphatic focus questions, along the lines we have already seen. Thus intonational information, in addition to the lexical semantic properties of strong negative polarity items, plays the central role in the grammaticization of emphatic focus questions as complex speech acts.[20]

7.9 Conclusions

The analysis of biased questions outlined above was formulated within the theory of the alignment of linguistic form and illocutionary force supplied by SDRT. Many diverse aspects of linguistic form contribute to the determination of discourse function, including, but not limited to, clausal syntax and semantics, specific lexical choices, and intonation and prosody.

[19] In some cases the intonational meaning might duplicate the meaning of the assertion, yielding an emphatic effect.

[20] There are some complexities to the story about intonation that our proposal doesn't address, as noted by one reviewer of this paper. It may be not only the F0 contour that is responsible for the complex speech act, but other factors as well—for instance, increased intensity, lengthening over the syllables. We suspect that such variables play an important role in distinguishing our emphatic focus questions from questions that may look quite similar with respect to the F0 contour, though even there we think that the contours are not exactly the same. Here are some examples noted by the reviewer.

(i) a. You spent several years working on a farm in Illinois.
b. Did you ENJOY working on a farm?

(ii) a. Falwell said that NYC had brought the attacks on itself.
b. Did he INTEND to be inflammatory?

(iii) a. A: Let's take Terri bowling when she comes to visit.
b. B: Oh, does Terri LIKE to bowl?

The interaction of intonation and prosody with lexical and compositional semantics in computing the rhetorical role of an utterance in a discourse or dialogue is especially interesting, as the discussion above we hope shows, in addition to being very complex. Intonational contours at the edge of phonological phrases, for example, signal an underspecified relation between the speaker, the propositional content of his or her utterance and the common ground or public commitments of various discourse participants (Gussenhoven 1984; Steedman 2000; Gunlogson 2003; Marandin, Beyssade, Delais-Roussarie, Rialland and Fornel 2005). Nuclear pitch accents contribute additional "information structural" information, marking information as given or new, in addition to introducing a set of alternative propositions.

All of this information is essential to determining the informational needs of the speaker alluded to in the introduction, and which we modeled above with rhetorical relations in SDRT. Questions that convey an informational need addressing the status of a proposition with respect to the common ground were modeled with the rhetorical relation $\textit{Confirmation}_q$; questions that request information about the cognitive state of a conversational partner were modeled with the relation $\textit{Acknowledgement}_q$. Of course, there may be other relations that address similar informational needs. The discussion above focuses on just biased questions as a means of getting a handle on the contribution to information structure of complete intonational tunes.

Part II

Cross-Linguistic Variation and Diachronic Change

8

Towards a typology of focus realization

DANIEL BÜRING

8.1 Introduction

This chapter presents a first attempt to formulate a cross-linguistic theory of *focus realization*, that is, of how different languages express focusing. It is common knowledge that by far not all languages mark focused constituents via pitch-accent placement in the way Germanic languages do. Rather, focusing is variously reflected in prosodic phrasing (i.e. boundary placement), constituent ordering, via special focus morphemes, and perhaps in some cases not at all. One might therefore suspect that there cannot be any interesting theory of focus realization cross-linguistically; languages just choose some aspect of their grammatical structure, prosodic, syntactic, or morphological, to realize focus.

While this may very well turn out to be the case, this chapter explores the hypothesis that there is still something systematic to be said about focus realization, that is, that there is a common analytical apparatus that can capture the cross-linguistic variation. I will take as my point of departure what I will call the *prominence theory of focus realization*, henceforth PTF, pioneered in Truckenbrodt (1995), among others. The idea behind this approach is that focus is always realized by structural prominence, and that the only way focus interacts with the formation of structure is through a constraint called FocusProminence:

Parts of this chapter are based on joint work I did in the summer of 2005 with Katharina Hartmann and Malte Zimmermann as a guest of the SFB632 project B2, Focus in Chadic Languages in Berlin. I would like to thank them for generously sharing their data and ideas with me. Thanks also to the participants at the workshop on focus at the National Institute of Informatics in Tokyo in November 2005, especially Chris Tancredi and Makoto Kanazawa, who made that wonderful meeting possible, and Katalin É. Kiss, Carlos Gussenhoven, Malte Zimmermann, and Caroline Féry, who provided comments on an earlier version of this chapter.

(1) *FocusProminence*
Focus needs to be maximally prominent.

In Sections 8.3–8.5 I will show how languages in which focus is realized by pitch accent (as in English), as well as languages in which it is realized by prosodic phrasing, constituent order variation, or a mixture of all of these can be analyzed under the umbrella of (1), understanding the notion of "prominent" there as a prosodic one (as originally intended by Truckenbrodt). In Sections 8.6 and 8.7 I will then go on to cases that appear to elude such an analysis, and I will speculate about ways to accommodate them. Section 8.7 in particular toys with the idea of parameterizing the very notion of prominence (in particular to include syntactically defined prominence) in (1) to try to bring these cases into the reach of a common analysis.

Obviously, this chapter cannot even begin to do justice to the arguments that have been put forward in favor (or, in some cases, against) the language specific analyses of focus realization I rely upon here, and the generalizations I report are certainly oversimplified in many cases. Likewise, the actual account of cross-linguistic focus realization I develop is a mere sketch. Yet I hope that this can serve to give the reader some idea of the overall project, and whether or not they deem it worthy of pursuit.

8.2 Basic notions

8.2.1 *The notion of focus*

Since we are interested in cross-linguistic differences in the way focus is realized, we have to provide some means of identifying focus independent of its particular realization. I assume, with much of the literature, that we can identify focus via the pragmatics. Concretely, certain (semantic or pragmatic properties of certain) contexts systematically trigger focus. While there is no common definition of focus-triggering contexts (or focus, for that matter) in the various analyses I draw upon, there is a common extensional core of contexts that people seem to agree trigger focusing. For example, all sentences in U1 will elicit an utterance of U2 with focus on the direct object, as does continuation of U2 with U3:

U1: a. What do you put in your pasta sauce?
b. What was it that you put in your pasta sauce?
c. Do you put tarragon or thyme in your pasta sauce?

d. Do you put tarragon in your pasta sauce?

e. Daniel puts tarragon in his pasta sauce.

f. First, I put tarragon in my pasta sauce, then . . .

U2: I put $[\text{thyme}]_F$ in my pasta sauce.

U3: . . . , not tarragon.

Though presently I do not see an alternative to giving a common core of focus-triggering context by simple ostension (rather than any more interesting definition) for a study like this, at least two shortcomings resulting from this bear mentioning. First, even if we agree that all of these contexts trigger focusing, we almost certainly lose valuable information by *restricting* our attention to them, since we cannot safely bring more "exotic" species of focus (such as second occurrence foci, see e.g. Rooth, this volume) to bear on the theoretical issues.

Second, we are stuck with the assumption that no matter which of the above context types a particular study used, they thereby elicited the same kind of focus, so that cross-linguistic comparison of its realization is warranted. While this strikes me as an extremely attractive hypothesis, it is also possible that an eventual theory of focus will have to grammatically distinguish different kinds of focus in some of these contexts (say, "answer focus" as opposed to "correction focus"), in which case it would be problematic (or at least uninstructive) to compare the realization of these foci across languages. For now I can only hope that not too much distortion will be created by the assumption that the differences between these "focus types" are purely pragmatic, not grammatical, and will therefore not influence the way the grammar realizes these foci.

8.2.2 *Prosodic prominence*

What does it mean for a focus to be prominent? As a starting point, I will define prominence in prosodic terms, using a hierarchy of prosodic units: one or more syllables form a *prosodic word* (PWd), one or more PWds form a bigger prosodic unit (variously called accent(ual) domain/phrase, intermediate phrase, major phrase etc. in the sources I rely on) called *phonological phrase* (pP) here; and one or more pPs form an *intonational phrase* (IP). Among the immediate constituents (or "daughters") of a given unit, one is its *head*. If you think of prosodic structure as a tree as in (2), this means that each node has exactly one strong daughter, which we can mark by +:

(2)

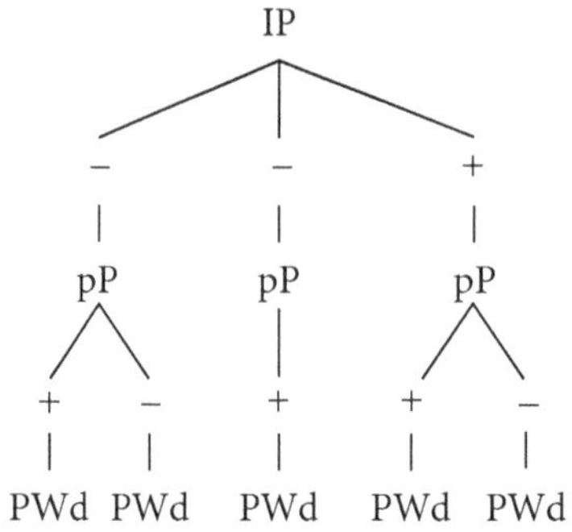

If you prefer to think of prosodic structures as bracketed strings, we can mark the head of a constituent by a grid mark at the pertinent level:

(3) $(\qquad\qquad\qquad\qquad\qquad * \qquad\qquad\quad)_{IP}$
$(\quad * \qquad\qquad)(\; * \;)(\quad * \qquad\qquad)_{pP}$
$(PWd)(PWd)(PWd)(PWd)(PWd)_{PWd}$

By definition, the head of any constituent (marked by +/*) is more prominent than any of its sisters (any other element within that constituent).

What this means in terms of the phonetic realization of the string, however, varies from language to language, and therein lies one source of cross-linguistic variation, as we will see.

Step 1 in formulating an account of focus realization in a given language is to determine its default prosodic structure, in particular:

(4) a. the standard mappings from syntax to prosody e.g. PWd $\approx X^0$, pP $\approx$ XP, IP $\approx$ sentence etc.;
b. the standard headedness for each prosodic phrase: left or right.

This chapter deals with two levels of prosodic structure, the topmost, intonational phrase, and the one below it, the phonological phrase.

Phonological phrases correspond roughly to maximal projections of lexical categories. Since syntax is recursive, but prosodic structure (normally) is not, a dilemma appears if one or more lexical XPs are contained in another, say objects within a VP. Descriptively, in a situation like (5), languages can choose between at least three strategies for their standard mapping:[1]

[1] I borrow the term "wrapping" from Truckenbrodt (1995), who gives an account of these three strategies in terms of re-rankings of a small number of constraints.

(5) $[\text{XP YP Z}]_{\text{ZP}}$

a. radical splitting: $(\text{XP})_{\text{pP}}$ $(\text{YP})_{\text{pP}}$ $(\text{Z})_{\text{pP}}$
each XP, as well as any remaining non-phrasal elements, gets its own pP;

b. moderate wrapping: $(\text{XP})_{\text{pP}}$ $(\text{YP Z})_{\text{pP}}$
each XP gets its own pP, but non-phrasal elements are "wrapped" with the structurally closest phrase;

c. radical wrapping: $(\text{XP YP Z})_{\text{pP}}$
the biggest XP, with everything it contains, gets wrapped into one big pP.

Next, the head of pP needs to be determined, in case it has more than one daughter. The two choices, obviously, are: the rightmost/leftmost PWd in a pP becomes its head. I will write this as pP–Head–R and pP–Head–L, respectively, meaning "the head of pP in this language wants to align with the left/right edge of pP."

Finally, the pPs join to form an IP, for which, again, the head-question arises, with choices being right- and leftmost (IP–Head–R/L).

8.2.3 *Prosodic structure amended for focus*

Suppose now an element A, which corresponds to a phrase at level L1 (an L1-phrase or L1P for short), is not by default the head of the next higher prosodic constituent (L2P) at level L2, e.g. A is left-peripheral in L2P, but L2 is right-headed (L2–Head–Right), yielding the unmarked structure in (6):

(6)

	$(\quad\ \ *\ )_{\text{L2}}$
	$(\ *\ )(\ *\ )_{\text{L1}}$
default:	A B

Now, if A is focused, then FocusProminence wants the prosodic structure to change so as to make A maximally prominent (in its L2-phrase). Generally, any one of three remedies can be used:

(7)

$(\ *\quad\ \ )_{\text{L2}}$	$(\ *\ )(\ *\ )_{\text{L2}}$	$(\ *\quad\ \ )_{\text{L2}}$
$(\ *\ )(\ *\ )_{\text{L1}}$	$(\ *\ )(\ *\ )_{\text{L1}}$	$(\ *\quad\ \ )_{\text{L1}}$
swap: A B	insert: A B	delete: A B

"Swap" the head at level L2 to be A: This violates the headedness of L2, but leaves the phrasing intact and makes A the head of its L2-phrase.

Insert an L2-boundary at the right edge of A: This makes A the head of its own L2-phrase, in which it is moreover (trivially) rightmost; note that this maneuver doesn't make A maximally prominent yet (it just does not make it less prominent than B). For that to happen, the L2-phrase that contains A has to become the head at the next higher level, call it L3, so this strategy will make sense only if the L3-head is, or at least can be (more easily than the L2-head), to the left.

Delete an L1-boundary at the right edge of A: This strategy will only make sense if A manages to become the head of the L1-phrase containing it and B. The "super-sized" L1-phrase will trivially be rightmost in the L2-phrase containing it, thus satisfying L2–Head–R. Note that even though the prominence mark in this option is as far from the right edge of the L2-phrase as in the swapped structure, its *head* in the relevant technical sense—the whole L1-phrase (*A B*)—is not.

The default, as well as the three amended structures, are given in tree notation again in (8), with a higher level L3 added for the insertion case (again L*n*P means "phrase at level *n*"):

(8)

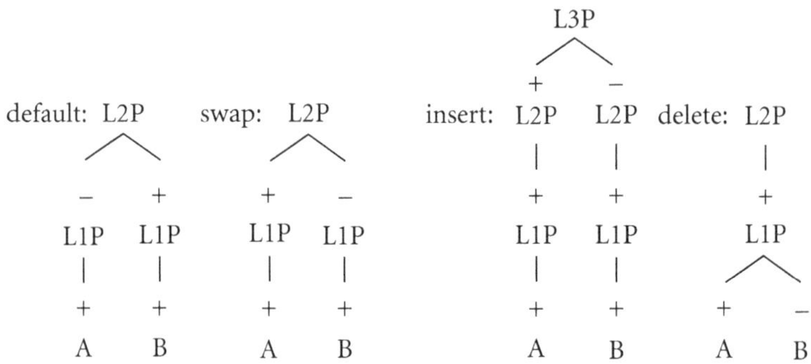

An instance of "swapping" the prosodic phrase head might be the change from *pleasant TRIP* to *PLEAsant*$_F$ *trip* in English.

Deletion of structure as a strategy to mark focus has been observed in many languages. In English, Greek, Bengali, German, and French, among others, focus causes the deletion of all accents to its right within the intonational phrase. In Japanese and Korean it forces deletion of all intermediate phrase/accentual phrase boundaries to its right within the intonational phrase, and in Shanghai Chinese deletion of all tones to the right within the major phrase.

Addition of structure is equally ubiquitous, found among others in Chichewa (addition of a phonological phrase break to the right of focus), English, German, Greek, French (addition of a phonological phrase break to the left), and Japanese (addition of an intermediate phrase boundary to the left).

8.2.4 *Preview*

We will now turn to a number of illustrations. Given what has been said so far, languages can differ in at least the following aspects: (i) default mapping, in particular the question whether and to what extent they split or wrap when mapping recursive syntactic structure to non-recursive prosodic structure; (ii) headedness at the various levels; (iii) the choice of adjustment strategies (head-swapping, boundary insertion, or deletion) at the various levels. This will lead us to a first typology of languages that realize focus prosodically (Section 8.3). An additional point of variation is (iv) how languages phonetically realize phrase boundaries and heads (again, at the various levels). Standard candidates for head-marking are stress and pitch accent (PA), and for boundaries boundary tones, lengthening, and pauses. In our model, every language has boundaries and heads at all levels. Descriptively, however, not all languages mark all of them (English, for example, realizes heads by stress and accent, but has no obvious phonetic marking of low-level boundaries; Korean has clear boundary markings, but no obvious phonetic marking of higher level heads), so an added option we need to consider is that heads or boundaries (at least at certain levels) are not marked at all.

Adjustment strategies need not be prosodic, however. If the prosodic system of a language does not allow *A* in the string *A B* to be prominent, two more options arise: no adjustment whatsoever (i.e. violation of FocusProminence); or *syntactic* adjustment to *B A* to get A into a prosodically prominent position. The latter case will result in a language that marks focus by constituent order variation (Sections 8.4–8.5).

In addition, I will suggest in Section 8.6 that in some languages, morphemes can serve to mark heads and/or boundaries. That is to say, so-called focus morphemes do not literally mark the focus, but rather the structural position in which focus is realized.

As I will show in Section 8.7, there are cases that still do not seem to fit an account along these lines. So as a last, and most stipulative, point of variation I will suggest that perhaps (v) prominence may, in some languages, be defined not over prosodic structures, but syntactic ones. Whether this is a useful move, or even one compatible with the spirit of a "prominence-based" account in general is very much unclear to me at present, and will, I hope, be the topic of future work.

8.3 Boundary languages

I will call a language a boundary language if it accords to (9):

(9) Focus is marked by insertion of a prosodic phrase boundary to the left or right of the focus.

8.3.1 *Chicheŵa*

The influence of focus on phrasing in Chicheŵa, a Bantu language, has been described in Kanerva (1990). The basic word order in Chicheŵa is SVO, and its default phrasing at the pP-level is (10) (as Kanerva shows, pP-boundaries are detectable because they block various segmental processes):

(10) (S)(V O1 O2 Obl)

Focus is marked by a pP-boundary to its right. If the default phrasing does not provide such a boundary, one is inserted (Kanerva does not discuss any additional effects of focus, say, on pitch or intensity):

(11) a. (**VP**)
What did he do?
([Anaményá nyumbá ndímwáála]$_F$)$_{pP}$.
hit house with rock
'He hit the house with a rock.'

b. (V OBJ **OBL**)
What did he hit the house with?
(Anaményá nyumbá ndímwáála$_F$)$_{pP}$.

c. (V **OBJ**)(OBL)
What did he hit with a rock?
(Anaményá nyumbá$_F$)(ndímwáála).

d. (**V**)(OBJ)(OBL)
What did he do to the house with the rock?
(Anaményá$_F$) (nyumbá) (ndímwáála)$_{pP}$.

(12) a. (**VP**)
What did they do?
([Anagóná mnyumbá yá Mávúuto]$_F$)$_{pP}$.
'They slept in Mavuto's house.'

b. (V **OBL**)
Where did they sleep?
(Anagóná mnyumbá yá Mávúuto$_F$)$_{pP}$.

c. (**V**)(OBL)
What did they do in Mavuto's house?
(Anagóná$_F$)(mnyumbá yá Mávúuto)$_{pP}$.

Following the terminology of Section 8.2.2, we can describe Chicheŵa as a radical wrapping language (since all objects are wrapped into one pP). The focus effect will follow if pPs are strictly right-headed: by creating an additional pP, the focus becomes rightmost in that pP. This violates the language's preference for radical wrapping, but satisfies FocusProminence. In other words, pP–Head–R dominates the constraints that regulate default phrasing.[2]

Kanerva does not discuss intonational phrases. For completeness, we will assume that the pP containing the focus also becomes the head of the IP, meeting FocusProminence at the higher level:

(13)			
	(	*	$)_{IP}$
	(	*) (	*$)_{pP}$
	(Anaményá)	(nyumbá$_F$)	(ndímwáála)$_{PWd}$
	hit	house	with rock

This would indicate that the head of IP is neither strictly leftmost nor strictly rightmost, but rather flexible in its alignment. I should emphasize, though, that any assumption about IP is made here on purely theoretical grounds.

8.3.2 *Bengali*

Bengali, as discussed in Hayes and Lahiri (1991), is an SOV language with unmarked phrasing (S)(O)(V). Similar to Chichewa, pP-boundaries are detectable through various prosodic processes; in addition, each pP has one L* pitch accent (on its head), and one final boundary tone (notated here as H−).[3] The head of IP can be H*, as in the declarative in (14), or L*, in other structures, including narrow focus; no PAs can follow the IP-head (L% is the IP-boundary tone):

(14) a. ami kágoǰola -ke dek^{h}lam
I newspaperman -OBJ saw

			H*	L%	
L* H−	L*	H−	|	|	
(| |	|	|	*	$)_{IP}$	
(*)	(*	)	(*	$)_{pP}$	
b. ami	kágoǰola-ke		dek^{h}lam		(neutral)

In present terms, then, Bengali is a strictly splitting language, the opposite of Chichewa. Assuming that the accent placement within pPs reflects the position of the pP-head, we furthermore see that pPs are strictly leftheaded. The placement of the IP-accent shows that the IP, on the other hand, is right-headed.

[2] Additional pP heads are inserted after every XP following the focus. Truckenbrodt (1995) provides an elegant account of this by assuming that, once it is impossible to wrap the entire VP (due to FocusProminence), lower-ranked constraints favor a one-to-one mapping of XPs to pPs.

[3] What I call pPs here are Hayes and Lahiri's P-phrases and Selkirk's (2007) Ma(jor)P(hrase)s.

It is then predicted that if Bengali uses pP-boundary insertion to realize focus, it will do so to the left of the focus, in order to make the focus leftmost in, and thus the head of, its pP. According to Selkirk (2007), this is indeed the case; she provides the examples in (15), where a boundary is inserted to the left of the focus *č^h obir*, "pictures", in (15b), which is not present in the neutral version (15a) (various other adjustments happen which cannot be discussed here):

```
                                             H*  L%
      L*  H-  L*                         H-   |   |
    (  |   |   |                          |   *   ) IP
    (  *   )(  *                          )(  *   ) pP
```

(15) a. Neutral: ami raǰar čʰobir ǰonno ţaka anlam
I king's pictures for money gave
'I gave money for the king's pictures'

```
                    L*                       L%
      L*       H-   |   H-          H-       |
    (  |        |   *   |            |       ) IP
    (  *        )(  *   )(           )(      ) pP
```

b. focused: ami raǰar čʰobir$_F$ ǰonno ţaka anlam

Hayes and Lahiri (1991), on the other hand, claim that the boundary to the left of the focus is optional, whereas a boundary to its right is obligatory (this boundary is evident, too, in (15b), contrary to what the PTF would lead us to expect, cf. the extensive discussion in Selkirk (2007)); Khan (p.c. and 2007), following Michaels and Nelson (2004), suggests that the alleged boundary tone is part of the pitch accent. More investigation is required here.

What about the IP level? As indicated in (15b), the IP-head aligns with the focus-pP (which, in this case, is reflected in a larger pitch excursion), and all subsequent accents are deleted. This is what we predict, given IP–Head–R.

Does the IP-head shift to the left, or are post-focal pP boundaries deleted so as to make the focal pP rightmost in IP? Hayes and Lahiri claim that while all pitch accents after the focus are deleted, the pP boundaries after the focus are present as before (i.e. segmental processes are blocked where they normally are, even post-focally, and H– boundary tones still occur). This would mean that the head of IP shifts to the left, but no pP-deletion occurs (as indicated by the boundaries and H-tones in (15b), which are not given in Selkirk's paper). If this is the correct analysis, it means that post-nuclear deaccenting is not necessarily the result of dephrasing, an issue we will return to in our discussion of English; either post-nuclear pPs do not have heads (as tentatively assumed in (15b)), or their heads cannot be associated with pitch accents.

8.3.3 *Japanese*

Japanese, like Bengali, is an SOV language, but unlike Bengali it shows some wrapping, its unmarked phrasing being (S)(OV) (the phrases indicated here are called "intermediate phrases" in Beckman and Pierrehumbert (1986), so I will mark them ip rather than pP; another common term is "Major-Phrase" (MaP)).

To detect ip-boundaries, we have to look at the next smaller prosodic phrase level, called the *accent domain* (AD). Within an AD, a lexical H* tone triggers downstep on all following H tones. The boundaries of ips are detectable because they reset this downstep, that is to say that the first H within a new ip is significantly higher than a downstepped one preceding it; in addition, ips have a final L% boundary tone and trigger lengthening of their final element.

Focus has a variety of effects on prosody in Japanese. First, the F-marked constituent is marked by an increased tonal pitch, even if it is not lexically accented. 'Second, and more importantly for our purposes, focus triggers an intermediate phrase boundary to its left, (compare (16b) to (16a)), like in Bengali, blocking downstep within the focus (Beckman and Pierrehumbert 1986, sec. 4):

(16)

a.
```
   H*L      !H*L
(   |          | )ip
(uma'i)(mame')ad
```

b.
```
   H*L      H*L
(   |  )(     | )ip
(uma'i)(mame'F)ad
```

Third, all ip-boundaries to the right of the focus are erased (i.e. downstep is not ever reset after a focus (see Truckenbrodt 1995 and references therein, but also Sugahara 2002 for conflicting findings):

(17) (Náoko wá)$_{ip}$ (niciyóobi)$_{ip}$ (Nágoya dé)$_{ip}$ (Mári ní átta)$_{ip}$
Naoko TOP Sunday Nagoya at Mari with met
'Naoko met with Mari in Nagoya on Sunday.' (unmarked)

a. (Náoko wá) (nichiyóobi) (Nágoya dé) (Mári ní)(átta$_F$)$_{ip}$

b. (Náoko wá) (nichiyóobi) (Nágoya dé) (Mári$_F$ ní átta)$_{ip}$

c. (Náoko wá) (nichiyóobi) (Nágoya$_F$ dé Mári ní átta)$_{ip}$

d. (Náoko wá) (nichiyóobi$_F$ Nágoya dé Mári ní átta)$_{ip}$

e. (Náoko$_F$ wá nichiyóobi Nágoya dé Mári ní átta)$_{ip}$

Truckenbrodt (1995) suggests a straightforward analysis for this pattern: the ip-boundary insertion tells us that ips in Japanese are strictly leftheaded (like the pPs in Bengali):

```
(18)      (*   )ip
     a.   A  BF      (bad: focus is not head of ip)
          (  *  )ip
     b.   A  BF      (bad: ip has its head on the right)
          (*) (*)ip
     c.   A    B     (ip a little small, but FocusProminence and ip-Hd ok)
```

The increased pitch indicates that the accent domain containing the focus is the head of the intonational phrase (IP). Finally, and this sets Japanese apart from both Chicheŵa and Bengali, the IP is strictly right headed. Therefore, the ip containing the focus can only become the head of IP by being rightmost in IP, which in turn requires deletion of all following ip-boundaries.

```
(19)      (               *)IP
          (*) (*      )(*)ip
     a.   A   BF C    D      (bad: focus ip not head of IP)
          (    *           )IP
          (*) (*      )(*)ip
     b.   A   BF C    D      (bad: IP head is not right)
          (    *           )IP
          (*) (*           )ip
     c.   A   BF C    D      (ip a little big, but FocusProm. and IP-Hd-R ok)
```

Similar patterns have been described for Korean (Jun 1996) and Greek (Condoravdi 1990; Baltazani and Jun 1999), among others.

8.3.4 *English*

Focus in English is first and foremost realized by pitch accents, whereas no consistent segmental or tonal effects of boundary insertion or deletion have been discussed in the literature. Yet, assuming that accents indicate heads of prosodic phrases, the analysis of English fits well into the mold of the analyses outlined so far, as we will see.

English has a basic (S)(VO) prosodic structure; it is a moderate wrapping language (though (S)(V)(O) is possible, too). Heads of pPs are realized by pitch accents, the IP-head being the nuclear pitch accent; unlike Bengali, English has a wide range of pitch accents and pitch-accent combinations, so that it is impossible to identify the IP-head by its tonal shape. The primary indicator of the IP-head is that, like in Bengali, no pitch accents can follow the one marking the IP-head, which yields the familiar effect that a non-sentence-final focus is marked by an early nuclear pitch accent, that is, by the absence of later pitch accents which would otherwise be present.

Whether this is the result of shifting the IP-head to a non-final pP, as in Bengali, or of deleting post-focal pPs, similar to Japanese, is at present unclear. I do not know of any segmental processes that indicate pP-boundaries in English. Jun and Fougeron (2000) mention in passing that postfocal material in English (unlike in French) is characteristically shortened, which they suggest may indicate the absence of pP-boundaries (which would induce pP-final lengthening). Hayes and Lahiri (1991), on the other hand, point out that the alignment of post-focal phrase tones may indicate the right boundary of the focus pP, which would suggest the presence of (accent-less) postfocal pPs, as in Bengali. I have to leave this question for further investigation.

What about pP-insertion? Note that in certain copular (as well as intransitive) structures, English wraps subject and predicate, as in (20a); narrow focus on the predicate yields a shift to a two-accent (non-wrapped) structure, (20b):

(20)	a.	(out of the blue:)	(*)IP
			(*)pP
		Your COAT's on fire.	[your coat's on fire]$_F$
	b.	('Why are you staring at my coat?')	(*)IP
			(*)(*)pP
		Your coat's on FIRE.	your coat's on fire$_F$

This indicates that the focus on the predicate *on fire* triggers the insertion of a pP-boundary to its left (rather than a shift of the pP head to the right), i.e. that the constraint pP–Head–L is stronger than the constraints governing default pP-formation.

This concludes the discussion of boundary languages. We have seen that various boundary languages can be analyzed using a small number of parametric choices. Boundary languages comprise languages in which pitch accent placement seems to be the main indicator of focus (English), as well as languages in which other prosodic or segmental cues are the main correlate of focus (Chicheŵa, Bengali, Japanese). Given the perspective of the PTF, these languages can be seen as variations of the same focus realization strategy.

8.4 Edge languages

Edge languages are characterized by (21):

(21) Focus is marked by non-standard constituent order, with the focus in left- or right-peripheral position.

On the face of it, edge languages display a fundamentally different strategy of focus marking, syntactic vs. prosodic (perhaps involving some specialized phrase-structural position such as a Focus Phrase).

At least for some languages, however, it has been argued that the focus position is truly defined in terms of being peripheral to a particular (prosodic) domain, but not in terms of being in a particular structural position. We will start by examining those, and then return to the question whether there are true "focus-position languages".

The basic analysis for edge languages uses the exact same ingredients as the analysis of boundary languages in the previous section. Like boundary languages, edge languages minimize the material between the focus and the relevant prosodic phrase edge, but unlike boundary languages, they do so via syntax. More precisely, these languages can not, or only to a very limited extent, amend the prosodic structure (they are *strict*, as opposed to *transparent*, in the terminology of Vallduví 1992).

8.4.1 *Spanish*

Zubizarreta (1998) provides the seminal discussion and analysis of an edge language, Spanish. The following discussion adopts her basic insight, but is closer in detail to the analysis in Büring and Gutiérrez-Bravo (2001).

The basic description of the facts is rather straightforward. Taking accents as indicative of pP-heads, Spanish shows a basic (S)(VO) or (S)(V)(O) phrasing (similar to English in that respect, the choice depending, among other things, on metrical factors). Unlike in English, however, this basic prosodic pattern does not change in response to focusing. Rather, constituent order is used to make sure that the focused constituent ends up clause finally. For example, in full-sentence answers, (non-corrective) subject focus is possible only with clause-final subjects, whereas broad-focus sentences show SVO order:

(22) Q: Qué pasó?/Qué hizo Juan?
'What happened?'/'What did Juan do?'

A: Juan compró ayer el peRIÓdico.
Juan bought yesterday the newspaper
'Juan bought the newspaper yesterday.'

(23) Q: Quién compró el periódico ayer?
who bought the newspaper yesterday

A: Ayer compró el periódico JUAN.
yesterday bought the newspaper Juan

A′: #JUAN compró ayer el periódico.
Juan bought yesterday the newspaper
'Juan bought the newspaper yesterday.'

The well-formed answers in (22) and (23) all have the same prosodic structure, by which I mean the following: all sentences above map the immediate constituents S, O, Adv, and V onto one pP each (or optionally wrapping V with O, which I do not discuss here). A transitive subject-focus sentence like (23) thus has the same prosodic pattern as an all new sentence like (22)—(XP) (V) (YP) (ZP)—which follows if we assume that Spanish is an optional (moderate) wrapping language that matches pPs with XPs. (24) illustrates for (22A):

```
(24) (                                  *      )IP
     (   *) (      *) (   *) (          *      )pP
     (Juan) (compró) (ayer) (el peRIÓdico)PWd
```

The focus effect will follow if we assume that both pP-formation and IP–Head–R are very strict. That is, a focused subject can not become prominent by shifting the IP-head to a non-final pP, (25a), or by deleting post-focal pPs, (25b). Short of violating FocusProminence itself, the language resorts to constituent order variation, (25c):

```
(25)     (   *                                   )
         (   *   )(      *) (   *) (      *      )
     a. *(JuanF) (compró) (ayer) (el periódico)        IP-head not right

          (  *)
          (  *)
     b. *(JuanF compró ayer el periódico)              pPs not properly built

         (                                  *  )
         (   *) (      *) (        *     ) (  *  )     non-standard order but
     c.  (Ayer) (compró) (el periódico) (JuanF)        happy prosody
```

Note that this analysis is different from saying that Spanish moves the focus into a right-peripheral position. There is no evidence that, say, a focused object as in (26A) occupies a position different from that of the object in neutral/all-new or VP focus sentences like (22), or that in the subject-focus VOS sentence (23)A/(25c) (analogously for focused adverbials, VPs, and sentences):

(26) Q: Qué compró Juan?— A: Juan compró el peRIÓdico.
what bought Juan Juan bought the newspaper
'What did Juan buy?'—'Juan bought the newspaper.'

On the other hand, Spanish does not allow VOS order in regular all-new sentences, (27), even though that results in the standard prosodic structure (V)(O)(S), and has the nuclear accent within the focus, (28):

(27) (*)
(*)(*)(*)(*)
Qué pasó?—*(Ayer) (compró) (el periódico) (Juan)

One could take this to mean that in an all-new sentence the main accent has to be on the O (by some principle like a syntactic Nuclear Stress Rule), and that that in turn is only possible in O-final order. I think it is advantageous, however, to simply derive this from a syntactic preference for SVO order, which will only be violated under pressure from FocusProminence.

But how could this be implemented? Given that FocusProminence makes reference to prosodic structure (by talking about prominence), does this mean that marked constituent order is established only after prosodic structure is built (and prosodic structure is subsequently "re-built")? This is more or less the position taken in Zubizarreta (1998).[4] Alternatively, one can reverse the causal chain and claim that each syntactic structure, canonical or inverted, is associated with the set of its possible focus markings (roughly the position of Reinhart 1995, 2006); the constituent order variation itself is then essentially optional. Finally, in a system that accesses syntactic and prosodic structure in parallel, questions of this sort do not even arise.

8.4.2 *Italian*

The basic pattern of data in Italian is the same as in Spanish: focused subjects appear in non-canonical, sentence final, position:

(28) Q: What happened?
(* $)_{IP}$
(*)(* $)_{pP}$
A: [(Gianni) (ha scritto) (una lettera$)_{PWd}$]F
G. has written a letter

[4] Zubizarreta's analogy to FocusProminence does not directly reference prosodic structure, but the notion of Nuclear Stress, which is crucially established after the syntactic structure has been completely assembled. In case of a mismatch between focus and Nuclear Stress, a late syntactic adjustment rule changes the constituent order.

(29) Q: Who laughed?

	(	*	)IP
	(	*	)pP
A:	(Ha riso) (Gianni$_F$)$_{PWd}$		
	has laughed G.		

Detailed investigation in Samek-Lodovici (2005) and Szendrői (2001, 2002) (as well as, from a different theoretical vantage point, Frascarelli 2000) shows some additional details that are worth pointing out, though.

First, in (30), the focused subject appears between the direct and the indirect object:

(30) Q: Chi ha mandato il vino a Marco?
who has sent the wine to Marco

	(						*)	(*	)IP
	(		*)	(	*)		(*)	(*	)pP
A:	(Gli	ha	mandato)	(il	vino)		(Gianni$_F$)	(a	Marco)$_{PWd}$
	to him	has	sent	the	wine		Gianni	to	Marco
	'Gianni sent the wine to Marco.'								

The subject is thus neither in its canonical preverbal position, nor peripheral, which may appear strange from the vantage point of the PTF. However, as indicated in (30A), in these structures the sentence-final, post-focal constituent is prosodically separated from, and perceptively less prominent than, the rest of the clause. The authors quoted interpreted this to mean that the indirect object in (30) has been syntactically right-dislocated, and prosodically forms its own IP. It is the latter fact that is important here. Within its own IP, the subject is right peripheral, so IP–Head–R is met.[5, 6]

Second, Samek-Lodovici and Frascarelli point out an interesting phenomenon that emphasizes the relation between edge languages and boundary languages. In certain cases it is syntactically impossible to make the focus truly peripheral within the main clause; for example, a focused numeral cannot occur post-nominally, as in (31b), forcing (31a) instead, a structure in which the IP-head is non-peripheral:

[5] If there is a next higher prosodic level, call it the "utterance phrase", we have to assume that the head of the UP is on the left IP (the main clause), i.e. that the UP is not as strictly right-headed as the IP.

[6] It is interesting to note that there *is* evidence that the post-focal material here is also syntactically moved. It thus seems as if syntactic extraposition is a prerequisite for "excluding" an argument from the main-clause IP.

(31) Q: Will you eat four sandwiches?/How many sandwiches will you eat?

a. Mangero TRE_F panini. b. *Mangero panini TRE_F.
eat-FUT-1SG three sandwiches

That is, there is small violation of either the headedness or the phrasing constraint, allowing the accent to shift slightly to the left, as it would in English.[7]

8.4.3 *Hungarian*

Let us now turn to Hungarian, which is perhaps the best-known language with a fixed focus position. The standard analysis for Hungarian describes it as having the constituent order in (32), where "Topic*" means "one or more topics":

(32) (Topic*) Focus V S O

Hungarian provides clear evidence for the structural reality of the focus position (in contradistinction to Italian, as discussed above), since the focus position is not obligatorily filled, and since placement of a constituent in that position is regularly accompanied by inversion of the verb and certain preverbal particles, glossed here as VM:[8]

(33) a. Mari fel hívta Pétert.
Mary-NOM VM rang Peter-ACC

b. MARI_F hívta fel Pétert.
Mary-NOM rang VM Peter-ACC

c. *MARI fel hívta Pétert.
Mary-NOM VM rang Peter-ACC
'Mary rang up Peter.' Szendrői (2003)

How does this fit the picture presented here? Szendrői (2001) argues that the movement to the pre-verbal position is prosodically driven (a position that is criticized in Horváth 2005). On her analysis, Hungarian is basically the mirror image of Italian: the IP is left-headed (an idea already proposed in Jacobs 1991/2b), drawing foci to the left periphery. Unlike in Italian, this involves

[7] There is a striking contrast with the cases discussed in Swerts, Krahmer, and Avesani (2002), which involve noun–adjective sequences (and arguably deaccenting, as opposed to narrow focus); they cannot be discussed here.

[8] É. Kiss (1998a and this volume) argues that foci can appear *in situ* (i.e. postverbally) as well as in the preverbal position, and that, rather, the structures involving a preverbal focus are akin to English clefts. If this is correct, Hungarian is not a strict positional focus language at all. According to Szendrői (2001), however, foci cannot generally occur *in situ*.

movement to a specifier position (simply called SpecF, for "functional"), rather than adjunction.

What of the pre-focus topics? Again, not unlike in Italian, topics are outside the main IP. Syntactically, they are adjoined to the projection hosting the focus, see (34b); prosodically, IP is recursive, and topics occupy the "extra-metrical" segments of IP, see (34c):

(34) a. A nö a KALAPJÁT vette le (nem a sálját).
the woman her cap-ACC took off
'It was her hat that the woman took off (not her scarf).'

b.

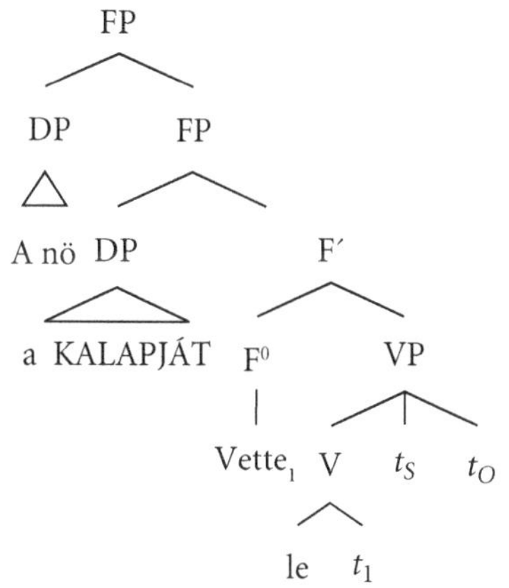

c. ((*))$_{IP}$
(*) (*) (*)$_{pP}$
(A nö) (a Kalapját vette) (le)$_{PWd}$

As mentioned before, IP is strictly left-headed in Hungarian. In the case of a recursive IP, the head will be left peripheral in the innermost segment of IP, as in (34c) (that way, each IP segment has a head, and none has two). As a result of this placement, the focus position is the most prominent one.

Unlike in Italian, by default the left-peripheral position in IP in Hungarian is filled by the verb rather than by an argument. Therefore, no argument is ever in a "focusable" position in the unmarked VSO order, making focus movement ubiquitous. Not only do these movements all target the same position, but they also all trigger particle–verb inversion, which is unlike in Italian. Nevertheless, according to Szendrői, there is nothing inherently focus-related about the preverbal position other than that it will eventually be left-peripheral within the main IP.

Szendrői (2001) presents additional evidence for this view, which, like in Italian, involves cases in which the focus is an element that is "naturally" left-peripheral. Since Hungarian has unmarked VSO order, this involves focus on V or a verbal projection. All these cases have an empty "focus position" and furthermore lack particle–verb inversion. In other words, just like in Italian object-focus sentences, there is no evidence of any "focus movement":

(35)

					*		
a.	A	kalapját	a	nö	[le	vette	az	elöszobában$]_F$.
	her	hat-ACC	the	woman	off	took	the	hall-in

'The woman took her hat off in the hall.' (**?** 2003: 12)

(((* $)_{IP}$

(*)(*)(*)(* $)_{pP}$

b. (a kalapját) (a nö) le vette)(az elöszobában$)_{PWd}$

topic — (a kalapját); topic — (a nö); VP_F — le vette)(az elöszobában)

8.4.4 *Relaxed-edge languages*

Let me briefly mention that in many languages, focus occurs *towards* an edge, but not exactly *at* the edge. Most common examples of this are strict V-final languages in which the focus appears pre-verbally (Hyun-Oak 1988):

(36) Gujarati (Indo-Aryan)

a. A chopli konne lidi?
this book who bought
'Who bought this book?'

b. (A chopli) me lidi.
this book I bought

c. #Me a chopli lidi.

(37) Telugu (Dravidian)

a. Sita ante evariki ishtam?
Sita ACC who like
'Who likes Sita?'

b. Sita ante Ramki ishtam.
Sita ACC Ram like
'Ram likes Sita.'

c. #Ramki Sita ante ishtam.

Similar patterns are documented for, among others, Turkish, Sherpa (Tibet), Mongolian, Hindi–Urdu (Hyun-Oak, 1988). Within the present analysis, these languages are basically like Italian and Spanish, except that they are head-final, and either completely disallow V–X structures or allow them only as right-dislocation-like structures akin to the Italian cases discussed above. Lacking the means to get the verb out of the way, as it were, and thereby making a focused argument truly right-peripheral, these languages minimize the IP–Head–R violation by bringing the focus as close to the right IP-edge as possible, and then wrapping it with the IP-final verb:

(38)
```
(            *            )
( *       )( *            )
 Sita ante  Ramki ishtam
```

Additional evidence for an analysis along these lines, as opposed to one that assumes a designated syntactic position for foci, may come from the fact that in many of these languages, it is not only verbs that can intervene between a focus and the left IP-edge. Often a whole range of arguments that are generally considered very "close" to the verb, such as locatives and indefinites, may separate the focus from the end of the clause. It seems promising to assume that the preverbal focus position is actually achieved by scrambling intervening phrases to the left, leaving only immobile elements to intervene between the focus and the IP-edge. Detailed studies of this language type within the present perspective have yet to be conducted.

8.4.5 *Strict-position languages*

To repeat, a strict-position language would be one that obligatorily puts focused constituents into a syntactically distinct position. The typological and theoretical literature abounds with claims that languages have "focus positions", for instance Armenian (Comrie 1984), Basque (Saltarelli 1988), Finnish (Vilkuna 1989), Georgian (A. C. Harris 1982), Hausa (Tuller 1986), Hungarian (Bródy 1990; É. Kiss 1987), Nupe (Baker and Kandybowicz 2003; Kandybowicz 2006), Turkish (Erguvanlı 1984); but as we have seen in our discussion of Hungarian, this claim can be open to re-evaluation. First and foremost, a language will qualify as a strict-position language only if it does not have an alternative *in situ* strategy; otherwise, the so-called focus construction is simply an information-structurally "loaded" construction, such as English clefts, for instance. That is to say, if a language has constructions in which foci typically occur in a specific position, that does not make it a strict-position language; it would only if *all* foci are realized in that construction. Upon closer inspection, Hausa, Nupe, and, ostensibly, Hungarian belong in the former category. (It is of course an interesting question what makes constructions such as clefting and dislocation information-structurally "loaded" in that way, but one that is orthogonal to the aims of the present chapter.)

Second, strict-position effects are generally amenable to a reinterpretation as edge effects. As discussed above, there are crucial cases to tease the two apart: do the focused constituents show any independent evidence for being in a structurally distinguished position, especially if their unmarked position linearly coincides with the focus position (as we saw, this does not seem to be the case for Italian or Hungarian)? And is the edge effect strict, or is there a structurally definable class of elements that intervene between the focus and the pertinent edge (most clearly illustrated in the relaxed-edge languages)?

Taking into consideration these tests and criteria, my cursory overview did not reveal any languages that clearly qualify as strict-position languages, though a more detailed investigation is needed here.

8.5 Mixed languages

In the previous section I argued that languages that appear to mark focus syntactically can be analyzed using the same framework used for languages that mark focus prosodically. We have also seen that many languages of the latter kind show prosodic effects of focus as well, especially where it is syntactically impossible to bring a focus all the way to the pertinent edge of the IP.

From the perspective of the PTF it is not only unsurprising, but expected that languages should mix these two strategies, or more precisely, mix them in a broader range of cases than the ones discussed so far. And indeed, examples of that language type abound, including most of the Slavic languages, German (and to a lesser extent other Germanic languages), Japanese, Korean, European Portuguese, and Finnish. In these languages, either prosodic or syntactic structure may be used to mark focus. In either case, the reward will be that the other "half" of structure is realized canonically; this leads to the following generalization:

(39) Generalization 1: Marked Word Order → Unmarked Prosody
Marked constituent order may be used for focusing X only if the resulting prosodic structure is less marked than that necessary to focus X in the unmarked constituent order.

Since the languages mentioned all are prominence-final (IP–Head–R), it furthermore follows that:

(40) Generalization 2: Marked Word Order°— Narrow Focus
Marked constituent order signals narrow focus on a right-peripheral element.[9]

I illustrate this with German. German is basic S IO DO V (with verb-second in root clauses yielding X V_{fin} S IO DO V_{inf}) with regular (S) (IO) (DO V) phrasing, that is, modest wrapping. Apart from its head-final VP, German behaves like English prosodically: the head of IP is right-peripheral, putting the nuclear pitch accent on the DO in neutral sentences. Shifting the head to the left yields post-nuclear deaccenting (and possibly destructuring), and unambiguously signals narrow focus on the constituent that corresponds to the IP-head.

Crucially, any focus pattern that is consistent with the unmarked order and unmarked prosody (i.e. DO_F, $[DO\ V]_F$, $[IO\ DO\ V]_F$, $[S\ IO\ DO\ V]_F$) *must* be realized with canonical word order. To focus, say, the IO, on the other hand, either the phrasing or the constituent order has to be amended:

(41)
(* $)_{IP}$
(* $)_{pP}$
a. (Er hat) (dem $Piloten_F$) (die Passagiere) $(vorgestellt)_{PWd}$.
he has to-the pilot the passengers introduced

(* $)_{IP}$
(*) (* $)_{pP}$
b. (Er hat) (die Passagiere) (dem $Piloten_F$) $(vorgestellt)_{PWd}$
'He introduced the passengers to the pilot.' (IO_F DO/DO IO_F)

[9] To be precise, this is true only if the marked constituent order results in a shifted nuclear accent. As Katalin É. Kiss (p.c.) reminds me, German allows fronting ("topicalization") of a nuclear pitch accented object while preserving VP focus:

(i) Was hat sie dann gemacht? (What did she do then?)
a. Sie hat [die ZEItung gelesen$]_F$.
she has the newspaper read
b. Die $ZEItung_F$ hat sie $gelesen_F$.
"She read the newspaper."

Though a detailed discussion would lead us too far afield here, it should be noted that crucially this is possible only if no accentable elements follow the preposed object, since that would yield a shifted nuclear accent:

(ii) (same question)
a. Sie hat einem Freund ein BUCH mitgebracht.
she has a friend a book brought
'She brought a book for a friend.'
b. #Ein Buch hat sie einem Freund mitgebracht. (any intonation)

Thus, German chooses the English strategy in (41a), but behaves like a relaxed-edge language in (41b).

It bears reiterating that sentences with marked constituent order have a strong tendency for sentence-final focus; that is to say, getting the focus into the unmarked nuclear accent position seems to be the primary motivation for constituent order variation in German. This is not the case in all mixed languages. According to Godjevac (2000), constituent order in Serbo-Croatian can vary independently of focus placement, defying Generalization 1 (it is not clear from the literature what determines constituent order in these cases). Yet, Serbo-Croatian does share two properties with German: early nuclear pitch accent unambiguously signals narrow focus (i.e. default prosody is only abandoned for special focus patterns), as does marked constituent order (Generalization 2).

To sum up this chapter up until here, we have seen that the PTF not only provides a unified account of edge, position, and accent-based strategies for focus, but can also account for a wide variety of mixed cases. In the next two sections, I will briefly look at less straightforward cases and offer some speculations as to how, if at all, these languages could be analyzed along similar lines.

8.6 Particle languages

I am interested here in languages that characteristically mark the focused constituent itself by a special morpheme (there are also languages in which special verbal or sentential particles mark a sentence containing a focus, which I will not discuss). Examples are ubiquitous, though detailed discussions are rare. A typical example is Chickasaw, a (Western) Muskogean language, which marks subject and object focus by suffixes *-akot*/*-akō* and *-ho:t*/*-ho* (Munro and Willmond 1994; Gordon 2007); there is no additional prosodic focus marking, hence no marking at all for verb or sentence focus:

(42) a. hat:ak-at koni(ã) pisa.
man-SUBJ skunk sees
'The man sees the skunk.'

b. hat:ak-akot koni(ã) pisa.
man-FOC.SUBJ skunk sees
'[The man]$_F$ sees the skunk.'

c. hat:ak-at koni-akõ: pisa.
man-SUBJ skunk-FOC.OBJ sees
'The man sees [the skunk]$_F$.'

It seems straightforward to analyze the focus morpheme as a direct spell-out of the syntactic feature F, in which case the prominence-based account has nothing to offer in the analysis of these languages. Alternatively, one could hypothesize that the focus morpheme marks prominence of *prosodic* units; that would make it the counterpart of positional head marking in the languages described so far.

To illustrate what kind of data may motivate such an analysis, I will briefly discuss Gúrúntúm, a Chadic SVO language spoken in Nigeria (Hartmann and Zimmermann in press; Zimmermann 2005; Fiedler *et al.*, this volume). Gúrúntúm has a focus marker *á*, which generally occurs before the focus:

(43) Neutral

Tí bá wúm kwálíngálá.
3SG PROG chew colanut
'He is chewing colanut.'

(44) Subject focus

a. **Á** kwá bá wúm kwálíngálá-í?
FOC who PROG chew colanut-the
'Who is chewing the colanut?'

b. **Á** fúrmáyò bá wúm kwálíngálá.
FOC fulani PROG chew colanut
'[The fulani]$_F$ is chewing colanut.'

(45) Object focus

a. **Á** kãèã mái tí bá wúmì?
FOC what REL 3SG PROG chew
'What is he chewing?'

b. Tí bá wúm-**á** kwálíngálá.
3SG PROG chew-FOC colanut
'He is chewing [colanut]$_F$.

There are two complications: first, with V and VP-focus, *á* occurs *after* the V:

(46) VP-focus

a. **Á** kãèã mái tí bá pí?
FOC what REL 3SG PROG do
'What is he doing?'

b. Tí bá ròmb-**á** gwéì.
3SG PROG gather-FOC seeds
'He is [gathering the seeds]$_F$.'

(47) V-focus

a. Á kãèã mái tí bá pí náa gwâì?
FOC what REL 3SG PROG do with seeds
'What is he doing with the seeds?'

b. Tí bá ròmb-**á** gwéì.
3SG PROG gather-FOC seeds
'He is gatheringF the seeds.'

Second, in sentences without overt objects, VP focus isn't marked by *á* at all:

(48) V-focus without overt object

a. **Á** kãèã mài tí bá pí náa dùsó-ì?
FOC what REL 3SG PROG do to car-the
'What is he doing to the car?'

b. Tí bá krí.
3SG PROG repair
'He is repairing$_F$ (it).'

In order to explain why *V à O* is ambiguous between V, VP, and O focus, we assume that V moves to a position outside of VP; *à*, even if marking VP, thus linearly follows the verb.[10] Furthermore, and more to the point at hand, assume that *à* actually marks the left edge of the prosodic phrase containing the (base position of the) focus. Since the verb has moved to a higher position, *à* relies on whatever is left in the VP to prosodically attach to. In case there is nothing there, *à*-marking is impossible, as in (48b).

This line of analysis, highly speculative though it is, extends the prosodic account of focus marking to morphological marking languages in an interesting way, in particular by proposing that the distribution of morphological markers may be determined by prosodic structure and in particular, the notion of "prosodic head".

Many questions remain open, though, such as why there is no *à*-marking on all-new sentences. We have been assuming that all-new sentences are all-focus sentences, but maybe this assumption should be called into question. Another possibility is that there are constraints regarding the entire focus realization paradigm, which block sentential *à*-marking since the result would look the

[10] This idea is less absurd than it may seem. Note that association with focus also disregards head-movement, as the following German example shows:

(i) (Wir lösen das Problem nicht.) Wir verLAGern$_F$ es **nur**.
we solve the problem not we relocate it only
'We don't solve the problem. We only relocate it.'

same as narrow subject focus (Malte Zimmermann, p.c.). Finally, this may suggest that, at least in Gúrúntúm, focus marking only takes place in order to mark a contrast in focus status; if everything is equally focused (as in an all-new sentence), there is no need to mark any asymmetries in F-marking. I have to leave these questions for further research.

8.7 Non-marking languages

Let me close this chapter with another puzzling case from another Chadic language, Hausa. Hausa has long been described as a language that marks focus by movement into a sentence-initial position, in certain tenses accompanied by a specific form of the auxiliary called the *relative form*, and raising of lexical high tones within the focus. It turns out, though, that alongside this *ex situ* strategy, Hausa can alternatively leave foci *in situ* (Green and Jaggar 2003; Hartmann and Zimmermann 2007*a*; Jaggar 2001). The two strategies do not appear to be correlated with any semantic or pragmatic distinctions such as exhaustivity and contrastivity.

(49) Q: Mèe su-kà kaamàa?
what 3PL-REL.PERF catch
'What did they catch?'

A: Sun kaamà dawaakii (nè).
3PL.PERF catch horses PRT
'They caught horses$_F$.' OBJ-NP

(50) Q: Mèe Tankò ya yi wà hàr̃aajì-n?
what Tankò 3SG.REL.PERF do to taxes-DET
'What did Tanko do with the taxes?'

A: Tankò yaa biyaa hàr̃aajì-n (ne).
T. 3SG.PERF pay taxes-DET PRT
'Tanko paid$_F$ the taxes.' V

(51) Q: Mèeneenèe ya fàaru?
what 3SG.REL.PERF happen
'What happened?' IP

A: Tankò yaa biyaa hàr̃aajìn (ne).
T. 3SG.PERF pay taxes PRT
'Tanko [paid the taxes]$_F$.'

This, in and of itself, is not novel. Two additional facts, however, make Hausa problematic. First, *in situ* focus appears to lack any acoustically measurable or perceivable marking; a sentence with basic SVO order is ambiguous between focus on either object, the verb, the whole sentence, or the verb phrase. So one might be inclined to think that focus is simply not relevant in Hausa at all. However, and this is the second crucial fact, focused subjects must occur in the focus position, and obligatorily trigger the relative form of the auxiliary (where applicable; all data from Hartmann and Zimmermann 2007*a* and p.c.):

(52) Q: Wàa ya-kèe kirà-ntà?
who 3SG-REL.CONT call-her
'Who is calling her?'

A1: Daudàa (nee) ya-kèe kirà-ntà.
D. PRT 3SG-REL.CONT call-her
'DaudaF is calling her.'

A2: #Daudàa ya-nàa kirà-ntà.
D.$_F$ 3SG-CONT call-her

The biggest problem with explaining this pattern within the prosodic account is that the subject in sentences without focus movement such as A2 is in the same prosodic position as focus in sentences with focus movement as in A1: left-peripheral. So even disregarding the differences between subjects and non-subjects, the prosodic theory cannot offer any reason why the subject should undergo essentially string-vacuous movement in order to be marked as focus (recall that the absence of any evidence for string-vacuous movement of peripheral constituents in Italian, Spanish, and Hungarian was considered a strong argument in favour of the prosodic account).

One way to incorporate a language like Hausa into the framework of the PTF is to allow for the notion of prominence to be alternatively defined in syntactic terms, along the lines of the hierarchies in (53):

(53) a. focus position $>_{Prom}$ rest
b. rest $>_{Prom}$ subject

That is to say, the syntactic subject position is inherently less prominent than the rest of the clause, and the syntactic focus position is inherently more prominent than the rest of the clause. FocusProminence requires action if the focus is less prominent than a non-focus; this would be the case if either a non-focus were moved to the focus position, or a focused subject were left *in situ*, exactly the two options excluded in Hausa. Non-subject foci are allowed to stay *in situ* since, absent a filled focus position, their position is (one of) the most prominent (ones) in the clause.

This analysis preserves the general architecture of the prominence-based theory of focus, but crucially parameterizes the notion of prominence as either syntax-based (Hausa) or prosody-based (English etc.). This doesn't mean that the prosodic (re)analyses of apparently syntactic focus strategies such as in Italian, Spanish or Hungarian should be abandoned. I still believe that the arguments given there are convincing. But it does mean that the strongest conceivable version of the prominence-based theory, which universally defines prominence in terms of prosodic structure, seems untenable. Of course it is very possible that future investigation of Hausa and other languages like it will reveal a way to reconcile it with the strongest theory. This question, too, will be left for future research.

8.8 Conclusion

This chapter presented, in very broad strokes, the outlines of a cross-linguistic theory of focus realization. I tried to show how a Prominence Theory of Focus realization affords a uniform and explanatory account of seemingly unrelated ways in which various languages realize focus, but also, where it fails to do so, at least in a straightforward way.

My intention in this chapter was to explain and illustrate the general logic of the PTF, indicate open theoretical and empirical questions, and thus invite further, and more detailed, research within this paradigm.

9

Focus in Aghem

LARRY M. HYMAN AND MARIA POLINSKY

9.1 Introduction

Aghem focus marking has long been noted by researchers as an unusual phenomenon, both descriptively and theoretically. The main descriptive generalization is that focus expression in Aghem is intimately connected with the immediate postverbal position, which, since the pioneering work of Watters (1979) and Hyman (1979a), has been known under the abbreviation IAV (immediately after the verb). To illustrate, compare the baseline sentence in (1), where the focus is either on the entire verb phrase or on the object (narrow focus), with (2a), where the narrowly focused subject appears in the IAV, and (2b), where the focused adjunct instead appears in the IAV:

(1)	tɨ́-bvʉ́	tɨ̀-bɨ̀ghà	mɔ̂	zɨ̀	kɨ́-bɛ́	↓nɛ́
	dogs	two	P$_1$	eat	D-fufu	today

'The two dogs ate the fufu today.'[1]

(2) a. Subject in IAV

à	mɔ̀	zɨ̀	<u>tɨ́-bvʉ́</u>	<u>tɨ̀-bɨ̀ghà</u>	bɛ́	↓kɔ́	nɛ́
ES	P$_1$	eat	dogs	two	fufu	D.OBL	today

'The TWO DOGS ate the fufu today.'

We are very grateful to Timothy Buo for sharing his knowledge of Aghem with us. We would also like to thank Mark Baker, Leston Buell, Ivano Caponigro, Lisa Cheng, Gennaro Chierchia, Laura Downing, Caroline Féry, Jeff Good, Laura Kertz, Shin-Sook Kim, Katalin É. Kiss, Beth Levin, Knud Lambrecht, Eric Potsdam, Hannah Rohde, Sharon Rose, Peter Sells, Anne Sturgeon, Malte Zimmermann, and the audiences at the Potsdam International Conference on Information Structure, 6–8 June 2006, and International Conference of the Association for Linguistic Typology in Paris 2007, for their helpful comments.

[1] *Fufu* is a traditional starchy accompaniment for stews or other dishes with sauce, made of corn or yams.

b. Adjunct in IAV

tɨ́-bvʉ́ tɨ̀ -bɨ̀ghà mɔ̂ zɨ̀ <u>nɛ́</u> ↓bɛ́ ↓kɔ́
dogs two P_1 eat today fufu D.OBL
'The two dogs ate the fufu TODAY.'

As seen in (3a), a matrix clause cannot end in a bare verb;[2] thus, intransitive verbs and transitive verbs that appear without a postverbal object have their IAV filled by the particle *nò*, which we will for now identify as the focus marker (FM):

(3) Bare verb: IAV has to be filled

a. ò mɔ̀ bvʉ̀ *(nò)
3SG P_1 fall FM
'He fell.'

b. ò mɔ̀ zɨ̀ *(nô)
3SG P_1 eat FM
'He ate (it).'

c. bvʉ́ ↓tɨ́ mɔ̀ bɛ́ ↓kɨ́ zɨ́ *(nô)
dogs D P_2 fufu D eat FM
'The dogs ATE the fufu.'

Unlike many other West African languages, Aghem does not have verb fronting, and verbs are focused *in situ*, often in combination with *nò*. The contrast between bare verbs and verbs followed by the particle *nò* is similar to the contrast between conjoint (conjunctive, weak, short) and disjoint (disjunctive, strong, long) forms in other Bantu languages (Hyman and Watters 1984; Creissels 1996; Güldemann 2003; Buell 2006): in all the relevant cases, the conjoint form cannot appear clause-finally. We will discuss the particle *nò* further in Section 9.4.

Finally, *wh*-words appear in the same position as the relevant focused material, thus invariably in the IAV. Compare *wh*-questions about object (4a), subject (4b), and adjunct (4c):

(4) a. bvʉ́ ↓tɨ́ mɔ̀ zɨ́ kwɔ̀ nɛ́ (à)
dogs D P_2 eat what today QM
'What did the dogs eat today?'

[2] Clauses with what Hyman and Watters (1984) call "auxiliary focus" do not require an overt IAV element (see also Hyman, to appear). Their analysis is beyond the goals of this paper.

b. à mɔ̀ zɨ̀ ndúghɔ́ ꜜbɛ́ ꜜkɔ́ nɛ́ (à)
ES P_1 eat who fufu D.OBL today QM
'Who ate the fufu today?'

c. tɨ́-bvʉ́ tɨ̀-bɨ̀ghà mɔ̂ zɨ̀ zɨ́n bɛ́ ꜜkɔ́ (á)
dogs two P_1 eat when fufu D.OBL QM
'When did the two dogs eat the fufu?'

While the empirical characterization of the IAV as being associated with focus is quite robust, how these facts fit in with the total analysis is less clear. Two sets of questions naturally arise. The first set concerns what the IAV is: is it the surface postverbal position that matters, or does the IAV represent a deeper structural association with the verb? Either way, is the IAV a dedicated focus/*wh*-position, as it superficially appears to be, or does it represent something else, either different from or more general than focus? If the IAV is indeed dedicated to focus, is it accidental that focus is in IAV and not somewhere else in the clause? If the IAV is not exclusively bound to focus, how does Aghem express focus in general?

The other set of questions concerns the contribution of the final marker *nò*: does it mark focus directly (as is assumed in Hyman and Watters 1984, Aboh 2006),[3] or is it simply a marker of constituency, thus only indirectly associated with focus, as has been argued for disjoint forms in Zulu (Buell 2006) or for focus in Naki (Good to appear)?

In the rest of the chapter we will present an analysis of focus expression in Aghem which addresses these questions. We will begin with a general discussion of the basics of Aghem grammar in Section 9.2, and we will then analyze two possible approaches to focus that have been proposed for Aghem and related languages: the representation of focus as a special projection in the verb phrase, and a movement analysis of focus (Section 9.3). We show that both approaches, although they offer considerable insights, are not supported by the empirical evidence. In Section 9.4, we present our own proposal concerning the grammar of focus in Aghem and analyze the status of *nò*. Our main conclusion is that exhaustive focus expressions and *wh*-words in Aghem have to occur in the verb phrase and receive the focus/*wh*-interpretation from an operator in the complementizer layer of the clause. Section 9.5 outlines our general conclusions and presents outstanding questions.

[3] Cf. similar analyses of conjoint/disjoint forms in other Bantu languages (Creissels 1996; Güldemann 2003).

9.2 Basics of Aghem structure

Aghem is a Western Grassfields Bantu language of the Ring subgroup spoken by about 27,000 speakers in Cameroon (Hyman 1979a, 1985, to appear; S.C. Anderson 1979; Watters 1979). Although more morphosyntactically analytic than the Narrow Bantu languages, Aghem maintains much noun-class morphology as well as some marking on verbs. Since choice of tone often reflects underlying syntactic structure in Aghem, we adopt the tone marking in (5):

(5) Aghem tones and tone marking

(´) high tone (ˆ) high-to-low falling tone (↓) downstep
(`) low tone (ˇ) low-to-high rising tone

The surface word order in Aghem is SVO, with the auxiliary verb always preceding the main verb, e.g. *mɔ̀* 'today past' (P$_1$) in (6a,b):[4]

(6) a. tɨ́-bvʉ́ tɨ̀-bɨ̀ghà mɔ̂ zɨ̀ kɨ́-bɛ́ ↓nɛ́
dogs two P$_1$ eat D-fufu today
'The two dogs ate the fufu today.'

b. fɨ́l á mɔ̂ fùo kɨ́-bɛ́ â bvʉ́ ↓tɔ́ á ↓ndúghó
friends D P$_1$ give D-fufu to dogs D.OBL LOC house
'The friends gave the fufu to the dogs in the house.'

9.2.1 *Noun phrase*

The structure of the DP is quite complicated, and in this chapter we will address only those aspects that are relevant to the general points made here; for a more detailed discussion, see Hyman (1979a, 1985, to appear). Unlike better-known Bantu languages, Aghem actually has a morphological distinction between direct and oblique cases, where the direct case is assigned by the verb (light verb for the subject,[5] and lexical verb for the object), whereas the oblique case is assigned by overt prepositions, as in (7), or by a suprasegmental preposition consisting solely of an underlying high tone, as in (8).

(7) Oblique case assigned by an overt preposition

a. â bvʉ́ ↓tɔ́
to dogs D.OBL
'to (the) dogs'

[4] Here and elsewhere, we cite individual morphemes with their underlying tones.

[5] In particular, we assume that assignment of direct case to the subject by the light verb takes place via an Agree/m-command relationship established between the specifier and head of the vP, prior to movement of the subject to Spec,TP.

b. à bvʉ́ ↓tɔ́
with dogs D.OBL
'with (the) dogs'

(8) Oblique case assigned by an underlying high tone

a. ò mɔ̀ kɔ̀ʔ wù
3SG P_1 see person
'He saw a person.' (NB: There is no high tone between 'see' and 'person')

b. à mɔ̀ kɔ̀ʔ wǔ ñɔ̀m 'a PERSON saw the animal'
ES P_1 see person animal
(/wù+ ′ +ñɔ̀m/)
person PREP animal

The morphological exponence of case is fused with the expression of the determiner, which follows the noun.[6] In the direct cases, the determiner procliticizes either to the next constituent, or to the noun, as in (9a), which shows the direct case of the subject, and (9b) illustrating the direct case on the object:

(9) a. bvʉ́ ↓tɨ́ mɔ̀ ñɨ́ŋ nô
dogs D P_2 run FM
'The dogs ran.'

b. bvʉ́ ↓tɨ́ mɔ̀ zɨ́ kɨ́-bɛ́
dogs D P_2 eat D-fufu
'The dogs ate the fufu.'

9.2.2 *Basic clause structures*

In analyzing Aghem clause structures, we adopt the standard assumptions concerning verbal and inflectional domains, namely, that a predicate's arguments are generated within vP. The highest XP in the vP domain then raises to Spec,TP; recall that verb-phrase raising to Spec,TP, attested in other West African languages, does not occur in Aghem. The structure of (9b) is shown in (10):

[6] Noun prefixes delete before all agreeing elements except numerals; see Hyman (1979a, 1985, to appear) on conditions on the realization of (9a) vs. (9b).

(10)

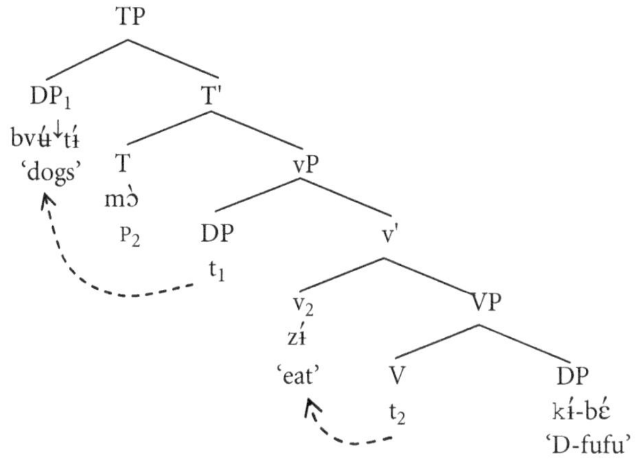

In this structure, the verb remains low in the verbal domain, raising to v^0 but not higher. Evidence for the low position of the verb comes from the placement of adverbials, which have to precede the verb when the object is in direct case, here *kɨ́-bɛ́* 'the fufu'.[7]

(11) bvʉ́ ↓tɨ́ mɔ̀ (á↓zɔ́ɔ́) zɨ́ (*á↓zɔ́ɔ́) kɨ́-bɛ́ (á↓zɔ́ɔ́)
dogs D P$_2$ yesterday eat yesterday D-fufu yesterday
'The dogs ate the fufu (yesterday).'

In addition to the SVO structure just shown, the object can also appear in the preverbal position. We assume that this preverbal position is adjoined to vP; thus, the sentence in (12a) has the structure in (12b).

(12) a. bvʉ́ ↓tɨ́ mɔ̀ bɛ́ ↓kɨ́ zɨ́ á↓zɔ́ɔ́
dogs D P$_2$ fufu D eat yesterday
'The dogs ate the fufu YESTERDAY[8].'

b. [$_{TP}$ bvʉ́ ↓tɨ́ [$_{T}$ mɔ̀ [$_{vP}$ ~~bvʉ́↓tɨ́~~ [$_{vP}$ bɛ́ ↓kɨ́ [$_{v}$ zɨ́ [$_{VP}$ [$_{VP}$ ~~zɨ́ kɨ́-bɛ́~~] á↓zɔ́ɔ́]]]]]]

Adjunction inside the vP results in typical middle-field effects, including defocusing (as in German or Slavic; cf. Reinhart 1996, Sturgeon 2006, among many others). However, there is no evidence for a dedicated Topic projection since the preverbal object does not show obligatory wide scope of the defocused constituent or reconstruction effects. In Section 9.4, we will return to

[7] In fact, temporal adverbials may also occur optionally in clause-final position, as shown in (11). The crucial point for our concerns is that the pre-verbal occurrence of the adverbial is obligatory.

[8] The reader will have noticed that the determiner is sometimes realized as a prefix on the noun, sometimes as a free-standing grammatical morpheme after the noun (or nominal modifier). We thus have glossed this variation as D-*fufu* in (11) and *fufu* D in (12). The reasons for the variation are orthogonal to the concerns of the present chapter, and we set them aside.

the differences between SVOX (11) and SOVX (12), but for now let us just emphasize that the preverbal position of the object (identified in Watters 1979 as immediately before the verb, IBV) is not tied to a unique interpretation.[9]

Next, Aghem has the inversion construction VS, illustrated in (13a), which contrasts with the regular SV structure in (13b):

(13) a. à mɔ̀ ñɨ́ŋ tɨ́-bvʉ́ á↓zɔ́ɔ
ES P_2 run dogs yesterday
'There ran dogs yesterday/The DOGS ran yesterday.'

b. tɨ́-bvʉ́ mɔ̀ ñɨ́ŋ á↓zɔ́ɔ
dogs P_2 run yesterday
'(The) dogs ran yesterday.'

Following extensive research on subject inversion in other languages (e.g. Kayne and Pollock 1978, 2001; Déprez 1990), we analyze (13a) as involving the movement of the verb to a higher head, with the subject staying in the base-generated position in Spec,vP. Unlike other languages with inversion where the verb moves to the inflectional head, in Aghem the verb stays lower, which is indicated by adverbial placement data: in inversion constructions, adverbs can still intervene between the tense marker and the verb, but not between the verb and the subject:

(14) à mɔ̀ (á↓zɔ́ɔ́) ñɨ́ŋ (*á↓zɔ́ɔ́) tɨ́-bvʉ́
ES P_2 yesterday run yesterday dogs
'There ran dogs yesterday/The DOGS ran yesterday.'

We hypothesize that the verb raises to a functional head in the extended verbal projection. This head is possibly aspectual, which would account for the episodic interpretation found in presentational contexts; we refer to it as Asp from now on. Aghem does not allow predicate fronting to satisfy the EPP,[10] nor is it a pro-drop language, hence inversion constructions invariably have an overt expletive subject in the preverbal position, thus (English glosses are substituted for Aghem words):

(15) $[_{TP}$ *expl* $[_{TP}$ P_2 $[_{AspP}$ $[_{Asp^\circ}$ run$[_{vP}$ dogs $[_{v^\circ}$ ~~run~~$]$ $[_{VP}$ ~~run~~$]]]]]]$

Although it may appear on the surface that Aghem allows inversion with transitive verbs, as seen in many examples in Hyman (1979b), closer inspection reveals that such clauses are in fact different from the uninverted ones.

[9] For a similar treatment of a related phenomenon, see Skopteas and Fanselow (this volume), who treat scrambling A-movement as a possible, though not obligatory, device for marking givenness in German.

[10] Cf. Alexiadou and Anagnostopoulou (1998) and Massam (2000) on predicate raising satisfying the EPP.

For example, the baseline sentence (16a) cannot be simply inverted, hence the ungrammaticality of (16b), where the object *kɨ́-bɛ́* 'the fufu' occurs in direct case:

(16) a. bvʉ́ ↓tɨ́ mɔ̀ zɨ́ kɨ́-bɛ́
dogs D P_2 eat D-fufu
'The dogs ate the fufu.'

b. * à mɔ̀ zɨ́ tɨ́-bvʉ́ kɨ́-bɛ́
ES P_2 eat dogs D-fufu
('There ate dogs the fufu.')

c. *à mɔ̀ bɛ́ kɨ́ zɨ́ tɨ́-bvʉ́
ES P_2 fufu-D eat dogs
('There ate dogs the fufu.')

For a two-place predicate to undergo inversion, the object has to be externalized outside the verb phrase. The descriptive generalization is as follows: the object cannot appear in the direct case and it has to be at the right periphery of the clause. This captures the ungrammaticality of (16b) and (16c) respectively.

In order to understand the structure as in (16a), we first need to address the status of the right-peripheral object.

The understanding of the right-peripheral position starts with the observation that direct arguments in Aghem are not dedicated topics—as (17) shows, sentences opening a narrative can have a preverbal subject in the direct case:

(17) wù lí fɨ́ kí tsɨ́ghá ↓kɔ́ wí
person certain once had HAB only wife
'A certain person once had a wife ...' (Hyman 1979b: 202)

More importantly, direct forms cannot appear in the right periphery, as shown by the ungrammaticality of (18):

(18) *bvʉ́ ↓tɨ́ mɔ̀ zɨ́ á↓zɔ́ɔ́, kɨ́-bɛ́
dogs D P_2 eat yesterday D-fufu
'The dogs ate yesterday the fufu.'

Oblique case-marked left-dislocated topics are possible but rare and dispreferred (Hyman 1979a: 70). Thus right-periphery seems to be the preferred place for topic expressions. Topic expressions on the right have to be in the oblique case, and they share this oblique cases marking with another set of peripheral expressions, vocatives (Hyman 1979a: 61); cf. Lambrecht (1996) on the systematic parallels between vocatives and base-generated topics in French.

Our approach to these right-peripheral expressions is that they are base-generated in a high adjoined position. In addition to the case marking, the evidence for their adjunct status comes from several quarters. First, such expressions can appear further to the right, following another adjunct, as in the following example:

(19) bvʉ́ ↓tɨ́ mɔ̀ zɨ́ á↓zɔ́ɔ́ bɛ́ ↓kɔ́
dogs D P$_2$ eat yesterday fufu D.OBL
'The dogs ate the fufu YESTERDAY.'

Next, this constituent is required to be marked by the high-tone preposition referred to earlier, which is realized on the surface only if it is both preceded and followed by a low tone, as in the following:

(20) a. ñɔ̀m mɔ̀ kɔ̀ʔ fɨ̀-kàa.
animal P$_1$ see squirrel
'An animal saw a squirrel.'

b. à mɔ̀ kɔ̀ʔ ñɔ̰̌m kàa fɔ̂
ES P$_1$ see animal squirrel D-OBL
/ñɔ̀m + ′ + kàa fɔ́/.
animal PREP squirrel D.OBL
'An ANIMAL saw a squirrel.'

We see in (20a) that the noun *ñɔ̀m* 'animal' has an inherent low tone. However, as seen in (20b), when 'animal' is followed by the oblique form of the noun *kàa fɔ̂* 'squirrel', it is realized with a low-to-high rising tone: *ñɔ̌m*. As indicated in the underlying representation to the right, the high tone that has been assigned to create the rise is the tonal preposition referred to above. Although underlyingly present to assign the oblique form of the following noun phrase, this high tone will only be realized if both the preceding and following tones are low (i.e. it will be "absorbed" into either a preceding or following high tone).

In sentences with rightward constituents, these expressions always take wide scope, as shown in (21a), where the clause-final locative takes wide scope over the subject. Compare that with example (21b), where such wide scope is optional:

(21) a. á-ghí á-tɨ́ghá mɔ̂ nùŋò (nò) án tɨ̀-mútù tɨ́-bɨ̀ghà
people three P$_1$ leave FM in cars two
'Three people went in two cars.'
(*two cars > three people*; **three people > two cars*)

b. á-ghí á-tɨ́ghá mɔ̂ án tɨ̀-mútù tɨ́-bɨ̀ghà nùŋò nò
people three P$_1$ in cars two leave FM
'Three people went in two cars.'
(*three people* > *two cars*—preferred; *two cars* > *three people*)

In addition to the wide scope, right-periphery expressions cannot include anaphors or NPIs, and they can be iterated and scrambled. All these properties are typical of topics. Given that right-periphery expressions appear in the oblique case, we therefore propose that they are base-generated in the right periphery of the clause, adjoined either to TP or to CP (assuming that CP may be projected optionally). As such, they have the status of external topics (cf. Aissen 1992, de Cat 2002). Thus, for (22a), the proposed structure is as in (22b).[11]

(22) a. bvʉ́ ↓tɨ́ mɔ̀ zɨ́ á↓zɔ́ɔ́ bɛ́ ↓kɔ́
dogs D P$_2$ eat yesterday fufu D.OBL
'The dogs ate the fufu YESTERDAY.'
(lit.: 'Of the fufu, the dogs ate YESTERDAY.')

b.

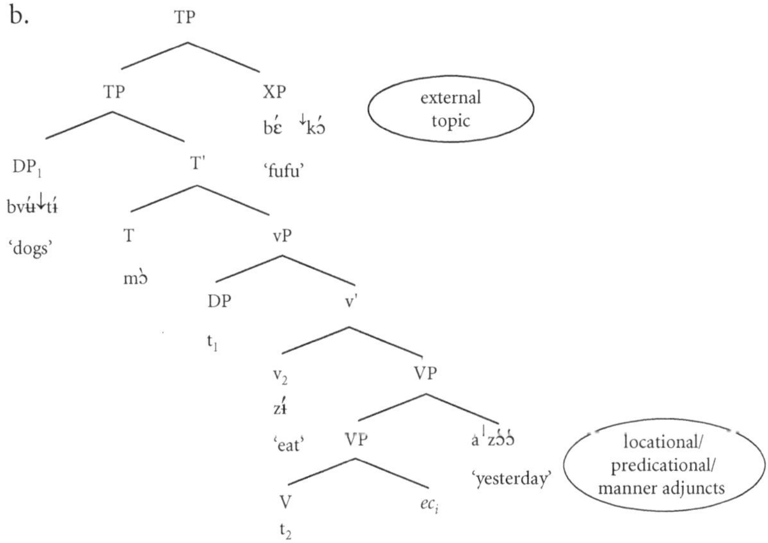

[11] In our approach, we do not assume the antisymmetric view, therefore, we allow right adjunction at various heights. We hypothesize that right adjunction at TP is the base-generated position of an external topic (cf. de Cat 2002 for a similar proposal for spoken French), and that right adjunction at VP is typically associated with manner or locative adverbials (cf. Ernst 2002, 2003, Shaer 2003 for similar structures in other VO languages).

TABLE 9.1 Basic Clause Structure in Aghem

Surface word order	Structural position of the verb	Structural position of the subject	Structural position of the object
SVOX	raises to v^0	Spec,TP	inside VP
SOVX	raises to v^0	Spec,TP	adjoined to vP (middlefield)
VSXO	raises to Asp^0	Spec,vP	inside VP (*pro*)

The next question has to do with the nature of the empty category bound by the right periphery topic. We hypothesize that this empty category in the argument position is a null pronominal; an outstanding question is why this null pronominal does not alternate with an overt pronominal.[12]

(23) bvʉ́ ꜜtɨ́ mɔ̀ [zɨ́ *pro*$_i$ nó] (ꜜbɛ́ ꜜkɔ́$_i$)
dogs D P$_2$ eat FM fufu D-OBL
'The dogs ATE (of) the fufu.'

The basic clause structures in Aghem are set out in Table 9.1.[13]

9.2.3 *Question formation*

Questions are marked by the particle *a*, which appears clause-finally, for example:

[12] Left-dislocated topics are in fact resumed by a pronoun, as shown in Hyman (1979a: 70, ex. (59)). The asymmetry between left-dislocation and right-dislocation with respect to resumption is reminiscent of the situation in spoken French (Lambrecht 1996; de Cat 2002), where two types of asymmetry are observed. First, there is an asymmetry between left- and right-dislocated expressions, where the left-dislocated ones are either very bad or incompatible with the null pronominal (i), but for the right-dislocation such a null pronominal is the norm (ii):

(i) a. */??ce petit con, *pro* mérite des baffes
b. ce petit con, il mérite des baffes

(ii) a. *pro* mérite des baffes, ce petit con
b. *il mérite des baffes, ce petit con
'Deserves a slap in the face, this little jerk.'

Second, the asymmetry in (i) and (ii) is particular to subjects; with objects, the null pronominal seems the norm (Lambrecht 1996).

[13] Some other facts of Aghem grammar deserve mention here. Aghem does not seem to have true complement clauses; the equivalent of familiar embedded declaratives and interrogatives is direct speech introduced by a quotative *ñɨ́ꜜá*. An intriguing phenomenon in its own right, this absence of complement clauses limits the standard syntactic diagnostics that can be used to test word order or extractions. Since there are no embedded complements, there is no embedded sluicing either.

(24) wò zɨ́á à-lɨ́m á
2SG eat yams QM
'Are you eating yams?

Wh-words are actually indeterminate expressions, and receive the interrogative interpretation due to the presence of a question particle.[14] Compare:

(25) a. bvʉ́ ꜜtɨ́ mɔ̂ kɔ̀ʔ kɨ̀-fɨ̀ghá
dogs D P$_1$ see thing
'The dogs saw something.'

b. bvʉ́ ꜜtɨ́ mɔ̂ kɔ̀ʔ kwɔ̀ à
dogs D P$_1$ see what QM
'What did the dogs see?'

Subjects cannot be questioned in the preverbal position:

(26) a. wù mɔ̀ ñɨ̀ŋ nô
person P$_1$ run FM
'Someone ran.'

b. *ndùghò mɔ̀ ñɨ̀ŋ nó à
who P$_1$ run FM QM

c. à mɔ̀ ñɨ̀ŋ ndúꜜghó á
ES P$_1$ run who QM
'Who ran?'[15]

Now that we have established the basic structures of Aghem, we are ready to address the status of the IAV and its relationship with focus.

9.3 IAV and focus: An overview of existing proposals

To recapitulate our starting point, it seems that focus expressions and *wh*-words in Aghem invariably appear immediately after the verb, in the IAV. In this section, we would like to address the question of whether the IAV is a dedicated focus projection. We will first address the analysis in terms

[14] It should be noted that the QM is often optional. While/ɛ/assimilates to the QM (e.g. /nɛ́/ 'today' +/a/ → [náà] in the examples in (4)) above, the more usual case is for the question marker to assimilate and lengthen other preceding vowels, e.g. /kwɔ̀/ + /a/ → [kwɔ̀:] 'what'. This sometimes makes it difficult to determine whether this marker is present.

[15] A reviewer points out that the notion of indeterminacy of *wh*-expressions, with a question particle doing the work of generating the interrogative interpretation, leads us to expect that the same forms should be used for the pronouns in the (a) and (b) cases in (25) and (26). We do not attempt to explain these facts here, but acknowledge the apparent redundancy.

of an articulated focus projection proposed for Aghem by Aboh (2006) and show that this analysis makes incorrect empirical predictions. In Section 9.3.2, we will examine another analysis of Aghem focus, one formulated in terms of movement with pied-piping/remnant movement, and show that it is also untenable.

9.3.1 *Focus projection in the vP*

Starting with Rizzi's seminal work (1997), many researchers have embraced the idea that information-structural categories are projected in the complementizer layer and may also be projected in the exploded vP layer, thus:

(27) Force P < **(TopicP) < FocusP < (TopicP)** < FinP < TP <
(TopicP) < FocusP < (TopicP) < vP < FocusP

Such clausal architecture is appealing because it allows linguists to associate certain positions with topic or focus interpretations and to constrain the range of possible parametric variation in topic/focus marking: some languages may house all their information-structural categories in the CP layers, others in the inflectional layer, and still others in the verbal layer. According to a recent proposal by Aboh (2006), Aghem instantiates the last kind, the low-focus projection type.

Aboh follows Belletti (2002) in assuming that the vP periphery includes a focus phrase, thus:

(28) vP < FocusP < (TopicP) < (TopicP) < VP

Languages where focus is in the complementizer layer are analyzed as high-focus languages, and the ones with focus in the verbal layer instantiate the low-focus type. Aboh proposes a principled difference between Kwa and Gur, which he shows to be high focus (Aboh 2004a, b, 2006), and Bantu, which he considers low focus. His argument for Bantu low focus relies heavily on Aghem data (Aboh 2006).

Aboh makes the following assumptions for Aghem: the verb invariably moves to T, and the focus projection follows. The head of the focus projection is the final marker *nò* which was introduced above; this marker can also alternate with a null element. The indication that *nò* is a phrasal head comes from the fact that it cannot be iterated, cf. the ungrammaticality of (29):[16]

(29) *fú kɨ́ mɔ̀ ñɨ́ŋ nô á kɨ́-↓bé nò
rat D P_2 run FM in compound FM
'The rat ran inside the compound.' (Watters 1979: 167)

[16] Aghem has verb serialization, and *nò* cannot be iterated even in serial-verb constructions.

Adding all these pieces together, the proposed syntax of focus in Aghem is as follows:

(30) bvʉ́ ↓tɨ́ mɔ̀ [$_{vP}$ bɛ́ ↓kɨ́ [$_{FocP}$ zɨ́ [$_{Foc}$ nô [$_{VP}$ ~~zɨ́ kɨ́-bɛ́~~]]]]
dogs D P_2 fufu D eat FM
'The dogs ATE the fufu.'

This account correctly captures the word-order facts and the distribution of *nò*. However, it makes a number of incorrect predictions and is incompatible with empirical data.

Firstly, the analysis predicts that focus phrases should be equally available in matrix and embedded clauses. However, the focus marker *nò* cannot occur in embedded clauses, despite the fact that a constituent inside the embedded clause can receive focus interpretation. For example, this marker cannot occur in a relative clause or in a purpose clause:

(31) a. wìzɨ́n wɨ́↓l [-á ò mɔ̀ tsɨ̀ghà bvʉ̀ (*nò)]
woman DEM-COMP 3SG P_1 pass fall FM
'the woman that fell'

b. ò lɨ̀ghà [ñɨ́↓á ǹ zɨ́ kɨ́-bɛ́ (*nò)]
3SG want COMP 1SG eat D-fufu FM
'He wants me to eat the fufu.'

Since the focus projection is low, in the verb phrase, such a restriction is unexpected.

Next, the focus-projection hypothesis assumes that *nò* is a phrasal head that takes the focused constituent as its complement. However, phonological phrasing evidence suggests that it does not have head properties. Consider the following example:

(32) à mɔ̀ nùŋò kɨ̀-kɔ̣̀ nô /kɨ́-kɔ̣̀ +′+ nò/
ES P_1 leave slave FM
'A slave left.'

In this example, the focus marker /nò/, which has underlying low tone, inexplicably appears with a high-to-low falling tone. Where is this extra high feature from? We suggest that it is exactly the same high tone that marks adjuncts in the right-peripheral position, as we saw in (20), repeated as (33a) below:

(33) a. à mɔ̀ kɔ̀ʔ ñɔ̠̌m kàa fɔ̂
ES P_1 see animal squirrel D-OBL
'An ANIMAL saw a squirrel.'

b. … ñɔ̱̀m + ′ + fɨ̀-kàa fɔ́ → ñɔ̱̀m+ fɨ̱́-kàa fɔ́ → ñɔ̱̀m +′̱kàa fɔ́ → ñɔ̱̌m kàa fɔ̂
animal PREP squirrel D.OBL

As seen in the derivation in (33b), the high is first assigned to the prefix of 'squirrel'. When the prefix falls out because there is an agreeing determiner /fɔ́/, the resulting freed high tone is then relinked to the preceding syllable to produce [ñɔ̱̌m]. In (32), the high tone is also assigned to the right, converting /nò/ to [nô]. The "oblique" tonal marking indicates that *nò* is not the head of a focus projection but instead a complement of a silent preposition. This makes *nò* similar to the adjuncts in verb phrases we have seen above, for example:

(34) bvʉ́ ꜜtɨ́ mɔ̀ zɨ́ áꜜzɔ́ɔ́ bɛ́ ꜜkɔ́
dogs D P_2 eat yesterday fufu D.OBL
'The dogs ate the fufu YESTERDAY.'

The analysis of postverbal adjuncts proposed above places the material introduced by a suprasegmental preposition in the right adjunction position in the VP, shown schematically in (35a). Accordingly, the particle *nò* could appear in a similar adjunct position, as shown in (35b):

(35) a. [$_{VP}$ [$_{VP}$ eat the fufu] [Adv]/[PP]]
b. [$_{VP}$ [$_{VP}$ eat] [$_{PP}$ Prp [*nò*]]]

Aghem has a variety of phrasal heads, most of them preceding the complement in surface structure (tense, mood), but also determiners, the question marker *á*, and suprasegmental hortative and imperative markers, which follow their complements on the surface (S. C. Anderson 1979: 106–9). None of these phrasal heads can occur with the tonal preposition, and in fact hortative and imperative markers combine with the preceding material conditioning tone change. Thus, *nò* behaves completely opposite from other comparable heads and would require a special analysis. When compared to other dependent forms, however, *nò* fits their profile.

Finally, as we already showed using adverbial placement data, Aghem does not have across-the-board V-to-T raising. The absence of this raising undermines one of the assumptions underlying Aboh's approach. If such raising were somehow available in spite of the adverbial placement facts, it is unclear how it could be reconciled with the occurrence of preverbal and postverbal objects. To accommodate structures where the object is preverbal (IBV) and postverbal (IAV), the low-focus analysis should presumably allow for verb raising (V-to-T) as well as for vP raising in (30) and in (36).

(36) tɨ́-bvʉ́ tɨ̀-bɨ̀ghà mɔ̂ [$_{vP}$ [zɨ̀ kɨ́-bɛ́]$_i$ [$_{FocP}$ t$_i$ [$_{Foc}$ nô [$_{VP}$ t$_i$]]]]
dogs two P$_1$ eat D-fufu FM
'The two dogs ate the fufu.'

Both verb-raising and VP-raising are generally available but it is unlikely for them to co-occur within a single language (Bobaljik 2002; Oda 2005).[17] If we were to follow the established contrasts between verb-raising and VP-raising languages proposed by Oda (2005), then Aghem would be closer to the profile of a verb-raising language, not VP-raising language: it does not allow nominal predicate fronting (VP-raising languages do), or any verb fronting for that matter, and it has the inversion alternation SV–VS, typical of V-raising languages. Unlike typical verb-raising languages, Aghem does not have obvious rich verb agreement; however, the cliticized determiners that we have glossed as D can arguably be identified as agreement markers (Hyman to appear). If this were the case, that would bring Aghem even closer to the profile of a verb-raising language.

All these considerations indicate that the low-focus projection in the verb phrase cannot account for Aghem focus.

9.3.2 *Focus movement*

In a number of African languages which are known to have articulated focus marking, focus has been analyzed on par with *wh*-movement as an A-bar movement of the relevant expression followed by pied-piping or movement of the remnant clause (cf. Nkemnji 1995, Koopman 2000, Sabel and Zeller 2006; see also Cheng and Downing 2006 for a different analysis employing A-bar movement). Nkemnji's analysis has been proposed for a closely related language, which makes it particularly relevant for discussion. The basic idea is that the focus expression/*wh*-word first moves to Spec,CP (alternatively, to the head of a high dedicated projection), and the remaining clause is then moved to the same position.

(37) A-bar movement and remnant movement

a. bvʉ́ ↓tɨ́ mɔ̀ zɨ́ kwɔ̀ à
dogs D P$_2$ eat what QM
'what did the dogs eat?'

b. bvʉ́ ↓tɨ́ mɔ̀ zɨ́ kɨ́-bɛ́ nò
dogs D P$_2$ eat D-fufu FM

[17] A reviewer points out, however, that German allows both V-to-C movement and VP-raising, even within the same clause. We would like to note that the issue of VP-raising in German is still controversial (cf. Taraldsen 2000).

c.

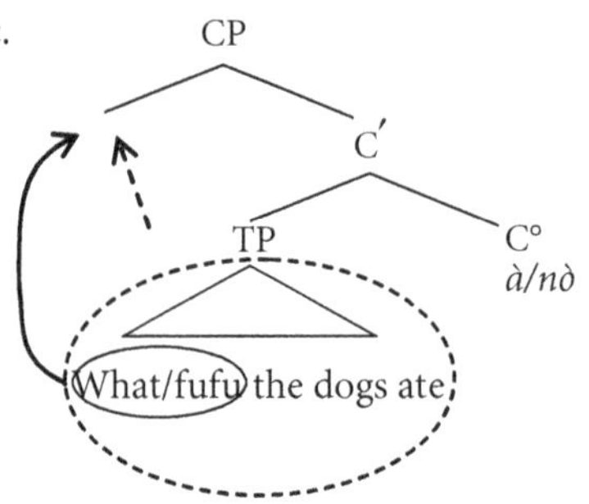

Under this analysis, the question marker *à* and the final marker *nò* can be identified as complementizer heads. That would presumably account for the absence of *nò* in embedded clauses (see (32a, b)), which was problematic for the low-focus projection account.

An argument in support of the movement analysis in Aghem comes from island effects, illustrated in (38) for a relative clause island:

(38) *m̀ mɔ̂ kɔ̀ʔ wɛ́ wɨ́l [↓á ò mɔ̀ nî kwɔ̀-kɔ̀] à
1SG P_1 see child DEM-COMP 3SG P_1 take what QM
('For what *x*, *x* a thing, did I see the child who took *x*?')

Surprisingly, island effects hold well only for *wh*-questions, not for focus; cf. the following example showing that focus in a relative clause is possible:

(39) m̀ mɔ̂ kɔ̀ʔ wɛ́ wɨ́l [↓á ò mɔ̀ nî bvʉ́ ↓tɔ́]
1SG P_1 see child DEM-COMP 3SG P_1 take dogs D.OBL
'I saw the child who took the DOGS.'

In addressing the movement analysis, we will set aside more general, conceptual arguments against movement for elements that on the surface appear *in situ* (cf. Reinhart 1998 for arguments against such an approach, and Heck 2004 for a general discussion of pied-piping). We remain agnostic as to whether such movement is justified or not; it is quite possible that languages differ with respect to it. Instead, we will concentrate on language-particular evidence which indicates that the movement analysis of Aghem focus is untenable.

First of all, the actual implementation of the two movement operations for Aghem would be difficult. One would need to move the *wh*-word to Spec,CP, and then front the remnant clause in front of that, presumably in some higher position (rather than to the same position). Even if the machinery of movement itself were to work, this analysis would still have to stipulate the IAV because the remnant would *have to* have the verb as the last part of the surface string. We have seen above that the postverbal area in Aghem is quite extensive, so this is a difficulty for this analysis.

The other difficulty of maintaining the movement analysis for Aghem questions has to do with the fact that many familiar manifestations of A-bar movement are absent. First, Aghem allows multiple *wh*-expressions but shows no superiority effects, which is again surprising under a movement analysis. Consider the following examples:

(40) a. à mɔ̀ zɨ̀ ndúghɔ́ kwɔ̀-kɔ̀ zɨ́n
ES P_1 eat who what when
'Who ate what when?'
'When did who eat and what?'

b. à mɔ̀ zɨ̌ kwɔ̀ ndúghɔ́
ES P_1 eat what who
'Who ate what?'
'What did who eat?'

Another argument against movement comes from an unusual asymmetry between arguments and adjuncts with respect to islands. In *wh-in situ* languages, arguments and adjuncts show different sensitivity to islands, with arguments escaping island constraints but adjuncts subject to them (cf. Huang 1982, Saito 1992, Lasnik and Saito 1992, Tsai 1995 for Chinese and Japanese). In Aghem, the situation is the opposite: as we saw in (38), arguments are impossible in islands, whereas adjuncts are accepted, if only marginally:

(41) ?à mɔ̀ tsɨ̀ghà bvʉ̀ ndúghɔ́ ↓zɨ́n
ES P_1 pass fall who when
'Who fell when?' (lit. 'when was it that who fell?')

Thus, the movement analysis of Aghem focus faces too many empirical challenges, and we will not pursue it further. This of course does not mean that the *wh*-movement analysis may not work for other languages, even closely related.

9.4 In search of an alternative

9.4.1 *Focus operator*

We have argued that proposals in terms of an articulated verb periphery and focus movement fail to capture the full range of Aghem data. In this section we will outline our own proposal on Aghem postverbal focus. Our main idea is that identificational (exhaustive) focus is sensitive to constituency and is assigned within the predicate phrase (VP); in particular, focus is located in the most deeply embedded consituent of the verbal projection. Conceptually, the association between focus and predicate is well established, so in that sense the idea that focus is linked to predication is not new (cf. Horvath 1986, Diesing 1992, van Geenhoven 1998, Reinhart 1998, É. Kiss 1998, Partee and Borschev 2003, Büring, this volume, for similar views).[18] The other main component of our proposal is that the focus/*wh*-interpretation is licensed by an alternative-activating operator in the complementizer projection.

The structure we propose is as follows: an interrogative or focus operator is in the specifier of the highest functional projection (CP or ForceP—nothing hinges on the actual category here) and unselectively binds the lowest XP in the clause.

(42) *Focus assignment in Aghem*

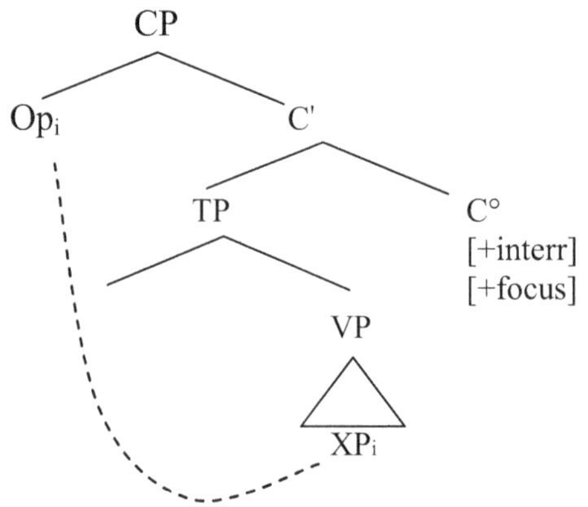

If our proposal is on the right track, the association between the IAV and focus is an artifact of a general focus-assignment rule. Focus is read off constituent structure, but is not directly projected.[19] Rather, it is associated with a particular projection, namely the verb phrase; the association between the verb (or predicate) phrase and focus is cross-linguistically quite common (see Lambrecht 1994: ch 5, Vallduví 1992, and many others).

9.4.2 *Deriving focus*

We will now show how this proposal works, starting with details of focus assignment. In an intransitive clause with no additional material after the

[18] A reviewer points out, on the other hand, that relating focus to predication is at odds with a structured-meaning approach to focus (von Stechow 1982, 1990; Krifka 2006), under which the predicate consists of the background component of an utterance.

[19] A reviewer drew our attention to the connection between this idea and that of Jacobs (1988) that even cases of apparent free focus involve binding by covert illocutionary operators in the left periphery of the clause.

verb, the focus is naturally interpreted with the VP, thus (only the relevant derivations shown):[20]

(43) a. ò mɔ̀ bvʉ̀ nò
3sg P$_1$ fall FM
'He fell.' (=(3a))

b. [$_{CP}$ Op$_i$ [$_{TP}$ he [$_{vP}$ fall [$_{VP}$ ~~fall~~]]]]

Turning to transitive clauses, we have seen that constituents in the IAV are interpreted as narrow focus or as part of the wide VP focus, thus in (44), repeated from (1) above, the focus is either *the fufu* or *ate the fufu*:

(44) tɨ́-bvʉ́ tɨ̀-bɨ̀ghà mɔ̂ zɨ̀ kɨ́-bɛ́ ↓nɛ́
dogs two P$_1$ eat D-fufu today
'The two dogs ate the fufu today.'

In this structure, the object is hierarchically the deepest XP, hence the narrow interpretation, and the well-known phrasal focus rule (Selkirk 1984: ch. 5; 1996) allows for focus to spread to the vP, yielding the wide scope interpretation.

(45) Phrasal focus rule.
A constituent may be in focus if (i) or (ii) or both hold:

(i) a constituent that is its head is in focus,
(ii) a subconstituent that is argument of the head is in focus.

Rule (45i) accounts for the wide focus on the verb phrase, and (45ii) for the narrow focus on the object. Back to the Aghem structure, the focus assignment rules work the following way (English glosses substituting for the Aghem words):

(46) [$_{TP}$ two dogs [$_T$ P1 [$_{vP}$ ~~two dogs~~ [$_v$ eat [$_{VP}$ ~~eat~~ the-fufu]]]]]
[narrow F]
[wide F]

In the subject-inversion structure, the focus is on the postverbal subject, with 'yesterday' in the right-periphery topic position:

(47) à mɔ̀ ñɨ́ŋ tɨ́-bvʉ́ á↓zɔ́ɔ
ES P$_2$ run dogs yesterday
'There ran dogs yesterday.'/'The DOGS ran yesterday.'

[20] We will turn to the category of *nò* in the next section; here, for ease of exposition we have not included it in (43b).

Since the unraised subject is not an argument of the verb, the presentational focus, which treats the entire verb phrase as focus, cannot be accounted for under the phrasal focus rule in (45). In other words, the wide focus in (47), represented schematically in (48), is unexpected:

(48) [$_{\text{TP}}$ *expl* [$_{\text{TP}}$ P$_2$] [$_{\text{AspP}}$ [$_{\text{Asp°}}$ run [$_{\text{vP}}$ dogs [$_{\text{v°}}$ ~~run~~] [$_{\text{VP}}$]]]]]
[narrow F]
[wide F]

Both focus readings are commonly found in inversion constructions cross-linguistically and have long posed a challenge to a derivational analysis of focus. Several approaches have been proposed, and here we would like to adopt the proposal advanced in Lambrecht (1994: 315–22), whose basic idea is that the wide-focus reading in inversion constructions is due to the recognition that the alternative could only be focus on the predicate. In avoiding this interpretation, speakers make an accommodation and recognize the entire inverted phrase as focus. Importantly, the problem identified here is not specific to Aghem.

If the object or subject position in these examples is filled with an indeterminate expression, this expression is interpreted as a *wh*-word when under the scope of an interrogative operator, thus:

(49) tɨ́-bvʉ́ tɨ̀ -bɨ̀ghà mɔ̂ zɨ́ kwɔ̀ à
dogs two P$_1$ eat what QM
'What did the two dogs eat?'

(50) à mɔ̀ ñɨ́ŋ ↓ndúghɔ́ à
ES P$_2$ run who QM
'Who ran?'

In all these examples, the narrowly focused constituent stays in the IAV on the surface.

In clauses with the object adjoined to the vP, thus appearing preverbally, focus is identified with the material remaining in the VP, for example, in (51a), where the focus is on 'yesterday', or in (51b), where the locative PP is in focus:

(51) a. bvʉ́ ↓tɨ́ mɔ̀ [$_{\text{vP}}$ [$_{\text{vP}}$ bɛ́ ↓kɨ́$_i$ [$_{\text{v}}$ zɨ́$_j$ [$_{\text{VP}}$ [$_{\text{VP}}$ t$_j$ t$_i$] á↓zɔ́ɔ́]]]]
dogs D P$_2$ fufu D eat yesterday
NARROW FOCUS
'The dogs ate the fufu YESTERDAY.'

b. fɨ̂l á mɔ̀ [$_{vP}$ [$_{vP}$ bɛ́ ꜜkɨ́ [$_{v}$ zɨ́ [$_{VP}$ [$_{VP}$ t_j t_i] án ꜜsɔ́m]]]]
friends D P_1 fufu D eat at farm
'The friend ate the fufu AT THE FARM.' (Watters 1979: 148)

Since the adverb inside the VP is not an argument of the verb, the focus remains narrow, on the adjunct.[21] Note however, that a focused adjunct is in a different structural position from that of a focused postverbal object:

(52) a. fɨ̂l á mɔ̀ zɨ́ áꜜzɔ́ɔ́ bɛ́ ꜜkɔ́
friends D P_2 eat yesterday fufu D.OBL
'The friends ate the fufu YESTERDAY.'

b. fɨ̂l á mɔ̀ [$_{vP}$ [$_{v}$ $zɨ́_i$ [$_{VP}$ [$_{VP}$ t_i] áꜜzɔ́ɔ́]]]

The difference in structure indicates that the IAV is a surface position only, derived in different ways. In that sense, the postverbal position in Aghem is no different from the preverbal focus position in other languages, which is uniform on the surface but can in fact mask a number of different configurations (see e.g. Horvath 1986, Kiss 1995, Kishimoto 2005, Hagstrom 1998).

In cases where there is a postverbal adjunct in focus and the object appears preverbally, in the middle-field, one may get an impression that the preverbal object is topicalized, or at least associated with "given" information. In fact, in the spirit of an articulated left periphery approach, it would be tempting to categorize the position occupied by the preverbal object (IBV in surface terms for Aghem) as a dedicated topic position, thus:

(53) [$_{vP}$ [$_{TopP}$ [$_{TopP}$ [$_{FocP}$ [$_{vP}$ [$_{VP}$]]]]]]

However, the preverbal position is not associated with known properties of topics: it does not have obligatory wide scope, can host non-specific expressions, and can be associated with contrastive or informational focus. For example, in (54), the postverbal subject 'friends' is identificational focus, but the preverbal (IBV) constituent is also interpreted as focus, this one being highly contrastive:[22]

[21] Incidentally, the fact that focus does not spread from the temporal or locative expression to the verb argues that the analysis of the verb phrase where the direct object is in Spec,VP and the locative/temporal expression is a complement of the lexical verb, as shown in (i), is untenable for Aghem.

(i) [$_{VP}$ fufu [$_{V'}$ [eat yesterday]]]

Baker and Collins (2006) and Kandybowicz (2006) propose this structure for several other Bantu languages. At this point, we do not have an explanation as to why such parametric variation in the structure of VP is possible.

[22] *Nzàŋ* is a kind of dance accompanied by singing.

(54) à mɔ̀ nzàŋ zɔ̩̀m á-fɨ̂n â bàʔtɔ̀m
ES P_2 Nzaŋ sing friends for chief
'The FRIENDS sang NZAŋ for the chief.'

We therefore conclude that whatever interpretive properties may be associated with this position, they do not follow from the presence of a dedicated topic projection.[23]

An outstanding problem for our proposal is that Aghem allows multiple *wh*-questions, as in the following example, which shows that the interrogative operator can take scope over the entire VP even with some material adjoined.

(55) à mɔ̀ [VP [VP zɨ̀ ndúghɔ́] kwɔ̀-kɔ̀ zɨ́n]
ES P_1 eat who what when
'Who ate what when?'
'When did who eat and what?'

We hypothesize that multiple *wh*-questions are licit for interpretive reasons, not for syntactic reasons. It is possible that the reading of the multiple indeterminate expressions as *wh*-words is forced in the presence of an interrogative operator.

Another problem for the proposed analysis has to do with the availability of focus in embedded clauses, for example, as in the relative clause in the following example. If the family of alternative-activating operators that we propose included focus and interrogative operators as equal members, then the availability of a focus reading contrasting with the impossibility of *wh*-questions in relative clauses needs to be accounted for.

(56) m̀ mɔ̂ kɔ̀ʔ wɛ́ wɨ́l [↓á ò mɔ̀ nî bvʉ́ ↓tɔ́]
1SG P_1 see child DEM-COMP 3SG P_1 take dogs D.OBL
'I saw the child who took the DOGS.'

The availability of focus in relative clauses is due to the difference between identificational and informational focus (É. Kiss 1998); while further work on this issue is needed, the focus reading in the relative clause in (56) could be accounted for under the informational interpretation, which is independent of an alternative-activating operator. Such an operator is needed only for the identificational (exhaustive) focus. With *wh*-words, a non-exhaustive reading is simply impossible, and this leads to the appearance of island effects shown here.

[23] See Sturgeon (2006) for a similar analysis of the middle-field in Czech, and Lee-Schoenfeld (2005) for a similar approach to the German middle-field.

To summarize what we have achieved so far, the identificational (exhaustive) focus in Aghem is located in the verb phrase, with the lowest XP in that phrase interpreted as focus. The interpretation of the entire verb phrase as focus also occurs under the focus-projection from object to the verb (accounted for under the phrasal focus rule), and in the presentational construction, where the explanation for wide focus is still outstanding.

9.4.3 *The content of CP*

Recall that we have proposed that the verb phrase is under the scope of an operator which we hypothesize to be in the complementizer phrase or, assuming an articulated complementizer layer, in the force phrase.

In this section, we will address two questions concerning the complementizer phrase that we associate with the focus and interrogative interpretations: the nature of the focus and interrogative operator and the head of this high structural phrase.

Our proposal is as follows: the specifier of the complementizer (force) phrase has an alternative-activating operator which binds the lowest constituent in the VP. For focus, this operator activates a set of alternatives forcing the selection of one element in a set to the exclusion of the others. This is a standard exhaustiveness operation, characteristic of identificational (exhaustive) focus (É. Kiss 1998; Rooth 1996a). For questions, the operator activates a set of alternatives that could constitute the answer to a given question (cf. Karttunen 1977, Ginzburg and Sag 2000). Thus, the proposed operators are as follows:

(57) a. Focus: Op [+exhaustive]
b. Question: Op [+interrogative]

Since this approach does not appeal to movement, a question arises as to why we still observe island effects associated with movement, for example in relative clauses:

(58) *m̀ mɔ̂ kɔ̀ʔ wɛ́ wɨ́l [↓á ò mɔ̀ nî kwɔ̀-kɔ̀] à
1SG P_1 see child DEM-COMP 3SG P_1 take what QM
'For what *x*, *x* a thing, did I see the child who took *x*?'

The ungrammaticality of such examples can still be accounted for, without reference to movement, under the rubric of standard intervention effects (de Swart 1992; Szabolcsi and Zwarts 1992–3; Beck 1996; Bošković 2000; Cheng and Rooryck 2001; Butler and Mathieu 2005). In this case, the relative-clause operator, which is hierarchically closer to the indeterminate expression inside the relative, blocks its binding by the interrogative operator:[24]

(59) [$_{\text{CP}}$ Op$_{[+\text{interr}]}$ [$_{\text{TP}}$ ... [$_{\text{vP}}$ [$_{\text{DP}}$ [$_{\text{NP}}$] [$_{\text{CP}}$ Op$_{[+\text{rel}]}$ ... indefinite]]] C°]]

As for the head of the complementizer (or force) phrase, we propose that it is filled by the interrogative particle *á* in questions and by a null head for focus. The association of interrogative particles with a high complementizer projection is well known (cf. Hagstrom 1998 and many others). For Bantu languages, Buell (2005) and Kandybowicz (2006) present arguments that interrogative markers are in the complementizer phrase. The usual evidence comes from the fact that such markers do not appear in embedded clauses and occur at the right periphery of root clauses. Genuine subordinate clauses embedded under bridge verbs or interrogative or resolutive predicates seem to be avoided in Aghem. Relative clauses are widely available but they do not allow interrogative markers anyway. Some evidence that the interrogative marker *á* is in a high structural position comes from its distribution with postverbal material: postverbal expressions, no matter how many, occur before *á*, for example:

(60) à mɔ̀ kɔ̀ʔ á-fɨ́n bvʉ́ à dzɨ́ ↓á
ES P$_1$ see friends dog and goat QM
'Did THE FRIENDS see a dog and a goat?'

However, in many cases the overt interrogative marker is omitted, thus alternating with a null particle. At this point we do not know what determines the optionality of *á*.

Turning to *nò*, recall Aboh's proposal that it heads a focus phrase (Section 9.3.1). While the evidence did not support the low-focus projection analysis, one might try to maintain that *nò* could be a complementizer head, with the feature [+focus], thus:

(61) [$_{\text{CP}}$ Op$_{\text{i}}$ [$_{\text{TP}}$ he [$_{\text{vP}}$ fall [$_{\text{VP}}$ ~~fall~~]]] [$_{\text{C°}}$ *nò*]]

Such an analysis would account for the fact that *nò* does not occur in relative clauses and would treat *nò* by analogy with the question particle *á*. However, there are at least three arguments indicating that this analysis is untenable.

[24] Recall that adjunct *wh*-words were marginally acceptable, even in the relative clause island. At present we have no explanation for such marginal acceptability. Peter Sells (p.c.) has suggested that this difference between arguments and adjuncts may be due to the fact that the adjunct is generated higher (on the right), which allows it to move out easier than an argument. This is an interesting possibility, which needs to be explored further.

First, *nò* has a different distribution from *á* with respect to right-periphery elements: it precedes right-periphery adjuncts and pronominal elements, for example:

(62) ò mɔ̀ fùò nó ꜜbɛ ꜜkɔ́ â bvʉ́ ꜜtɔ́
3SG P$_1$ give FM fufu D.OBL to dogs D.OBL
'He GAVE the fufu to the dogs.' (e.g. he didn't TAKE it from them)

(63) ò lɨ̀ghá sùghɔ̀ nó ꜜwɨ́n
3SG like also FM 3SG.OBJ
'He also likes him/her.'

Second, *nò* co-occurs with *á*, which indicates that they represent different categories. On the standard assumption that yes–no questions also have an exhaustivity operator (akin to the one found in *wh*-questions) and are incompatible with identificational focus (É. Kiss 1998), it is surprising to see *nò* in such questions:

(64) ò mɔ̀ wí kɨ́-kɔ̩̀ ꜜá ŋ́ꜜkáá ò mɔ̀ bò nó ꜜá
3SG P$_1$ kill slave QM or 3SG P$_1$ beat FM QM
'Did he kill the slave, or beat (him)?'

Finally, recall that *nò* may occur with the high-tone preposition documented in Section 9.3.1. The question marker *á* (and its variant *à*) does not occur with this preposition; it always undergoes tonal association with the preceding word. If *nò* and *á* belonged to the same category, such difference in distribution would be quite surprising.

The evidence assembled here points to the conclusion that *nò* is not a head and that its place in clause structure is much lower than that of *á*. Under the structure we propose, it cannot be a lower focus marker since the focus marker should head the higher projection. We hypothesize that *nò* marks off the right edge of the verb phrase. If non-verbal constituents on the right edge are present, it is optional; when the verb appears "bare", *nò* is required. We are not sure what accounts for this optionality.

The proposed approach to *nò* has two main components: first, it places *nò* in the lowest verb phrase, not higher, and second, it removes a direct association between *nò* and focus. This approach is very similar to the proposal made by Buell (2006): according to Buell, the conjoint–disjoint alternation in Zulu (and possibly in other Bantu languages) can be accounted for without direct reference to focus marking. As to the absence of *nò* in embedded clauses, we can offer only a preliminary observation. The particle *nò* is not the only constituent that cannot occur in embeddings. Embedded clauses (for instance, infinitival clauses) do not allow postverbal adjuncts either, which suggests that the verb in these clauses does not raise to the highest light verb as it raises in the matrix clause. Different structural positions of the verb phrase in matrix and embedded clauses are a well-known phenomenon (cf. den Dikken 1995, and many others), so it would not be surprising if this turns out to be the case in Aghem as well. More work on the word order in Aghem embedded clauses is needed, and these outstanding data may also be helpful in understanding the general constraints on distribution of conjoint and disjoint forms across Bantu.

9.5 Conclusions and outstanding questions

In this chapter, we have analyzed the expression of focus in Aghem, focusing primarily on the postverbal (IAV) position, which is associated with exhaustive or identificational focus. Previous accounts of Aghem focus have relied on a highly articulated left periphery of the verb phrase (so-called low-focus analysis, as in Aboh 2006) or an analysis in terms of focus/*wh*-movement and subsequent movement of the remnant. We found that both approaches face significant empirical challenges and are untenable for Aghem. We would like to underscore that their failure in Aghem does not render a more general verdict on accounts in terms of fine structure of left periphery or A-bar movement because languages may well vary in their expression of information structural categories. Nevertheless, Aghem data show that one has to proceed with caution and examine language-internal data carefully before extrapolating analyses from other languages to a new one.

Our approach to Aghem focus takes us back to the more traditional accounts, which associate focus with the verb phrase (predicate phrase) and derive focus interpretations indirectly from constituent structure, without dedicating special projections to focus (or topic for that matter, although we did not discuss it in this chapter). We propose that the focus and interrogative interpretation is derived without movement, via a relationship between the alternative-activating operator in the highest complementizer projection and the lowest XP in the verb phrase. The presence of lower operators, for example, as in relative clauses, creates an intervention effect breaking the binding chain between the high operator and VP-internal material; this accounts for the island effects in relative clauses, in the absence of A-bar movement.

The approach in terms of an alternative-activating operator captures the similarities between focus and *wh*-movement and allows us to analyze Aghem as a focus *in situ* language, where the focus interpretation is achieved scopally, relying on the syntactic structures made independently available by Aghem grammar.

The IAV, which appears uniform on the surface, actually masks different structural positions and is structurally non-uniform. In that sense it can be compared to the preverbal position used for focus in head-final languages—the latter is typically associated with focus but different constituents appearing in it may occupy different structural positions. The structural ambiguity of the IAV in Aghem points to the need to examine similar postverbal positions in other languages.

10

Subject focus in West African languages

INES FIEDLER, KATHARINA HARTMANN, BRIGITTE REINEKE, ANNE SCHWARZ, AND MALTE ZIMMERMANN

10.1 Introduction

This chapter provides an overview of how the information-structural category of focus is grammatically marked in three West African language groups, namely, Gur, Kwa, and (West) Chadic. The languages under discussion display a wide variety of grammatical focus-marking strategies, both within and across language groups. Some languages mark focus syntactically, for instance as clefts, or by means of focus movement. Others mark focus morphologically, where the morphological markers come from different categories, such as copulas, functional heads, or affixes. Yet others mark focus by prosodic means, for instance by prosodic phrasing. The primary goal of this chapter is to provide a comparison of the realization of subject focus from a cross-linguistic (West African) perspective (for a more comprehensive cross-linguistic overview of focus realization, see Büring, this volume). In doing so, we will mostly eschew a specific theoretical analysis of the focus-marking devices employed in the various languages. For instance, in Section 10.2.2 on syntactic focus marking, we remain neutral on the question of whether the positioning of the focus constituent in the left periphery of a clause is the result of clefting, of focus movement, or whether it involves some other syntactic configuration. A choice between the various analytical options would require a careful investigation of the syntactic structures involved (cf. Adger and Ramchand 2005 for an illustrative discussion of this point in Scottish Gaelic and Irish), and thus be well beyond the scope of the chapter.

We are grateful for valuable comments on an earlier version of this chapter by Mara Frascarelli, Larry Hyman, Caroline Féry, and Knud Lambrecht. This chapter is the result of a cooperation between projects B1 and B2 of the SFB 632, which is funded by the German Science Foundation (DFG).

Despite the observed differences in the realization of focus, the languages under discussion show a surprising degree of similarity at a more abstract level: the large majority of them display a subject–non-subject asymmetry when it comes to focus marking, irrespective of which language group they belong to. As will emerge, subject focus always requires a special linguistic coding, which either indicates the focus status of the subject directly, or else marks the entire clause containing the focused subject as thetic. Non-subject focus, in contrast, can often be unmarked (*marking asymmetry*). In addition, focused subjects are marked differently from focused non-subjects in many languages (*structural asymmetry*). In our view, the reason for the observed asymmetries is as follows. Focused subjects must be marked, often in a special way, in order to avoid a default interpretation of grammatical (preverbal) subjects as topics. This strict requirement on the marking of subject focus manifests itself differently in the languages under discussion, which show an interesting typological variation with regard to how structures with focused subjects are realized grammatically.

The geographic distribution of the three language groups is as follows: Kwa languages are found in Nigeria, Benin, Togo, Ghana, and Ivory Coast. Gur languages are spoken in Nigeria, Benin, Togo, Ghana, Ivory Coast, Burkina Faso, and Mali. Chadic languages are spoken in the areas surrounding Lake Chad in Northern Nigeria, Niger, Cameroon, and Chad. The discussion here is based on a language sample comprising the languages in (1):[1]

(1) a. Kwa: Aja, Akan, Awutu-Efutu, Ewe, Fɔn, Foodo, Lelemi

b. Gur: Buli, Byali, Dagbani, Ditammari, Gurene, Konkomba, Kɔnni, Nateni, Yom

c. Chadic: Bade, Bole, Duwai, Guruntum, Hausa, Ngizim, Tangale

All languages in the sample are tone languages, with tone taking over both lexical and grammatical functions. The basic word order of all languages is SVO with full NPs and there is no morphological case marking.

The chapter is organized as follows. Section 10.2 introduces by means of examples the different grammatical means for marking non-verbal focus that are employed by the languages in the sample. Section 10.3 discusses the two kinds of subject–non-subject asymmetry, namely, marking and structural asymmetry. Section 10.4 discusses the special status of focused subjects in more detail, and shows that there is a major typological split in our language sample concerning the way in which focused subjects are marked. While one

[1] For most of the languages we rely on data elicited by ourselves during field research. For reasons of space, not all languages are equally represented.

subgroup (West Chadic) treats focused subjects on a par with other focus constituents, there is another subgroup (within the Oti-Volta branch of Gur) that realizes sentences with focused subjects as thetic utterances with special structural properties. Section 10.5 concludes.

10.2 Linguistic means of expressing non-subject focus

This section introduces the various grammatical means of marking focus employed by the West African languages under consideration. For ease of exposition, the discussion is based on examples involving non-subject term focus (henceforth: NSF), that is, focus on nominal arguments and adjuncts. Before we turn to the different linguistic ways of coding focus, however, a few remarks are in order concerning our understanding of the notion of *focus*.

We understand focus in a very general sense, namely, as that information-structural component of a clause that is most important or salient relative to a given discourse situation, see for instance Dik (1997) and Jackendoff (1972). Relative importance or salience can be achieved in various ways, for example by introducing new information into the discourse (*information focus*), or by standing in explicit or implicit contrast to a set of comparable alternatives (*contrastive focus*) (Rochemont 1986; Rooth 1985). *Focus marking*, then, implies the linguistic realization of the information-structural category focus by way of special grammatical means. Notice that the term focus marking, when used in this general way, does not imply that the formal devices employed in the marking of focus are exclusively found in focus constructions. In many cases, the *focus marker* may serve other grammatical functions as well; see Sections 10.2.1 and 10.2.2 for relevant discussion.

Even though we concentrate on instances of explicit focus marking in what follows, it is worth pointing out that NSF need not be marked at all in many languages in the sample. All Kwa languages (Aja, Akan, Awutu-Efutu, Ewe, Fɔn, Foodo, Lelemi—see also Fiedler 2007, Schwarz 2009b, Fiedler and Schwarz 2007), as well as some of the Chadic languages (Bole, Hausa, see also Zimmermann 2006), can express NSF focus simply by using the canonical word order SVO (*in-situ focus*)[2] and without making use of special prosodic patterns or special morphological markers. To give a concrete example from Fɔn, an appropriate reply to the object question in (2Q), in fact the most common reply, is the answer in (2A) with canonical word order and without any additional morphological or prosodic marking. The focus constituent is

[2] Of course, it is also fine to answer the given questions with just a noun phrase, which can also be accompanied by a morphological focus marker in some languages. Also notice that the marking of multiple foci within one sentence is generally impossible in the languages under discussion.

set in bold face in the target language; in the English translation, it appears in small capitals. For reasons of space the focus-triggering questions appear in English throughout.

(2) Fɔn (Kwa, Gbe)

Q: What did the woman eat?

A: é ɖù **àyìkún**.
3SG eat bean
'She ate BEANS.'

We take such data to indicate that the postverbal position of a canonical SVO sentence constitutes the default position for NSF in these languages, making additional focus marking by syntactic movement and/or morphological marking and/or prosodic marking superfluous. Furthermore, as will emerge, the preverbal subject position in canonical SVO clauses is often associated with a default topic interpretation in these languages, so that such sentences are normally interpreted as categorical statements with a topic–comment structure.

Nonetheless, whenever NSF is overtly marked, we find a wide variety of focus-marking devices that are not necessarily expressed on the focus constituent itself (*in-focus part*), but may (also) occur in other parts of the clause (*out-of-focus part*), for instance in the form of special aspect markers on the verb. In the following subsections we introduce the various strategies of grammatical focus marking in turn.

10.2.1 *Morphological focus marking of non-subjects*

Morphological focus marking without any additional syntactic changes in the canonical SVO order is found in several West African languages, in particular in the Gur group, but also in some Chadic languages: it is reported for Buli, Byali, Dagbani, Ditammari, Gurene, Kɔnni (Gur), as well as for Bole and Guruntum (Chadic). The morphological focus markers in these languages precede the focus constituent in some of the languages while they follow it in others. As mentioned in the introduction, the class of morphological focus markers is not homogeneous but comprises at least the following list of formal elements, many of which also occur independently in non-focus contexts: (i) invariant information-structural particles; (ii) particles agreeing in gender with the focused NP/DP; (iii) copulas; and (iv) nominal affixes.

In the Gur language Buli in (3), the focus marker is the morpheme *ká*, which precedes the focus constituent under certain conditions.[3] It can change its

[3] In the subjunctive and under negation, the focus marker is less regularly applied. Furthermore, it cannot precede the focal verb (cf. Schwarz, to appear and 2009a).

surface tone because of Low-tone spreading, and it may be subject to heavy segmental erosion when—as is possible in some environments—it cliticizes phonologically on the preceding word (Schwarz 2005, 2009a).

(3) Buli (Gur, Oti-Volta, Buli-Kɔnni)

Q: What did the woman eat?

A: ɔ̀ = ŋɔ̀b **kà** **túé**.
3SG=eat FM bean:PL

'She ate BEANS.'

In parallel fashion, the morphologically invariant focus marker *a* precedes the focus constituent in the West Chadic language Guruntum, cf. (4) (Hartmann and Zimmermann, in press).

(4) Guruntum (West Chadic)

Q: What is he chewing?

A: Tí bà wúm **á** **kwálíngálá**.
3SG IPF chew FM colanut

'He is chewing COLANUT.'

In the Gur language Ditammari in (5), in contrast, the focus marker, which appears in the morphologically complex form N-CL with instances of NSF, follows the focus constituent and displays gender agreement (Reineke 2006a).

(5) Ditammari (Gur, Oti-Volta, Eastern)

Q: What did the woman eat?

A: ò dī **yātũ̀rà** **ǹyā**.
3SG eat PL:bean:PL FM:CL

'She ate BEANS.'

The fact that the languages discussed require a pre- or postfocal marker with instances of NSF while the focus constituent is placed in the postverbal default position for NSF indicates that morphological focus marking forms an integral part of their grammar.

10.2.2 *Syntactic focus marking of non-subjects*

Syntactic focus marking refers to the fact that a focus constituent is reordered relative to the other elements in the clause. This effects a deviation from the canonical word order such that the focus constituent is no longer realized in its base position (*ex-situ strategy*). There are two subcases depending on whether there are additional changes in the out-of-focus part of the clause (10.2.2.2) or not (10.2.2.1). This difference aside, an additional focus marker must precede or follow the focus constituent in some of the languages, while this is just optional in others.

10.2.2.1 Ex-situ *strategy without additional out-of-focus marking* This occurs in Aja, Fɔn, Foodo, Lelemi (all Kwa), and Konkomba (Gur). The Fɔn example in (6) exemplifies the *ex-situ* strategy without additional marking on the out-of-focus part of the clause. The focus constituent is compatible with the same information question as in (2), but, unlike in (2), it is placed in left-peripheral position and is optionally marked by the focus marker *wɛ̀* (see also Ameka 1992, Höftmann 1993, Fiedler 1998, 2008, Lefebvre and Brousseau 2002, Aboh 2004*a* for discussion of this phenomenon in various Gbe languages).

(6) Fɔn (Kwa, Gbe)

Q: What did the woman eat?

A: **àyìkún** (**wɛ̀**) é ɖù.
bean (FM) 3SG eat

'She ate BEANS.' ~ 'It is BEANS that she ate.'

Notice that the answer in (6) is not necessarily interpreted contrastively—nor is its counterpart in (2) used uniquely for information focus. This shows that there is no strict one-to-one correlation between the syntactic realization of a focus and its interpretation in this language.

10.2.2.2 Ex-situ *strategy with additional out-of-focus marking* This occurs in Buli, Byali, Dagbani, Ditammari, Kɔnni, Nateni, Yom (all Gur),[4] Akan (Kwa), and Hausa (Chadic). The Ditammari example in (7) represents the *ex-situ* variant of the *in-situ* strategy from (5) above, which exemplified morphological focus marking. In (7), the focus constituent is fronted to the left-peripheral position and is followed by the FM *ǹya*, which was also present in (5). Notice that an additional morpheme *mà*, which is not found in (5), is required in the out-of-focus part of the clause in sentence-final position.

(7) Ditammari (Gur, Oti-Volta, Eastern)

Q: What did the woman eat?

A: **yātṹrà** **ǹyā** ò dī *(**ma**).
PL:bean:PL FM:CL 3SG eat MA

'She ate BEANS.' ~ 'It is BEANS that she ate.'

[4] See also Fiedler (2006) for instances of this kind in Yom.

In (8) from Hausa (Chadic), the focus constituent is also fronted to the left-peripheral position, where it is optionally followed by a gender-agreeing particle *nee/cee*.[5] In addition to focus fronting, the person-aspect marker must appear in a special relative form (*tá-kee*), which is also observed with other kinds of A-bar-fronting, such as *wh*-question formation and relativization (Tuller 1986; Newman 2000; Hartmann and Zimmermann 2007a).[6]

(8) Hausa (West Chadic)

Q: What is Kande cooking?

A: **Kíifíi** (nee) Kandé tá-kee dáfaawáa.
fish (FM.M) Kande 3SG.F-IPF.REL cooking
'Kande is cooking FISH.'

Finally, in the Buli example in (9), the NP *tú-mɔ̀àntàŋā* ("the red beans"), which contains a pragmatically focal adjective, is located in the left-peripheral position of the clause, where it is optionally preceded by the focus marker *ká*. The characteristic out-of-focus features in this construction are (i) the clause-initial conjunction *tè* (*àtè* after a prosodic break),[7] and (ii) a predicate that blocks certain verb tone patterns and morphemes, such as the verbal suffix *-ya* and the postverbal occurrence of the focus marker *ká*, cf. example (3) (Schwarz and Fiedler 2007; Fiedler and Schwarz 2005; Schwarz to appear).

(9) Buli (Gur, Oti-Volta, Buli-Kɔnni)

S: The woman ate the black beans.

A: **(ká) tú-mɔ̀àntàŋā** tè wà=ŋɔ̀b.
(FM) bean-red:PL:DEF CNJ 3SG=eat
'She ate the RED beans.' ~ 'The RED beans is what she ate.'

[5] This particle is analysed as a focus-sensitive exhaustivity marker in Hartmann and Zimmermann (2007c).

[6] The occurrence of special forms of the person-aspect marker is restricted to the perfective and imperfective aspects (Newman 2000). Furthermore, left-dislocation of the object NP in the imperfective clause in (8) also affects the shape of the verb, with *dáfaawáa* replacing the short form *dáfaa*. See n. 9 for more discussion.

[7] The following example illustrates that *tè* functions as a clausal conjunction:

(i) Context: A dog is chasing a man.

àtè nùrùwá=á chālī, **àtè** bīākā=á ví=wà, **àtè** tìisàŋā à yāā
CNJ person:DEF=IPF run CNJ dog:DEF=IPF follow=3SG CNJ tree:PL:DEF then
chɔ̀gsì míiká...
grasp rope:DEF

'... and the man is running, and the dog is chasing him, and then the trees held the rope...'

Finally, syntactic focus marking is not automatically triggered by a *wh*-question, which induces information focus in the answer. Rather, it seems that the application of the *ex-situ* strategy is restricted to specific contexts and often—though not necessarily—involves an element of contrast, unexpectedness, etc. In many languages, then, the *ex-situ* strategy exists next to an *in-situ* strategy. In some languages, the first is reported to show a tendency to realize contrast, and the second to realize information focus (cf. É. Kiss 1998a, this volume), without there being a strict one-to-one correlation; see, for instance, (2) and (6), above, as well as Hartmann and Zimmermann (2007a) and **?** (2008) for discussion.

10.2.3 *Prosodic focus marking of non-subjects*

Apart from morphological and syntactic focus marking, there is also prosodic focus marking in the form of prosodic boundaries (cf. Féry and Ishihara, this volume). In example (10) from Tangale (West Chadic), a phonological phrase (φ-)boundary is inserted before the focus constituent. The presence of the φ-boundary results in the blocking of certain prosodic processes, such as e.g. *vowel elision* (*VE*), which would otherwise apply. The non-application of VE prevents the derivation of the canonical surface form *way-ug* from underlying *wai-gó* (Kenstowicz 1985; Tuller 1992; Hartmann and Zimmermann 2007b).[8]

(10) Tangale (West Chadic)
Q: What did Laku sell?
A: (Lak wai-**gó**)$_\varphi$ **lánda** vs. Lak way-ug lánda. (all-new)
Laku sell-PERF dress   Laku sell-PERF dress
'Laku sold a DRESS.'   'LAKU SOLD A DRESS.'

Even though a prosodic strategy proper is only attested in Tangale (and more marginally in Bole, see Zimmermann 2006; see also Schwarz 2009a concerning its absence in Buli), it completes our overview of how NSF is marked in the West African languages discussed.

[8] The presence of a prosodic boundary between a verb and a focused object following it is reminiscent of the widespread conjoint–disjoint distinction found in many Bantu languages, see Creissels (1996). The presence of a prosodic boundary between the focus and the rest of the sentence is also in line with recent analyses of a number of non-African languages; see, for instance, Kenesei and Vogel (1990) for Hungarian and English, Kenstowicz and Sohn (1996) for Korean, and Frascarelli (2000) for Italian. We would like to thank two reviewers for bringing these points to our attention.

10.2.4 *Conclusions*

From the findings so far, we draw the following conclusions:

(i) The languages under discussion show considerable variation with respect to how focus is marked (even within one and the same language group).
(ii) All languages looked at exhibit *in-situ* focus (either unmarked or morphologically marked) in postverbal position. We consider this configuration to constitute the default configuration for NSF.
(iii) Nearly all of the languages in the sample exhibit *ex-situ* focus constructions as well. In many languages, the *ex-situ* focus gives rise to additional semanto-pragmatic effects such as emphasis, contrast, or exhaustivity (cf. É. Kiss, this volume).
(iv) The focus-marking systems of Gur and Kwa differ typologically: The primarily agglutinating Gur languages tend to exploit morphology to a higher degree (cf. verb morphology and tone) than the more isolating Kwa languages.

Having shown the basic strategies of marking NSF in the languages under discussion, we now turn to the marking of subject focus (SF).

10.3 Asymmetries between subject focus and non-subject focus

This section investigates the grammatical marking of subject focus in West African languages. The main result of this study is summarized as follows: all languages under discussion exhibit at least one of two kinds of asymmetry, or both, between the marking of subject focus (SF) and non-subject focus (NSF). We distinguish two kinds of asymmetry: (i) a marking asymmetry, which pertains to the fact that SF must be marked whereas NSF need not, or even cannot be marked (Section 10.3.1); (ii) a structural asymmetry, which pertains to the fact that SF is often marked differently from NSF (Section 10.3.2).

10.3.1 *Marking asymmetry*

All languages in the sample exhibit a marking asymmetry, at least concerning the syntactic marking of focus:

(11) *Marking asymmetry*

(i) NSF cannot or need not be marked syntactically.

a. NSF is restricted to *in-situ* positions (Bole, Duwai, Bade, Ngamo (all Chadic))

b. NSF is not restricted to *in-situ* positions (Gur; Kwa; Hausa (Chadic))

(ii) SF must be marked.

Condition (11i) concerns the syntactic marking of NSF and divides into two separate subclauses: languages satisfying condition (ia) never mark NSF syntactically. Furthermore, with the exception of Bole, NSF is not marked by alternative means in these languages either (prosodic or morphological; Schuh 1982, p.c.). In languages satisfying condition (ib), syntactic marking of NSF is optional, as illustrated for Hausa in (12). Condition (11ii) is even more general and holds for all languages under discussion. It says that SF is special in that it must be grammatically marked, be it by syntactic or morphological means, or both.

The marking asymmetry is illustrated by means of the Hausa examples in (12) and (13). Example (12) shows that NSF can be realized either *in-situ*, syntactically unmarked (12a), or *ex-situ* (12b). As a matter of fact, Zimmermann (2006) and Hartmann and Zimmermann (2007a) show that *in-situ* NSF in Hausa is not marked prosodically either. Thus, (12aA) is fully identical to the canonical sentence:

(12) a. Hausa (West Chadic), *optional object marking*

Q: What is Kande cooking?

A: Kandé tá-naa dáfa **kíifíi**. — Unmarked NSF
Kande 3SG.F-IPF cooking fish
'Kande is cooking (A) FISH.'

b. A: Kande is cooking meat.

B: **Kíifíi (nee)** Kandé tá-kee dáfaawáa.[9] — Marked NSF
fish (FM.M) Kande 3SG.F-IPF.REL cooking
'It is (A) FISH that Kande is cooking.'

SF, in contrast, must be marked in Hausa by vacuous syntactic movement of the subject to a left-peripheral focus position. SF movement in Hausa is

9 Notice that left-dislocation of the object NP in the imperfective clause in (12b) has a reflex on the shape of the verb, with *dáfaawáa* replacing the short form *dáfaa*. The descriptive generalization is that the long form occurs in the imperfective aspect when there is no overt object NP following the verb, for instance, after left-dislocation (Newman 2000). We will have to leave it open whether the *waa*-extension indicates a verbal gerund (Tuller 1986), or whether it is a resumptive pronoun marking the base position of the dislocated object.

witnessed by the special relative form of the person-aspect marker *tá-kee*. In contrast, the canonical sentence in (13A′) with the person-aspect marker *tá-naa* is infelicitous as an answer to (13Q):

(13) Hausa (West Chadic), *obligatory SF-marking*

Q: Who is cooking (the) fish?

A: **Kandé** tá-**kee** dáfa kíifíi. Marked SF
Kande 3SG.F-IPF.REL cooking fish
'KANDE is cooking (the) fish.'

A′: #**Kandé** tá-**naa** dáfa kíifíi.

While the marking asymmetry is quite robust from a typological point of view, as either (11ia) or (11ib) and (11ii) hold for all of the languages discussed, it is considerably weakened by the restriction of condition (11i) to instances of syntactic focus marking. Nonetheless, even though languages with morphological focus marking—such as several of the Gur languages in the sample—do not show a marking asymmetry in the strict sense, as both SF and NSF are obligatorily marked, they show a structural asymmetry when it comes to how focus on SF and NSF is marked.

10.3.2 *Structural asymmetry*

The structural devices for marking SF and NSF differ in many languages. Crucially, these differences show up even if we compare sentence-initial instances of SF and NSF. Our language sample is heterogeneous with respect to the observed differences, but nevertheless, the languages can be organized on a scale of increasing complexity depending on quantity and quality of the structural differences between SF- and NSF-marking. At one extreme of the scale, there are languages with no structural differences between SF-marking and *ex-situ* NSF-marking (Guruntum, Byali; see Reineke 2007). At the other end of the scale, we find languages in which SF-marking and NSF-marking differ in several respects, or are even marked in a totally different fashion (Buli, Dagbani, Ditammari, Gurene, Kɔnni, Lelemi). In intermediate positions on the scale, there are languages that show only minor structural differences in the realization of SF- and NSF-marking (Aja, Akan, Awutu-Efutu, Ewe, Fɔn, Foodo, see Fiedler 2007, Konkomba,[10] Nateni, Yom, Bole, Hausa). Often, these

[10] In Konkomba, the encoding of focus depends on the position of the focus constituent within the sentence, rather than being directly determined by its grammatical status as subject or non-subject (cf. Schwarz 2009b).

minor differences concern the presence of the focus marker with instances of NSF-marking.

A representative example of such an intermediate language is Fɔn (Kwa). Comparing the *ex-situ* focus structures for SF and NSF in (14a, b), the only discernible difference consists in the occurrence of the focus marker: it is obligatory with SF, which would otherwise be formally indistinguishable from the canonical sentence, but it is optional with NSF. There are no further differences in the out-of-focus part of the clause.[11]

(14) Fɔn (Kwa, Gbe)

a. Q: Who ate the beans? — SF:

A: **nyɔ̀nú ɔ́** *(**wɛ̀**) ɖù — obligatory FM *wɛ̀*
woman DEF FM eat
àyìkún.
bean
'THE WOMAN ate the beans.'

b. Q: What did the woman eat? — NSF:

A: **àyìkún** (**wɛ̀**) nyɔ̀nú — fronting + optional FM *wɛ̀*
bean (FM) woman
ɔ́ ɖù.
DEF eat
'The woman ate BEANS.'

The Chadic language Bole exhibits an additional complexity in the marking of SF and NSF. As in Fɔn, the IS-particle *ye* is optionally inserted with instances of NSF, cf. (15b), but it is obligatory with SF (at least with transitive verbs), cf. (15a). In addition, the realization of SF in (15a) involves reordering of the subject to a postverbal position (cf. Section 10.4.2):

(15) Bole (West Chadic)

a. Q: Who is planting the millet? — SF:

A: (Àn) jìi kàppà mòrɗó *(**yé**) — inversion + particle *yé*
(3SG) IPF planting millet PRT
Léngì.
Lengi
'LENGI is planting the millet.'

[11] Thus, Fɔn differs from the closely related Ewe, which displays a special pronominal form in the out-of-focus part of the sentence after focal non-subjects (cf. Fiedler and Schwarz 2005: 114f, Schwarz and Fiedler 2007).

b. Q: What is Lengi planting? — NSF:

A: Léngì à jìi kàppà (**yé**) **mòrɗó.** — optional particle *yé*

Léngì	à	jìi	kàppà	(**yé**)
Lengi	AUX	IPF	planting	(PRT)

mòrɗó.
millet

'Lengi is planting MILLET.'

As for the languages at the top end of the scale, which exhibit a high degree of structural asymmetry in the realization of SF and NSF, we find that it is mostly Gur languages that belong to this group. Consider the following examples from Buli in (16a, b):

(16) Buli (Gur, Oti-Volta, Buli-Kɔnni) (cf. 9)

a. Q: Who ate the beans?

A:	(**ká**)	**Mary**	**àlē**	ŋɔ̀bī.	SF (rare with NSF):
	(FM)	Mary	LE[12]	eat.ASS	optional FM *ká*,
	'MARY ate them.'				morpheme *lē*, $verb_1$

b. S: The woman ate black beans.

A:	(**ká**)	**tú-mɔ̀àntàŋā**	**tè**	wà=ŋɔ̀b.	NSF:
	(FM)	bean-red:PL:DEF	CNJ	3SG=eat	optional FM *ká*,
	'She ate the RED beans.'				conjunction *tè*, $verb_2$
	~ 'The RED beans is what she ate.'				

In Buli, the preposed focus marker *ká* is not obligatory with a sentence-initial focus constituent, be it a subject or a non-subject. The two kinds of focus marking exhibit more fundamental differences, though, which concern (i) the category of the morpheme introducing the out-of-focus part, that is, the particle *lē* with SF,[13] and the conjunction *tè* with NSF; and (ii) the predicate. After *tè*, the sentence-final verb cannot take the assertive suffix *-ya*, while it tolerates the suffix in reduced form after *lē*, where it also displays tonal peculiarities under certain conditions.

[12] The particle *lē* (allomorph *nē*, with initial vowel, *àlē*, after a prosodic break) represents a connective preposition related to the preposition *lè* "with, and". Its occurrence is restricted to predicative constituents, such as VPs and predicative NP/DPs (cf. Schwarz 2009a), as illustrated in (i):

(i)	wá	**lē**	nààwā.
	3SG.DJ	LE	chief:DEF

'He is the chief.'

[13] Notice that the asymmetry in Buli is not absolute, as this particle is obligatory after focal subjects while rare, but not completely excluded from occurring after sentence-initial focal non-subjects (cf. Fiedler and Schwarz 2005: 119).

The only Kwa language in the sample that can be considered as representing a high degree of structural asymmetry is Lelemi. In Lelemi, the realization of SF, cf. (17a), requires the use of a special verb form of the "Marked Paradigm", which lacks the verb-internal subject agreement marker required elsewhere (cf. Schwarz 2008b). This form is also called *relative form* by Allan (1973), even though it is not constitutive for all types of relative clause. Instances of NSF, on the other hand, co-occur with the regular verb form, which follows the conjunction *nà* and is not found with SF, cf. (17b).

(17) Lelemi (Kwa, Na-Togo)

a. Q: Are the boys eating oranges?

A: ɔ̀nàabì **ùmwì pɛ́** mɔ̀-dí | SF:
boy one only REL.IPF-eat | special verb form
kùtú.
orange
'Only ONE boy is eating an orange.'

b. S: The boy is eating a banana.

A: **kùtú nà** ɔ̀nàabì ɔ́mɔ̀ ɔ̀ɔ̄-dì. | NSF:
orange CNJ boy DEM 3SG.IPF-eat | conjunction *nà* and
'The boy is eating AN ORANGE.' | regular verb form

From a more abstract perspective, we find a general tendency for languages that do not exhibit a marking asymmetry in the strict sense (see Section 10.3.1) to be located at the top end of the structural asymmetry scale (see Table 10.1).[14] SF is singled out by the grammatical systems of most languages as being peculiar, or special in some way. In Section 10.4 we turn to the reason behind the special status of subject focus. Before we do so, we provide a schematic summary of the asymmetries observed in the languages used for illustration.

10.3.3 *Summary*

While the marking of asymmetry is a general feature of our language sample, possibly with cross-linguistic implications for a wider range of languages, the structural asymmetry has language-specific traits. Table. 10.1 gives an overview

[14] This is only a tendency, and not a correlation, as witnessed by the fact that Guruntum (Chadic) and Byali (Gur), which mark focus morphologically, exhibit no structural asymmetries; see Table 10.1.

TABLE 10.1 Overview of the realization of SF and NSF in the languages illustrated

Language	NSF (term focus)		SF	
	NSF *in-situ*	NSF *ex-situ*		
Byali (Gur)	FM *le*	FM *le* + out-of-focus relative form	FM *le* + out-of-focus relative form	No difference
Guruntum (Chadic)	FM *á*	FM *á*	FM *á*	↓
Hausa (Chadic)	No marking	Optional FM *nee/cee* + relative TAM	Optional FM *nee/cee* + relative TAM	
Fɔn (Kwa)	No marking	Optional FM *wɛ̀*	FM *wɛ̀*	
Tangale (Chadic)	Phrase boundary	N/A	Subject inversion + phrase boundary	
Bole (Chadic)	Optional FM *yé*	N/A	Subject inversion + FM *yé*	
Lelemi (Kwa)	No marking	Optional CNJ *nà*	Relative TAM	
Buli (Gur)	FM *ká*	Optional FM *ká* + out-of-focus conjunction *tè* + $verb_2$[a]	Optional FM *ká* + out-of-focus *lē* + $verb_1$	
Ditammari (Gur)	FM *N-CL*$_1$	FM *N-CL*$_1$ + out-of-focus PRT *mà*[b]	FM *CL*$_2$	
Kɔnni (Gur)	FM *-wÁ*	Out-of-focus PRT *di* or special pronoun + $tone_2$	Out-of-focus verb suffix *-nÀ* + $tone_1$	High degree of structural difference

[a] In Buli, different verb forms due to affixes and tone (discriminable only under certain conditions) in NSF *ex-situ* and SF constructions are here labelled as $verb_1$ and $verb_2$, respectively.
[b] In Ditammari, the focus marker for NSF *ex-situ* constructions consists of N + Class marker of set 1, whereas for SF the focus marker only consists of the class marker of set 2 (cf. Reineke 2006b).

of the realization of SF/NSF in the languages discussed in this section, where the arrow on the right indicates an increasing degree of difference in the formal realization of SF and NSF. The table also shows that NSF may be realized *in-situ* or *ex-situ* in most languages. The Chadic languages Bole and Tangale are the only languages in which non-subjects must be realized *in-situ*. In Fɔn, Lelemi (Kwa) and Hausa (Chadic), *in-situ* NSF is not marked by additional means, in contrast to several Gur and Chadic languages. The comparison between the default *in-situ* strategy for NSF in column (2a) and the various marking strategies for SF in column (3) shows that *in-situ* NSF is generally less marked than SF.

10.4 On the special status of focused subjects

Having established that there is a robust asymmetry between subject and non-subject focus marking in a variety of West-African languages (for a subgroup of Gur and Kwa languages, see also Fiedler and Schwarz 2005, Schwarz and Fiedler 2007), we now discuss the possible reasons behind this asymmetry. The guiding hypothesis is that the special status of focused subjects is conditioned by information–structural factors: we assume that subjects in their canonical sentence-initial position are prototypically interpreted as topics (cf. Li and Thompson 1976) in our language sample. Consequently, if the context establishes a subject as focus, this conflicts with its primary information–structural function as non-focal topic (on the notion of topic, see Ebert and Hinterwimmer, this volume, and Tomioka, this volume). In order to resolve this conflict, the focused subject will have to be realized in a non-canonical structure, for instance, by means of special morphological markers and/or syntactic reorganization.

In our sample, we find at least three different strategies for a subject to avoid the canonical interpretation as topic that it would receive in a canonical sentence with topic–comment structure (Lambrecht 1994: 132). The first strategy is found among the Gur languages and consists in the formal concatenation of subject and predicate, which is the hallmark of thetic statements in these languages (Section 10.4.1). A second strategy, subject inversion, is found in some of the West Chadic languages (Section 10.4.2). Finally, there is a third strategy, which is found in many Kwa languages, where focused subjects remain in their canonical preverbal position and are marked for focus by a morphological focus marker. This third strategy will not be further discussed in this section.

10.4.1 *Marked subject–predicate concatenation in thetic statements*

The investigation of the marking of subject focus in Gur gives rise to the empirical generalization in (18), which holds for a subgroup of Gur languages of the Western Oti-Volta branch (Buli, Kɔnni, Dagbani, Gurene).

(18) *Empirical generalization I (subgroup in Gur)*
Whenever a subject is not to be interpreted as topic, but as focus, it occurs in a special construction, in which subject and predicate are concatenated in a special way and form a single unit from the semanto-pragmatic perspective (Sasse 1995). The concatenated structure serves to express an internally unstructured thetic statement.

While there is a clear grammatical focus strategy for non-subjects, which are morphologically marked in their default postverbal focus position in canonical categorical utterances, at least in the affirmative indicative, this position is unavailable for focused subjects. Rather, subjects obligatorily precede the verb (stem). Furthermore, a sentence with a focused subject always has a special structure, representing a *thetic statement*. In this construction, the topic–comment split is voided by the formal concatenation of subject and predicate (see below). As a result of the concatenation, subject and predicate are assigned equal pragmatic status. In all the languages cited above, the thetic construction is expressed on the predicate, albeit in different ways. Kɔnni, for instance, employs a verbal suffix *-nÀ* (*-nà*, *-nè* due to vowel harmony, see Cahill 2007), which triggers a special tone pattern on the verb. This is illustrated in the following examples, where (19a) illustrates a categorical statement with object focus and with no concatenation features on the verb. Example (19b) illustrates subject focus with the concatenation features on the verb.

(19) Kɔnni (Gur, Oti-Volta, Buli-Kɔnni)

a. **ʊ̀=nìgì-wá** ʊ̀= bʊ́à. — Object Focus: canonical verb + FM
3SG =hit-FM 3SG=child
'She hit HER CHILD.'

b. Q: Who hit Peter?
A: Mary **nígí-nà** wà. — Subject Focus: suffix *-nÀ* + verb tone$_1$
Mary hit-NA 3SG
'MARY hit him.'

In other languages, the concatenation of subject and predicate in the thetic construction is expressed by means of preverbal elements. This is illustrated for Buli in (20), where—in contrast to the categorical statement in

(20a)—the answer in (20b) contains the preverbal particle *lē* (cf. Schwarz 2009a, and fn. 12).

(20) Buli (Gur, Oti-Volta, Buli-Kɔnni)

a. nípōōwá **fɔ̀b kà Peter.** — Object Focus:
woman:DEF slap FM Peter — canonical verb + FM
'The woman hit PETER.'

b. Q: Who hit Peter?
A: (ká) **Mary àlē fɔ̀b=wā.** — Subject Focus:
(FM) Mary LE slap=3SG — particle *lē* + verb$_1$
'MARY hit him.'

The claim that subject-focus constructions with the formal concatenation of subject and predicate, such as (19b) and (20b), actually represent thetic statements in this language group is supported by the occurrence of sentences with the same structural features in environments in which pragmatically unstructured thetic statements can be expected.

First, utterances with focus on the whole sentence are expressed in the same way as subject foci. There is thus an isomorphism between sentences with subject focus and sentences in which the entire clause is focused, resulting in focus ambiguity in the absence of context. This isomorphism is fairly widespread within the Gur languages and is illustrated in (21) for Kɔnni. Apparently, the lack of any sentence-internal topic–comment structure serves the expression of sentence focus in the same way as subject focus. It is therefore also typically employed with event reporting sentences.

(21) Kɔnni (Gur, Oti-Volta, Buli-Kɔnni)

a. Q: Who hit Peter?
A: Mary **nígí-nà** wà. — Subject Focus:
Mary hit-NA 3SG — suffix -*nÀ* + verb tone$_1$
'MARY hit him.' (= 19b)

b. Q: What happened?
A: Mary **nígí-nà** Peter — Sentence Focus:
Mary hit-NA Peter — suffix -*nÀ* + verb tone$_1$
'MARY HIT PETER.'

Certain peculiarities are observed with pronominal subjects in thetic statements, as briefly illustrated in the following for Kɔnni. In the case of sentence focus, a conflict may arise when the thetic encoding of subject and predicate requires a (pronominal) subject expression that typically refers to topical entities. This situation is not uncommon in languages without passive constructions, where a passive sentence like *A child was born* must be expressed actively by using a third-person-plural class pronoun as dummy subject.

(22) Kɔnni (Gur, Oti-Volta, Buli-Kɔnni)

Q: What happened?

A: **bà**=mììrì-wá bʊ̀ànyààlíŋ.
3PL=give.birth-FM baby:CL
'A CHILD WAS BORN.' (lit.: 'They gave birth to a child.')

If uttered as a reply to the question *What happened?*, as in (22), the whole sentence can be taken to be in focus from a semantic–pragmatic point of view. Consequently, one would expect it to be realized by a thetic construction with suffix *-nÀ*, as was observed with other instances of sentence focus, such as (21b) from above. The use of the disjunctive subject pronoun *bámíŋ* "they" is impossible, however, as this would immediately yield a referential reading implying focus on the subject rather than on the entire sentence (#***bámíŋ bà***=*mɪɪrɪ*-***na*** "THEY gave birth to the child."[15]) On the other hand, the use of the common proclitic subject in combination with the *-nÀ*-marked verb would result in the formation of a backgrounding clause that needs to be followed up by a main clause (#***bà***=*mɪɪrɪ*-***na*** *bʊ̀ànyààlíŋ*... "When they gave birth to the child, ... "). Hence, even though the answer in the Kɔnni example in (22) expresses a sentence focus pragmatically, the thetic encoding cannot apply and the entire clause has to be construed as a categorical statement with topic–comment structure.

The second argument supporting the thetic analysis is based on the fact that this construction is typically found in text-initial position. Text-initially, discourse topics are rare or even absent and have to be first introduced, for instance by means of a thetic statement, before any topic–comment structuring can apply. Generally, such text-initial thetic statements locate an entity in time or space, as illustrated in (23), again for Kɔnni.

(23) Kɔnni (Gur, Oti-Volta, Buli-Kɔnni)

gbáníŋ **díísí-nè** tébùlì-ké síkpèŋ.
book lie-NA table-DEF top
'There is a book on the table.'

[15] Though this is still a matter of investigation, the available data suggests that a disjunctive pronoun like *bámíŋ* cannot immediately function as subject of the *-nÀ*-marked verb form and is therefore supported by a verb-initial proclitic subject, here ***bà***=.

Parallel findings showing the use of thetic statements under similar focus conditions are also reported for some European and other languages (see Sasse 1987, 1995). They corroborate our analysis of the subject and sentence focus constructions as based on a pragmatically unstructured thetic utterance. It can therefore be concluded that, in these languages, the unmarked topical status of the subject in canonical categorical statements blocks its focalization. If the subject represents the unmarked topic, focus remains within the comment. However, if the focus is not restricted to the comment, but concerns either just the subject or the whole sentence (see also Schwarz 2007: 135), a topicless (thetic) construction is used. The thetic construction involves a marked concatenation of subject and predicate and is realized in different ways in each of the languages concerned. Common to all languages, however, is the formal manipulation of the predicate in such a way that it is inseparably linked to the subject from a semanto-pragmatic point of view.

10.4.2 *Subject inversion*

Another strategy to strip the subject of its default topic interpretation is exhibited by a subgroup of geographically related West Chadic SVO languages that mark narrow subject focus by means of subject inversion (Bole, Tangale, Bade, Ngizim, Duwai): the focused subject does not occur in its canonical preverbal position, where it would be interpreted as a topic, but is placed in the postverbal domain in which focus is typically realized.[16] In other words, focused subjects and non-subjects behave alike from a structural point of view. Example (24), repeated from (15), illustrates subject inversion in Bole (cf. also the discussion of inversion constructions in Aghem, Hyman and Polinsky, this volume). Recall that the inverted subject must be preceded by the particle *yé* in Bole. Example (25) from Tangale shows that the positional restriction on focused subjects extends to *wh*-subjects, which must also appear postverbally.[17]

[16] Unlike in Aghem (Hyman and Polinsky, this volume) and other Bantu languages, focused subjects in Bole and Tangale are not placed immediately after the transitive verb but the direct object frequently intervenes between the verb and the focused subject, as in (24) (cf. Tuller 1992). Nor are postverbal focused subjects restricted to sentence-final position. This is shown in examples (27a, b), where the postverbal focused subjects are followed by a temporal and a locative adjunct, respectively. For this reason, we tentatively propose identifying the postverbal focus position of objects and subjects with the right edge of VP.

[17] In (25), subject inversion is accompanied by the insertion of a prosodic phrase boundary before the inverted subject. In contrast to (10), the underlying verbal form *wai-go* surfaces as *way-ug* because of the application of two segmental processes, vowel elision and [u]-epenthesis. Recall from the discussion of (10) that these processes apply within the phonological phrase (φ) only, which shows that verb and object must form a prosodic unit in (25). The existence of a φ-boundary before the inverted subject is evidenced by the non-application of *left-line delinking* (LLD), see Kidda (1993: 118), another

(24) Bole (West Chadic)

Q: Who is planting the millet?

A: (An) jìi kàppà mòrɗó **yé** **Léngì**.
(one) IPF planting millet PRT Lengi
'LENGI is planting the millet.'

(25) Tangale (West Chadic)
(Way-ug land-í)$_{\varphi}$ **nóŋ**?
sell-PERF dress-DEF who
'WHO sold the dress?'

At first sight, the sentences in (24) and (25) appear to resemble predicate inversion structures as found in the English sentence [*The one planting the millet*] *is Lengi*, where the DP containing an (empty-headed) relative clause is the predicate that moves across the subject of the predication (see e.g. Frascarelli 2007 for a recent analysis along these lines). However, the following data show that an analysis of SF in Bole and Tangale in terms of predicate inversion (or pseudo-clefting) cannot be correct. First, the particle *yé* in Bole is not identical to the relative marker *la*, which furthermore precedes the relative clause. This is shown in (26):

(26) Bole (West Chadic)

In gomu ga memu [**la** Bamoi essungo yê].
1SG met with man REL Bamoi called DEF
'I met the person that Bamoi called.'

We conclude that there is no empty-headed relative clause in (24). Second, the inverted subject in postverbal position can be followed by additional background material in both languages. This is shown for Tangale in (27a), and for Bole in (27b):

(27) a. Tangale (West Chadic)

wa patu ayaba **nuŋ** ta luumo (Tuller 1992: 307, ex. (4b))
FUT buy banana who at market
dooji.
tomorrow

'Who will buy bananas at the market tomorrow?'

tonal process that does not apply across φ-boundaries. LLD dissociates tonemes from their original tone-bearing units (TBUs) after rightward spreading onto the following TBU. In (25), the presence of the high tone on the second vowel of *landí* shows that LLD is blocked, thus indicating the existence of a φ-boundary that separates the object from the inverted *wh*-subject.

b. Bole (West Chadic)

Q: Who is planting the millet at the farm?

A: An jìi kàppà mòrɗó **yé** **Léngì** ga ga
3SG IPF plant millet PRT Lengi in inside
kori *(yê).
farm PRT
'LENGI is planting the millet at the farm?'

The fact that backgrounded material can both precede and follow the SF in (27a, b) provides further evidence against an analysis of subject inversion in terms of a predicative cleft-like construction in which the backgrounded material forms a single constituent, e.g. a relative clause.[18]

The behaviour of focused subjects in this subgroup of West Chadic is captured in the form of the empirical generalization in (28):

(28) *Empirical generalization II*
Whenever a subject is not to be interpreted as topic, but as focus, it must occur in the prototypical focus position, that is, in a postverbal position at the right edge of VP.

The generalization in (28) holds not only for certain West Chadic languages. A similar requirement has been observed for some Romance languages, including Italian (Frascarelli 2000; Samek-Lodovici 2005) and Spanish (Zubizarreta 1998), for Old High German (Hinterhölzl, this volume), as well as for Bantu (Demuth and Mmusi 1997; Hyman and Polinsky, this volume). It has been argued that some languages have a fixed position to which nuclear stress is assigned. As focused constituents must associate with main stress, they must appear in this position. In the case of focused subjects, this requirement thus triggers prosodically motivated movement. To illustrate, consider the Spanish example in (29) (Zubizarreta 1998: 125f). The non-canonical VOS order is only compatible with an information focus interpretation of the subject.[19]

[18] Taking up a discussion from Section 10.3, the obligatory occurrence of a second instance of *yé* (+ final low tone) in final position suggests that these particles should not be treated as genuine focus markers, but as background markers that mark the material to their left as presupposed (Schuh 2005). Given the observable tendency for focus constituents in Bole and Tangale to occur at the right periphery of the clause, background marking on postfocal material is obligatory.

[19] Contrastive focus on the subject is realized clause-initially in Spanish:

(i) **María** me regaló la botella de vino, no Juan.
Maria to.me give.PERF the bottle of wine not Juan
'MARIA gave me the bottle of wine, not JUAN.'

The asymmetry between information focus and contrastive focus has two interesting implications. First, it shows that the postverbal position is not the unique stress position in Spanish. Second, information focus and contrastive focus on subjects are realized differently, which corroborates arguments

(29) Spanish (Romance)

Q: Who gave you the bottle of wine?

A: [Me regaló la botella de vino]$_i$ **María** t_i.
to.me give.PERF the bottle of wine Maria
'MARIA gave me the bottle of wine.'

Zubizarreta (1998: 127) analyzes (29) in terms of leftward adjunction of the complex phrase originating immediately below the subject. Since leftward adjunction is motivated by the need to place the subject in the stress position, movement in (29) is prosodically driven.

The Romance data, as well as the Chadic data, suggest that there is a strong requirement for *all* focus constituents, and not only non-subjects, to be placed in a prototypical, and in some sense prominent, focus position behind the verb. The marking of SF in this group of subject-inverting languages is thus subject to a stronger requirement than the mere need to mark a subject as non-topic, as postulated in the initial hypothesis: in these languages, focused subjects need not only be marked as non-topics, but they must also be marked as focus.

This conclusion has three immediate consequences for the discussion of focus in West Chadic. First, there is a structural analogy between SF and NSF as both must be in the same linear relation with the verb. Second, the obligatory placement of focused subjects in the prototypical focus position at the right edge of the VP shows that focus *is* coded in the grammatical system of these languages after all, in spite of the fact that explicit focus marking is optional with non-subjects (see Section 10.2). Third, sentence focus cannot be expressed by means of inversion in these languages, as it is impossible for the entire focused clause to occur in the prototypical postverbal focus position. There is thus no isomorphism of structures with subject focus and sentence focus in the West Chadic languages, unlike in the Gur languages discussed in Section 10.4.1. Instead, sentence focus is realized in the canonical SVO order with no additional marking. As a result, the realization of sentence focus is formally identical to information focus on non-subjects in the West Chadic languages. Interestingly, an analogous isomorphism is found in many intonation languages such as English and German, which mark focus prosodically by a pitch accent on the focused constituent. Example (30a) illustrates information focus on the direct object; (30b) is an instance of all-new or sentence focus. In both cases, the main accent is realized on the direct object.

to the effect that elements in clause-initial position often receive a special pragmatic interpretation (e.g. ? 2008).

(30) a. To which country did the president travel?
The president traveled **to BhuTAN**.

b. What happened?
The president traveled to BhuTAN.

10.4.3 *Summary*

To summarize, this section discussed two different strategies for focused subjects to escape their canonical interpretation as topic, employed by subgroups of the Gur and West Chadic languages. In a subgroup of the Gur languages, SF is not marked by a specific linguistic encoding on the focused subject alone. Rather, sentences with subject focus behave on a par with all other thetic sentences, which are characterized by the concatenation of subject and predicate. In such cases, the subject remains in preverbal position, but the clausal construction changes, resulting in an isomorphism between subject focus and sentence focus. In contrast, in the West Chadic languages discussed here, focused subjects do not appear in their canonical preverbal position, but must invert to the postverbal default focus position at the right edge of VP.

10.5 Conclusion

The following points emerge from the present discussion of (term) focus marking in a range of West African languages of the Kwa, Gur and (West) Chadic group. First, focus in these tone languages is realized in a variety of grammatical ways, that is, by syntactic, morphological, or prosodic means. Second, the majority of the languages in the sample show a subject vs. non-subject asymmetry when it comes to the realization of focus. Third, the special status of focused subjects in Kwa, Gur, and (West) Chadic follows from the cross-linguistically well-attested fact that the subject in sentence-initial position is assigned a default interpretation as unmarked topic. Fourth, if there is no match between subject and topic, for instance if only the subject is focused, the languages of our sample show parametric variation with respect to the grammatical realization of subject focus: (i) some languages (among Gur) concatenate subject and predicate to mark theticity; (ii) some languages (many West Chadic) mark non-topical, focused subjects by placing them in the prototypical postverbal focus position; and (iii) a third group of languages (many Kwa) simply put a focus marker on the non-topical subject, which remains in its prototypical sentence-initial position.

Further research on focus in non-European languages will show whether these properties are peculiar to the focus-marking systems of West African languages, or whether they reflect more general properties of the languages of the world when it comes to the marking of focus.

11

Information structure and OV order

ÞORBJÖRG HRÓARSDÓTTIR

11.1 Introduction

This chapter presents an empirical study of the loss of OV orders in the history of Icelandic, showing that change in word order cannot always be explained in terms of changes in morphology (loss of case, as has been proposed by various authors for English). A modified version of Hróarsdóttir's (2000) analysis of OV orders in Older Icelandic (OI) illustrates how mixed OV/VO word-order patterns are due to (a) information structure and prosodic parameters and (b) that the pertinent change in the history of Icelandic can be adduced to one rule, namely the extraction of non-finite VPs out of complex finite VPs.

Morphological triggers, or the absence thereof, have been related to clustering of syntactic properties in both diachronic change and acquisition. The correlation between overt verb movement and "rich" agreement paradigms has, for instance, been proposed by several researchers in recent years (see Kosmeijer 1986, Rohrbacher 1994, 1999, Holmberg and Platzack 1995, Vikner 1997, Koeneman 2000).[1] I.G. Roberts (1997) argues that English lost overt object movement due to the loss of morphological case in Middle English (ME). As seen from the English viewpoint, inflectional morphology may express the relevant cue for parameters and so the loss of inflection may (but does not have to) lead to a grammar change. This analysis does not carry over to Icelandic as the loss of OV in Icelandic took place despite rich case morphology. We assume that the relevant cue may be expressed differently among the languages: while it may have been expressed through morphology in Old English (OE), it was expressed through prosodic properties in OI. In

[1] But see Hróarsdóttir, Hrafnbjargarson, Wiklund, and Bentzen (2006), Wiklund, Hrafnbjargarson, Bentzen, and Hróarsdóttir (2007), and Hróarsdóttir, Wiklund, Bentzen, and Hrafnbjargarson (2007) for a critical view of this correlation.

both cases, external effects led to fewer expressions of the relevant cue and a grammar change took place. If we are correct, then the relationship between strong morphology and OV order has been overestimated. We claim that this relation is only one-directional, in the sense that languages that lose strong morphology also tend to lose overt OV order (English) because of structure sandwiching, but not vice versa. Therefore, languages may have overt OV orders despite weak morphology, and languages may lose overt OV structure regardless of whether there have been any changes in the morphology or not (Icelandic). Moreover, many modern languages have rigid word order despite their case morphology, and vice versa (cf. Siewierska 1998).

Finally, the loss of OV orders in OI may not be explained in structural linguistic terms only, but in connection to sociolinguistic terms: a certain type of word order became unfashionable. We will address this possibility in Section 11.6. We discuss what might happen when a population changes rapidly, for example by epidemics.

11.2 Loss of OV in Icelandic grammar

11.2.1 *Introduction*

Let us now look at the loss of OV word order in the history of Icelandic. Modern Icelandic has pure VO order within the VP, as shown in (1). The word order in (1) with auxiliary verb–main verb–object is the only possible order of these elements in Modern Icelandic (abstracting away from topicalization and stylistic fronting). The two non-finite verbs and their objects must follow the negation and sentential adverbs.

(1) Þeir munu aldrei **hafa lesið bókina**.
they will never have read book.the
'They will never have read the book.'

Unlike in Modern Icelandic, several orders of the VP-internal arguments were possible at earlier stages in the history of Icelandic, including both OV and VO word-order patterns (cf. Sigurðsson 1988, Rögnvaldsson 1996, Indriðason 1987, Hróarsdóttir 1996, 2000). The attested OV word-order patterns were lost at the beginning of the nineteenth century. Two examples of OV word order in OI are shown in (2) below.

(2) Pure OV word order

a. að hann hafi **hana drepið**. (Álf)
that he had her killed
'that he had killed her.'

b. að þú...hafir **það bréf fengið**. (letters)
that you...have that letter received
'that you have received that letter.'

Icelandic has had rich subject–verb agreement morphology and case morphology throughout its history. It is also generally assumed that both Old and Modern Icelandic have an obligatory overt movement of the finite verb out of the VP (to IP/CP), in both main and subordinate clauses.

11.2.2 *The diachronic aspect*

With regard to the order of objects and non-finite verbs, OI differs from Modern Icelandic in three major aspects, as shown in (3). All these three word-order patterns are ungrammatical in Modern Icelandic (ignoring negative phrase constructions).

(3) a. V_{fin} ... object–V_{aux}–V_{main}

b. V_{fin} ... V_{aux}–object–V_{main}

c. V_{fin} ... (object)–V_{main}–(*object)–V_{aux}–(object)

Hróarsdóttir (2000) studied the frequency of OV and VO patterns in various texts dating from the fourteenth to the nineteenth centuries, in addition to personal letters dating from throughout the nineteenth century. Here, we have built on this database, making it approximately double in size based on a wider extraction of the same texts. A list of the 16 texts used for this study is given in Appendix A, together with bibliographical information. These texts are literary works, all in reliable editions based directly on the original composition. Approximately 55 pages were extracted from each text, where possible, until a corpus of approximately 8,500 sentences each containing at least one non-finite verb had been reached, exhibiting either OV or VO word order. Nineteenth-century letters by 75 individuals were also studied. Bibliographical information for the nineteenth-century letters are given in Appendix B, together with an explanation for the abbreviations in parentheses in the examples. Some OI examples are shown below.

(4) V_{fin} ... object–V_{aux}–V_{main}

a. að hon mundi eigi **barn mega eiga**. (Finn)
that she would not child may own
'that she would not be allowed to have a child.'

b. ef hann hefði **það viljað fága**. (Guðm)
if he had it wanted clean
'if he had wanted to clean it.'

(5) V_{fin} . . . V_{aux}–object–V_{main}

a. að hann skyldi aldrei **mega sól sjá**. (Árm)
that he should never be-allowed sun to.see
'that he should never be allowed to see the sun.'

b. og ekki skal faðir minn **geta þér hjálpað**. (Álf)
and not shall father mine can you helped
'And my father will not be able to help you.'

(6) V_{fin} . . . V_{main}–V_{aux}–object

a. at hann mun **raða vilia ferðum sínum**. (Finn)
that he will decide want journeys his
'that he wants to decide his own journeys.'

b. þu munt **haft hafa harða landtoku**. (Finn)
you will had have hard landing
'You have had a rough landing.'

(7) V_{fin} . . . object–V_{main}–V_{aux}

a. og enginn þóttist **þvílíkan veikleika séð hafa**. (Álf)
and no-one pretended such weakness seen have
'And no one pretended to have seen such weakness.'

b. að eg mundi **hann sigrað geta**. (Árm)
that I would him defeat could
'that I would be able to defeat him.'

As mentioned, all these word-order patterns are ungrammatical in Modern Icelandic and the striking fact is that they all disappeared from the language at the same time in the history of Icelandic, namely, in the beginning of the nineteenth century.

11.3 Information structure

11.3.1 *Information structure and OV order in Older Icelandic*

Let us now turn to the role of information structure as a cue for the OV/VO parameter in Older Icelandic. The cause of the grammar change must lie in some alteration to the primary linguistic data (PLD). Here, we address the variation in the PLD that led to the grammar change. We will illustrate the gradual loss of the OV word-order patterns that took place in the centuries prior to the grammar change, focusing on the changes in the PLD that must have paved the way for the parameter change when OV word order was lost.

We propose that the cue for the OV/VO word-order variation in OI was expressed through prosodic properties associated with syntactic positions.

If we now focus only on the nominal objects in pre- and postverbal positions in the attested OI texts, then we see a gradual increase of VO word orders with nominal objects from the seventeenth century on. Examples (8) through (11) show some simple cases of pre- and postverbal nominal objects in OI.

(8) Full DPs: OV word order

a. efftir þad þeir høffdu **eplid** eted. (Dín)
after that they had apple.the eaten
'after they had eaten the apple.'

b. þä skilldu þeir **lijffed** missa. (Dín)
then should they live.the lose
'Then they should die.'

(9) Full DPs: VO word order

a. að hann hefði etið **kjötið**. (Munn)
that he had eaten meat.the
'that he had eaten the meat.'

b. hvört hann vilji ei kaupa **þræla**. (Árm)
whether he wanted not buy slaves
'whether he didn't want to buy slaves.'

(10) Pronouns: OV word order

a. að eg hafi **hana** beðið. (letters)
that I have her asked
'that I have asked her.'

b. at hann mundi **þat** eigi gera. (Finn)
that he would it not do
'that he would not do that.'

(11) Pronouns: VO word order

a. að ég hafi aldrei borgað **þér**. (letters)
that I have never paid you
'that I have never paid you.'

b. að guð vilji brúka **þig** lengi. (letters)
that God wants use you long
'that God wants to use you for a long time.'

TABLE 11.1 Full DPs versus pronouns

Texts	Full DPs			Pronouns		
	OV	VO	% OV	OV	VO	% OV
14th c.	75	186	28.7	92	33	73.6
15th c.	66	110	37.5	54	28	75.0
16th c.	69	112	38.1	34	7	82.9
17th c.	94	245	27.7	162	120	57.4
18th c.	25	125	16.7	27	58	31.7
19th c.	99	1,603	5.8	258	1,050	19.7

Consider Table 11.1 which illustrates the total frequency for the texts from each century. Pronouns were preferred in preverbal position more often than full DPs in OI. The overall difference is not great, but it is clearest in the earliest texts. There is also a grammatical ordering with respect to old information preceding new information in the OI texts, as shown in Table 11.2. This means that objects previously mentioned within the five preceding sentences have a tendency to occur in preverbal position, more than objects introduced for the first time in the context.[2] The reason for why we do not see a clear rule here, with OV structures always involving old information and VO structures always involving new information, is that the word order is also subject to a stylistic rule in terms of weight. This means that light DPs (including pronouns) are preferred in preverbal position and heavy DPs are preferred in postverbal position. Hence, a relatively heavy DP with old information would normally occur in postverbal position. The heaviness factor is illustrated in Table 11.3.

Some examples of the division of full DPs according to their heaviness are shown below. Only sentences with OV word order are exemplified.

(12) DPs with one word

a. Þar er kaleikur sem álfafólk hefur **kirkjunni** gefið. (Álf)
there is chalice that elves have church.the given
'There is a chalice there, that some elves have given to the church.'

b. at ek mun **sæmd** af þier hliota. (Vikt)
that I will honor of you get
'that I will get honor from you.'

[2] Objects with old information here mean personal pronouns, and full DPs already mentioned in the preceding sentences.

Table 11.2 Old versus new information

Texts	New information			Old information		
	OV	VO	% OV	OV	VO	% OV
14th c.	75	204	26.9	156	36	81.3
15th c.	48	136	26.1	110	24	82.1
16th c.	50	132	27.5	77	17	82.0
17th c.	79	314	20.1	215	104	67.4
18th c.	20	98	16.9	39	43	47.6
19th c.	84	1,312	6.0	261	1,072	20.0

(13) DPs with two words

a. hafer þu **þinu lidi** jatat þeim. (Vikt)
have you your assistance promised them
'if you have promised them your assistance.'

b. at hann muni **eitthvert ráð** til leggja. (Guðm)
that he would some advice to put
'that he would offer some advice.'

(14) DPs with three words

a. að prestur þessi hefði **sína fyrri konu** misst. (Álf)
that priest this had his first wife lost
'that this priest had lost his first wife.'

b. því hann vildi **þann saklausa mann** til (Morð)
because he wanted this innocent man to
dauða dæma.
death sentence
'because he wanted to sentence this innocent man to death.'

Table 11.3 Number of words within full DPs

	1 word % OV	2 words % OV	3+ words % OV
14th c.	72.5	36.8	10.9
15th c.	66.3	32.7	10.3
16th c.	60.0	31.8	8.0
17th c.	46.0	23.5	2.8
18th c.	43.8	20.6	1.7
19th c.	13.7	4.6	1.6

No preverbal DPs in the corpus contained more than three words. On the other hand, postverbal DPs containing more than three words were common (163 examples found). Some of these examples are shown in (15).

(15) DPs with more than three words: VO word order

a. Ungbarna veiki hefur sópað burt **miklum** (letters)
baby sickness have swept away big
þorra þessa ungviðis.
majority these children
'Diphtheria has taken most of these children.'

b. eg heffe feinged **eina fräbæra edur** (Dín)
I have got one distinguished or
faheyrda sött.
outrageous sickness
'I have got a distinguished or outrageous sickness.'

c. Vil eg nú upphéðan heita yður **minni** (Munn)
want I now from-now-on promise you my
fullri og fastri vináttu.
complete and solid friendship
'From now on I will promise you my complete and constant friendship.'

Single DPs (DPs containing only one word) occurred as frequently in preverbal position as in postverbal position in the earliest texts (dating from the fourteenth through the sixteenth centuries). In texts dating from the next two centuries (the seventeenth and the eighteenth centuries), single DPs started to occur in postverbal position more often, as illustrated in Table 10.3. We take this to reflect language change; that is, change in usage during the seventeenth and eighteenth centuries. Finally, in the nineteenth-century texts and letters, preverbal single DPs have become very rare, as a consequence of an abrupt grammar change in the beginning of the nineteenth century. Single DPs thus lost their tendency to occur to the left of the main verb. DPs containing two words clearly are preferred in postverbal position, throughout the period studied. However, these DPs could also occur preverbally. DPs consisting of three or more words were very rare in preverbal position in all the texts and letters studied. It is therefore clear that light DPs more often occurred in preverbal position than heavy DPs in OI. The heaviness factor seems to have been stronger than the information factor: Light objects with old information (unfocused) are always preverbal and heavy objects with new

TABLE 11.4 Objects occurring before the negation

	Pronouns	DPs	PPs
All centuries	27	5	3

information (focused) are always postverbal. Light objects with new information are also usually preverbal, while heavy objects with old information are usually postverbal. DPs with two words seem to have been more or less neutral with regard to the heaviness factor; hence, unfocused (old information) DPs with two words are usually preverbal, while focused (new information) DPs with two words are usually postverbal. Hence, pronouns and other light categories are preferred in preverbal position, while (heavy) full DPs and objects with new information are preferred in postverbal position. In other words, there is a prosodic ordering associated with syntactic positions with respect to:

- full DPs are more likely to occur in VO structures than pronouns;
- new information is more likely to occur in VO structures than old information;
- heavy DPs are more likely to occur in VO structures than light DPs;
- quantified and negated objects occur in preverbal (OV) position.[3]

When we take all these prosodic orderings into consideration, we can account for the OV/VO word-order pattern in the OI corpus. In other words, it is more or less predictable which objects occur preverbally and which occur postverbally.

11.3.2 *Further restrictions*

Note that there also was a restriction in OI, similar to the one of Object Shift in Modern Icelandic, which applies to pronominal objects and definite DPs. In the OI data, where a sentential adverb/negation intervenes between the preverbal complement and the non-finite main verb, there are no instances of indefinite objects. That is, the pattern [$DP_{indefinite}$–sentence adverb–V_{main}] seems to have been absent. Pronouns also had a much stronger tendency to occur to the left of an adverb/negation than full DPs or other types of argument, as illustrated in Table 11.4. The distinction is very striking: 27 pronouns occurred to the left of the negation, only five full DPs, all definite, and three arguments of another type (PP or a small clause predicate). Some of the OI sentences are shown in (16) to (18).

[3] Quantified and negated objects always occurred preverbally in OI, and they still do in the modern language. These are the remaining OV orders in Modern Icelandic.

(16) Pronoun to the left of a sentence adverb

a. Og vér fáum **það** ei þekkt. (Álf)
and we get it not recognized
'And we cannot recognize it.'

b. at hann mundi **þat** eigi gera. (Finn)
that he would it not do
'that he wouldn't do it.'

c. að þeir hefðu **þau** aldrei út gefið. (Morð)
that they had them never out given
'that they had never published them.'

(17) Full DP to the left of a sentence adverb

a. þui ath hann willdi **leoninum** giarna ekki (Sig)
because that he wanted lion.the readily not
grand giora.
damage do
'because he didn't want to harm the lion.'

b. Enn hier med a ofan hef eg **greind** (Afs)
but here with in addition have I mentioned
Kugillde alldrei teked.
cow-values never taken
'But, in addition, I have never taken these cows.'

c. Hefi eg **þessa kapla** aldri aptur fengið. (Skál)
have I these cables never back got
'I haven't got these cables back.'

(18) Complemental adverbial/PP to the left of a sentence adverb

a. þegar hann hefur **þar** ei lengi verið. (Álf)
when he has there not long been
'when he hasn't stayed there for a long time.'

b. því þeir hafa **hér** aldrei fyrri komið. (Árm)
because they have here never before come
'because they have never been here before.'

c. Hefr síðan **um þann Haug** ekki (letters)
has since about that grave-mound not
grenslast verid.
inquired been
'Since then, no one has inquired about that grave mound.'

Furthermore, at most one argument could precede the sentence adverb at a time. In VO word order, on the other hand, two or three arguments frequently followed the sentence adverb.

In the patterns [object–V_{aux}–V_{main}] and [object–V_{main}–V_{aux}], the preverbal argument could be either a DP, a PP, or a small clause predicate. However, only the DP occurred with any significant frequency. In almost all the examples found, the DP here is a pronoun. Two examples are shown in (19) and (20).

(19) [DP–V_{aux}–V_{main}]

ef hann hefði **þat viljað fága**. (Guðm)
if he had it wanted clean

'if he had wanted to clean it.'

(20) [DP–V_{main}–V_{aux}]

að eg mundi **hann sigrað geta**. (Guðm)
that I would him defeat could

'that I would be able to defeat him.'

The pattern [V_{main}–V_{aux}–object], on the other hand, occurred with full DPs only, but never with pronouns. The sentences in (21) show some examples of this pattern.

(21) [V_{main}–V_{aux}–DP]

a. at hann mun **raða vilia ferðum sínum**. (Finn)
that he will determine want journeys his
'that he will want to determine his own journeys.'

b. þu munt **haft hafa harða landtoku**. (Finn)
you will had have hard landing
'You have had a rough landing.'

11.4 Theoretical implications

In this section, partly following Hróarsdóttir (2000), we make use of remnant-movement of various kinds of predicative phrases to provide an analysis of OV orders, and, correspondingly, a proposal as to which aspect of Icelandic syntax must have changed when VO word order became the norm: the essential change is loss of extraction of the embedded VP from the matrix VP (so-called VP-extraction). The claim is that the crucial difference between OV and VO languages is simply that OV-languages lack the VP-preposing Modern English and other VO languages have. The picture is slightly more complicated for mixed OV/VO languages such as OE, OI, and Afrikaans. Our hypothesis here has three main transformations:

- The derivation of OV obligatorily involves VP-extraction.
- Unfocused DPs move to Spec,AgrOP, while focused DPs do not.
- Obligatory preposing of the remnant vP, containing at least the finite verb, at all stages of Icelandic.

There are four points to be made. First, the derivation of OV necessarily (obligatorily) involves VP-extraction. This is the crucial difference between OV and VO orders, and therefore also the key for the grammar change. Second, DPs behave differently according to their type: light DPs (i.e. DPs consisting of less than three words) and DPs with old information (unfocused DPs) have to move to a special position, outside the VP-domain, and higher than the negation. We call this position Spec,AgrOP. However, this is now a position for specific DPs, rather than a case-checking position. Heavy DPs (i.e. DPs consisting of three words or more) and DPs with new information do not have to check this specificity feature, hence they do not move to Spec,AgrOP but are left within the VP-domain, lower than the negation. Third, the remnant vP, containing at least the finite verb, obligatorily moves to Spec,FP (in the CP-domain, across all complements).[4] Finally, VP/vP-movement cannot take place with the DP inside. Along with Mahajan (2003), Biberauer and Richards (2006), Hróarsdóttir *et al.* (2006), and Wiklund *et al.* (2007), we assume that some material (crucially always the DP and never the verb) has left vP/VP prior to their relevant movements.

Let us start by illustrating the derivation of a simple VO order.

(22) *The initial structure*

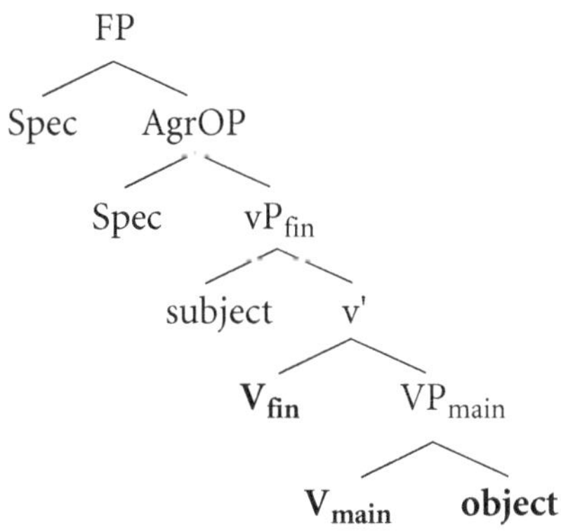

[4] This leaves open the question of how the subject ends up in the initial position. Either it is generated in Spec,vP and moves along when the finite vP is preposed, or it is merged higher in the structure. In the following, we simply assume the former.

(23) VO order: unfocused (light/old information) DPs
First step: DP moves to Spec,AgrOP
Second step: The remnant vP moves to Spec,FP

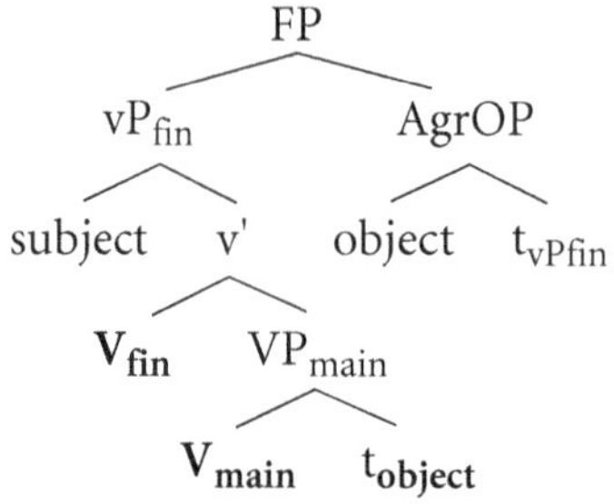

In cases where the DP is unfocused (light/old information), it obligatorily moves to Spec,AgrOP. Then the remnant finite vP obligatorily moves to Spec,FP, masking the DP movement. This derives the surface VO word order [V_{fin}–V_{main}–DP].

(24) VO order: focused (heavy/new information) DPs
First step: The DP is evacuated from the vP
Second step: The remnant vP moves to Spec,FP

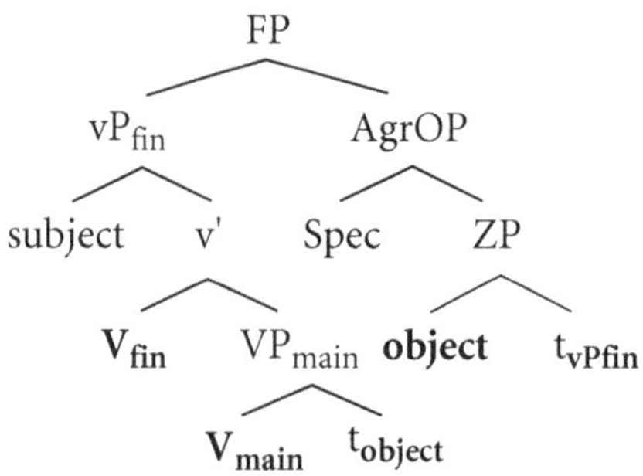

As shown in (24), the DP does not move to Spec,AgrOP when it is focused (heavy/new information). However, as we have mentioned, no vP/VP-movement can take place before the DP has been evacuated. Thereafter, the remnant finite vP is obligatorily fronted, as before, to Spec,FP, deriving again the surface order [V_{fin}–V_{main}–DP].

Let us now turn to the derivation of OV word-order patterns, illustrated here by the simple case with only one non-finite verb: V_{fin}–DP–V_{main}. As in the case of a VO order, the DP in this simple construction could be of different types in OI, indicating two different derivations, as shown in (25) and (26). In OV word order, the crucial fact is that the preposing of the finite vP does not mask the object movement, because of the additional step in OI involving VP-movements. Hence, it now becomes important exactly where the object ends up (in Spec,AgrOP or in the evacuation position just above the vP). This can now be seen overtly by the presence of the negation (NEG), that we assume is situated just below AgrOP.

(25) OV order: [V_{fin}–DP–V_{main}]: unfocused (light/old information) DPs
First step: DP moves to Spec,AgrOP
Second step: VP-extraction (V_{main} moves out of vP)[5]
Third step: The remnant vP moves to Spec,FP

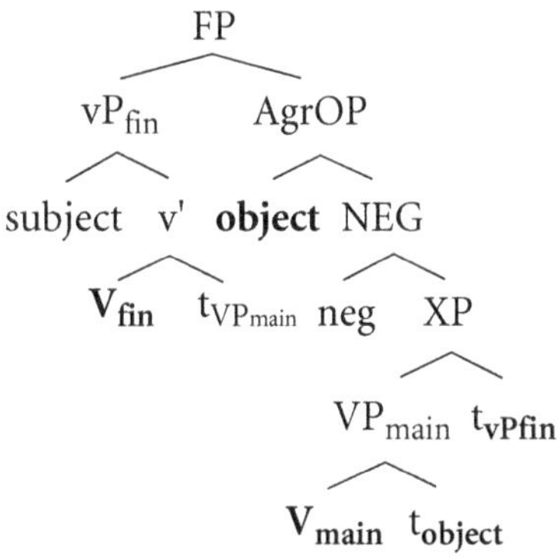

As in the derivation of a VO order, unfocused (light/old information) DPs must move to Spec,AgrOP. Importantly, the derivation of OV orders always includes an additional movement, VP-extraction, meaning an extraction of the non-finite VP from the finite vP. Finally, as before, the remnant vP obligatorily moves to Spec,FP. As shown in (25), this derives the OV order [V_{fin}–DP–Neg–V_{main}]. As we discussed, DPs occurring higher than the negation are indeed usually unfocused (light/old information) in OI. This was the dominant OV order in OI.

The derivation of OV order with focused (heavy/new information) DPs is shown below.

(26) OV order: [V_{fin}–DP–V_{main}]: focused (heavy/new information) DPs
First step: The DP is evacuated from the VP
Second step: VP-extraction (VP_{main} moves out of vP, pied-piping the DP (YP))
Third step: The remnant vP moves to Spec,FP

In cases in which the DP in OI was focused (heavy/new information), it does not move to Spec,AgrOP but is evacuated from the VP prior to the

[5] For simplification, we assume that VP_{main} moves to some XP-position immediately above the finite vP. However, this is presumably more complex, as shown in Hróarsdóttir (2000), where each verb has its own VP-projection and PredP-projection. In order to obtain successive cyclic application of VP-extraction, VP-extraction may then be taken to be PredP-extraction, where PredP is immediately above the VP. Hence, $PredP_{main}$ in fact raises to Spec, $PredP_{fin}$. We do not need to assume this more complex structure here to illustrate our main point.

VP-extraction. We assume that the DP moves to some evacuation position, YP, immediately above VP_{main}. Then, the additional OI step takes place, moving the non-finite VP_{main} from the finite vP. Finally, the remnant vP moves to Spec,FP. These derivations are shown in two steps in (27) (DP-evacuation) and (28) (after VP-extraction and vP-movement).

(27) The DP is evacuated from the VP

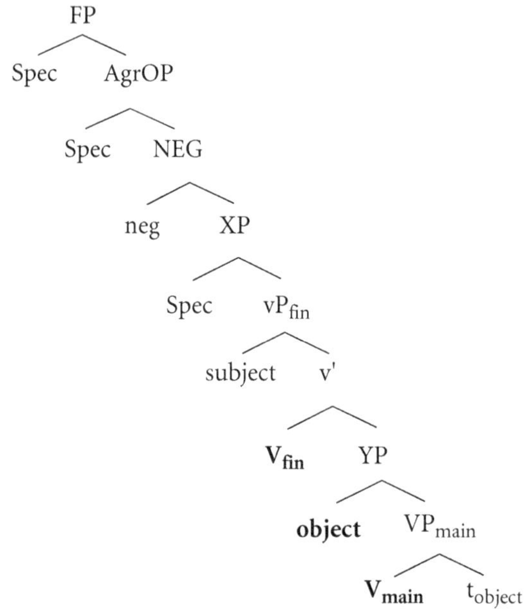

(28) The final result

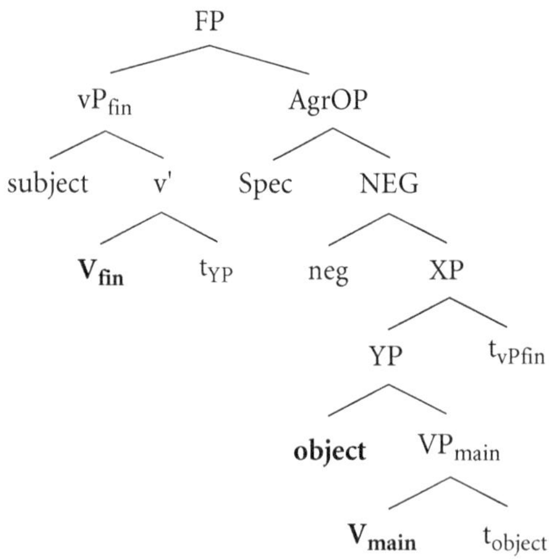

This derives the OV order [V_{fin}–NEG–DP–V_{main}], with the DP lower than the negation, and indeed, this was the position of focused (heavy/new information) DPs in OI.

The other OV options (listed in (3)) can be derived in a similar way with the same three main transformations: DP-movement, VP-extraction, and remnant vP-preposing. The preposing of the remnant finite vP will always mask the object movement, deriving VO word order only, as long as no VP-extraction has taken place. This hypothesis accounts for all the various attested pure and mixed word-order patterns in Icelandic, besides providing an account for the diachronic development, explaining the grammar change in question in terms of a single parameter change. Given this remnant hypothesis, all the various OV word orders would disappear as soon as the possibility of VP-extraction was lost. In other words, the remnant finite vP will always necessarily contain all the non-finite verbs in addition to the finite verb, and the preposing of this remnant vP will always mask the object movement, deriving VO word order only. However, this proposal does not provide an explanation for why the third leftward movement process, the VP-extraction, was lost. The remaining question is what the trigger for VP-extraction was and why it was lost, or became inoperative, in the language.

We argue that the strong feature that must have led to the additional (overt) movement of VP-extraction in OI entails a more complex representation for the child-learner, and hence, that it is marked. These conclusions are in accordance with I. G. Roberts' (2001) claim that simplicity is a function of movement relations, since the old language here has an additional movement over the modern language. Hence, the old grammar is more complex for the child-learner than the new grammar not involving this additional movement. At a certain point in the history of Icelandic, the marked value of this parameter, that is, the strong feature leading to VP-extraction, was no longer expressed (highly enough) in the PLD and the unmarked value became to be triggered by default.

The loss of this additional movement, the option of VP-extractions, might be further related to information structure: that is, VP-extraction and its loss was associated with the difference in behavior stated between unfocused (light/old information) and focused (heavy/new information) DPs. So far, we have noted both an empirical and a theoretical difference with regard to focused versus unfocused objects. First, unfocused (light/old information) DPs more often occurred in a preverbal (OV) position in OI than focused (heavy/new information) DPs. As we have seen in Table 11.1 in Section 11.3, the frequency of preverbal pronouns (which are both light and encode old information) drops spectacularly from the seventeenth century onwards. At the same time, as illustrated in Table 11.2, there is a major drop in the frequency of preverbal DPs in general associated with old information. Second, by hypothesis, unfocused (light/old information) DPs move to a different position than focused (heavy/new information) DPs. While the unfocused DPs must move out of the VP-domain (to Spec,AgrOP in the functional domain), focused DPs are left within the VP-domain where they must interact with VP-extractions in the derivation of OV word order.

11.5 Object Shift

In this section we briefly compare OV orders in OI and Modern Icelandic Object Shift. Note that although Modern Icelandic is generally assumed to have SVO surface word order, as already mentioned, there still exist three constructions exhibiting some sort of SOV word order: negative (and quantified) phrase construction, Object Shift, and Stylistic Fronting. We will leave Stylistic Fronting aside here, and take a look at a correlation between OV word-order patterns in OI and Object Shift in the modern language.[6] Consider example (29):

(29) Object Shift

a. María las **ekki** **bókina**.
Mary read not book.the
'Mary did not read the book.'

b. María las **bókina** **ekki**.
Mary read book.the not
'Mary did not read the book.'

c. *María las **ekki** **hana**.
Mary read not it
'Mary did not read it.'

d. María las **hana** **ekki**.
Mary read it not
'Mary did not read it.'

Object Shift in Icelandic is generally assumed to involve a leftward movement of definite objects and pronouns across a sentence adverb in clauses containing

[6] Stylistic Fronting rarely leads to an OV order, but often to an order where a complement, or part of a complement, precedes the finite verb (see Maling 1990 for the first explicit discussion of Stylistic Fronting in Icelandic, together with the discussions in Jónsson 1991, 1996, Holmberg 1997, 2000, and Hrafnbjargarson 2004).

no non-finite verbs, that is, in sentences where verb movement has occurred (cf. Holmberg's Generalization 1986). Thus, the employment of Object Shift does not lead to OV word order in a strict sense, but to a word order where the object has moved out of the VP, since it shows up to the left of the sentence adverb, marking the left periphery of the VP. An important element in connection to OV orders in OI is the analysis of these possible OV orders still found in Modern Icelandic. This has been the subject of much recent work, and it would indeed be interesting to see what the empirical and analytical differences between the earlier stages and the present-day stage are. In particular, it is interesting to compare obligatory Object Shift of pronouns and optional Object Shift of definite DPs in Modern Icelandic. For instance, how did whatever changes that have taken place give rise to the differences between pronouns and non-pronouns that we observe today? It is very likely that English has undergone a parallel development; Late Middle English probably had optional Object Shift for non-pronouns and obligatory Object Shift for pronouns, whereas sixteenth-century English was like the Modern Mainland Scandinavian languages; pronouns moved whenever the verb moved but no other complements moved (see I. G. Roberts 1995).

Bobaljik and Thráinsson (1998) discuss the connection between Object Shift in Modern Icelandic, the remains of OV orders still existing, and information structure, giving independent support to our claim that the cue for OV order in OI indeed was expressed through information structure. They give the following example:

(30) a. Ég las **þrjár bækur** ekki.
I read three books not

b. Ég las ekki **þrjár bækur**.
I read not three books
'I didn't read three books.'

They claim that the alternation between (a) and (b) here depends on information structure, with a contrast between specific and non-specific, or given versus new information. Hence, in order to be able to shift, the DP object must represent old information, while a DP object representing new information cannot shift and must remain internal to the VP. They conclude that AgrOP has thus nothing to do with case theory, but instead with information structure. This is exactly what we have been arguing for OI in Section 11.2.4 about the theoretical implications: objects behave differently according to their type. Light DPs and DPs with old information (unfocused DPs) have to move to Spec,AgrOP, while heavy DPs and DPs with new information (focused DPs) do not have to move to this position but are left within the VP-domain lower than NEG.

11.6 View to a sociolinguistic change

In this section we argue that the loss of OV in Icelandic might be connected to social/historical changes in Icelandic at the time of the diachronic change. We discuss what might happen if the population changed rapidly, for example by epidemics.

The eighteenth century marked the most tragic age in Iceland's history. In 1703, when the first complete census was taken, the population was approximately 50,000. Between 1707 and 1709 the population sank to about 35,000 because of a devastating smallpox epidemic. Twice again the population declined catastrophically, both during the years 1752–1757 and 1783–1785, owing to a series of famines and natural disasters. The summer of 1782 marked the beginning of the presumably most tragic period ever, the so-called *Móðuharðindi*, where a tremendous natural catastrophe took place with a great volcanic eruption that led to increased cold weather and hard times, including starvation and different diseases. It is usually assumed that a fifth of the population, or 10,000 people, died because of the *Móðuharðindi*. After the population had been severely decimated by these epidemics, the surviving population found themselves in a new situation. First, the age distribution changed, and the average life expectancy decreased. Second, the language would be more sensitive to other sources of noise, for instance, language contact. Historical and social changes in the form of increased intervention, administration, and the power of the Danes led to an increase in foreign influence in Iceland at a time when the stability of the language was less secure due to the fact that quite a lot of the older bearers of the language had perished. Note that we expect the child-learner to be sensitive to changes in the surrounding language environment, since that language environment is what she uses as a model. The language environment providing the PLD had changed so that there was more insecurity about the correct word-order patterns due to an increased influence from Danish and the loss of older individuals, followed by an increase in the proportion of children. It is obvious that we can expect the language to change when the transmission of the tradition is disturbed. When a severe plague hits a population causing the sudden death of older speakers, much of the conservatism of the language (use) may be lost. If we take VO word order to be the more innovative structure, it is not surprising that the frequency of the use of the old OV structure drops when the language environment is affected in this way.

We can assume that the devastating smallpox epidemics between 1707 and 1709 and other natural disasters in Iceland throughout the eighteenth century are a plausible example of the kind of population change that can cause a language change, which may then lead to a parameter change such as the loss of OV word-order patterns. Hence, the case we are studying here is a language that had a variation (OV and VO word order) before the epidemic, and lost it after the epidemical years during the eighteenth century. As can be seen in the tables in Section 11.3, the proportion of OV word order remained surprisingly stable from the fourteenth to the seventeenth century. The first notable drop in frequency takes place during the eighteenth century, during the epidemic. This reflects a language change, where constructions with OV word order came to be used less and less, until the frequency of OV had dropped below a certain threshold to be useful as cues (Lightfoot 1999). This in turn led to a parameter change in the beginning of the nineteenth century where OV orders suddenly disappeared.

To summarize, after the smallpox epidemics in the very beginning of the eighteenth century and the natural catastrophic *Móðuharðindin* in the middle of the century, the age distribution changed due to a loss of older individuals, followed by an increase in the proportion of children. Hence, the language environment had changed and we assume this to have led to more insecurity about the correct word-order patterns. At the same time, there was an increase of foreign influence from Danish, a language that already had pure VO word order at the time.

Finally, another social/historical change in Icelandic at the time of the language change supports our findings. Note that the frequency of OV/VO word-order patterns showed a remarkable stability from the earliest texts until the eighteenth century. More interestingly, so did the Icelandic society: from the time of the settlement until the eighteenth century, it was a rather stable (old-fashioned) country of farmers, without any important formation of cities or towns. It was a society of big families living together. The elderly were not sent away to old people's homes and the children were not sent away to nursery schools or kindergartens. The big family stayed together, and worked together, on the farm. Older children took care of their younger siblings, and, most noticeable perhaps, the children grew up with their parents and their grandparents and their language (including language use and tradition). It was not until the eighteenth century that we have the formation of big towns. Between 1760 and 1770, a great disease hit Icelandic livestock, killing as many as half of the sheep. As a consequence, people had to flee the country side and move to the coast to survive. The fishing industry increased and fishing towns thrived. Reykjavík gradually became the capital of Iceland. This also meant that the big families were split up and children went to school with other children and hence their language acquisition was now, to a much greater extent than before, affected by the language of other children. This is another example of a change in language environment that has had consequences for language use leading to a parameter change. Together these historical and social changes during the eighteenth century in Iceland explain why the language change took place—there was a change in the language environment, affecting the use of the language—paving the way for the grammar change.

There is also some direct evidence that OV and VO word order might have correlated with style; the frequency of OV word order can sometimes be associated with a more elevated or formal style (and formal style is typically more conservative). For instance, not only personal letters were studied dating from the nineteenth century, but also more formal letters written by Icelandic priests at the request of a special committee for archaeology in Copenhagen. When the proportions of OV orders in these two types of letter are compared, it appears that on average, the rate of OV word order is higher in the more formal letters than in the personal letters.

11.7 Conclusions

In this chapter, we have studied the change from object–verb order to verb–object order. A number of European languages have undergone such a change and we have focused here on the change in Icelandic. Morphological triggers, or the absence thereof, have often been related to clustering of syntactic properties in both diachronic change and acquisition. As seen from the English viewpoint, inflectional morphology may express the relevant cue for parameters, and hence the loss of inflection may lead to grammar change. We have shown here that this analysis does not carry over to Icelandic, as the loss of OV in Icelandic took place despite rich case morphology. Instead, we have argued that the relevant cue may be expressed differently among languages: while it may have been expressed through morphology in English, it was expressed through information structure in Icelandic. In both cases, external effects led to fewer expressions of the relevant (universal) cue and a grammar change took place. We have also shown how Object Shift in Modern Icelandic supports the claim that the placement of the object in OI is connected to information structure.

Let us conclude by briefly mentioning two additional arguments in favour of our hypothesis. First, Westergaard (2004a, b) has recently shown that a UG principle of information structure may indeed interact with input in the acquisition of word order. The Tromsø-dialect of Norwegian has both SVO and V2 word order in *wh*-questions, and the child language shows a weak tendency for topicalization. There, the order XVS is preferred when the subject conveys new information, and XSV when the subject conveys given information. Hence, in the early stages of first language acquisition, verb movement is more likely to occur in sentences with a subject that conveys new information. In other words, certain word-order patterns in the Tromsø-dialect that are dependent on information structure (V2 and V3) are acquired very early, and they appear in child language even in structures where there is no evidence for them in the input. Westergaard (2004b) shows that the children seem to be guided by a natural word order with respect to information structure in their very early production of topicalization constructions, which may support the claim that this reflects an underlying word order principle given by UG. Westergaard (2004a) also shows that similar patterns can be seen in Old and Middle English, where V2 and XSV (V3) structures co-exist, and this can be used to explain the loss of V2 in English. As the subject normally conveys given information, the XSV word order will be very frequent in the input (PLD) for the child-learner, and hence this order may be grammaticalized. Westergaard argues that a change in the statistical frequencies in the input to children, derived by patterns of information structure, may be responsible for the development from V2 to XSV order in the history of English. Hence we have a grammar change from a language allowing several word-order patterns based on information structure to a consistent non-V2 (XSV) grammar.

Second, Hróarsdóttir *et al.* (2006), Wiklund *et al.* (2007), and Hróarsdóttir *et al.* (2007) have independently argued that the role of morphology in syntax has been overestimated. They present new data from Regional Northern Norwegian (ReNN) dialects and varieties of Icelandic that show that (a) verb movement in non-V2 contexts in ReNN is possible rather than impossible despite weak morphology (see also Bentzen 2005, 2007), and that (b) verb movement is optional rather than obligatory in the same contexts in varieties of Icelandic (see also Angantýsson 2001). This means that Icelandic has no independent verb movement at all to the inflectional domain, contrary to standard assumptions. These new findings weaken the Rich Agreement Hypothesis which proposes a correlation between the absence and the presence of rich verbal morphology and independent V-to-I movement (cf. Vikner 1995, Rohrbacher 1999, Bobaljik and Thráinsson 1998, Bobaljik 2002).

Appendix A: Primary texts

[Finn]. *Finnboga saga ramma.* Edited by Hugo Gering. Verlag der Buchhandlung des Waisenhauses, Halle, 1879. Heroic epic. Date of composition: 1330–70.

[Guðm]. Saga Guðmundar Arasonar, Hóla-biskups, eptir Arngrím ábóta. *Biskupa sögur.* Second volume, pp. 1–220. Hið íslenzka bókmenntafélag, Copenhagen, 1878. Story of bishops. Date of composition: 1350–65.

[Árn]. *Árna saga biskups.* Edited by Þorleifur Hauksson. Stofnun Árna Magnússonar in Iceland, Reykjavík, 1972. Story of bishops. Date of composition: 1375–1400.

[Dín]. *Dínus saga drambláta.* Edited by Jónas Kristjánsson. Riddarasögur I. Háskóli Íslands, Reykjavík, 1960. Chivalric romance. Date of composition: 1375–1400.

[Sig]. *Sigurðar saga þögla.* Edited by M. J. Driscoll. Stofnun Árna Magnússonar in Iceland, Reykjavík, 1992. Icelandic romance. Date of composition: early fifteenth century.

[Vikt]. *Viktors saga og Blávus.* Edited by Jónas Kristjánsson. Riddarasögur II. Handritastofnun Íslands, Reykjavík, 1964. Chivalric romance. Date of composition: *c.*1470.

[Afs]. *Morðbréfabæklingar Guðbrands biskups Þorlákssonar, 1592, 1595 og 1608, með fylgiskjölum.* Sögufélagið, Reykjavìk, 1902–6. Afsökunarbréf Jóns Sigmundssonar. Document/formal letter. Date of composition: 1502–6. Transcript made by Bishop Guðbrandur Þorláksson, 1592.

[Morð]. *Morðbréfabæklingar Guðbrands biskups Þorlákssonar, 1592, 1595 og 1608, með fylgiskjölum.* Sögufélagið, Reykjavík, 1902–6. Morðbréfa-bæklingar Guðbrands biskups. Document. Date of composition: 1592.

[Skál]. Sögu-þáttur um Skálholts biskupa fyrir og um siðaskiptin. *Biskupa sögur.* Second volume, pp. 235–65. Hið íslenzka bókmenntafélag, Copenhagen, 1878. Story of bishops. Date of composition: late seventeenth century.

[Árm]. *Ármanns rímur eftir Jón Guðmundsson lærða (1637) og Ármanns þáttur eftir Jón Þorláksson,* pp. 91–121. Edited by Jón Helgason. Íslenzk rit síðari alda, first volume. Hið íslenzka bókmenntafélag, Copenhagen, 1948. A short narrative story. Date of composition: late seventeenth century.

[Munn]. *Munnmælasögur 17. aldar.* Edited by Bjarni Einarsson. Íslenzk rit síðari alda, volume 6. Hið íslenzka fræðafélag í Kaupmannahöfn, Reykjavík, 1955. Folk tales, in oral tradition. Date of composition: 1686–7.

[J.Ey.]. *Ferðasaga úr Borgarfirði vestur að Ísafjarðardjúpi sumarið 1709, ásamt lýsingu á Vatnsfjarðarstað og kirkju.* Eptir Jón Eyjólfsson í Ási í Melasveit. Blanda II. Fróðleikur gamall og nýr, pp. 225–39. Sögufélagið, Reykjavík, 1921–3. Travelogue; a story from a journey. Date of composition: 1709.

[Bisk]. *Biskupasögur Jóns prófasts Haldórssonar í Hítardal.* Með viðbæti. Skálholtsbiskupar 1540–1801. Sögufélagið, Reykjavík, 1903–10. Story of bishops. Date of composition: 1720–30.

[Próf]. *Æfisaga Jóns prófasts Steingrímssonar eptir sjálfan hann.* Sögufélagið, Reykjavík, 1913–6. Biography. Date of composition: 1785–91.

[Álf]. *Íslenzkar þjóðsögur og ævintýri.* Nýtt safn. Volume VI, pp. 1–39. Collected by Jón Árnason. Edited by Árni Böðvarsson and Bjarni Vilhjálmsson. Bókaútgáfan Þjóðsaga, Reykjavík, 1961. Álfarit Ólafs í Purkey. Folk tale, fairy tale. Date of composition: 1820–30.

[Esp]. *Íslands Árbækur í söguformi.* Af Jóni Espólín fyrrum Sýslumanni í Skagafjarðar Sýslu. Hið íslenzka bókmenntafélag, Copenhagen, 1843. Jón Espólín. Annual stories, in epical form. Date of composition: first half of the nineteenth century.

Appendix B: Bibliographical information for the nineteenth century letters

Biskupinn í Görðum. Sendibréf 1810–53. Edited by Finnur Sigmundsson. Íslenzk sendibéf II. Bókfellsútgáfan, Reykjavík, 1959.

Bjarni Thorarensen, Bréf. First volume. Edited by Jón Helgason. Safn Fræðafélagsins um Ísland og Íslendinga XIII. Hið íslenzka Fræðafélag í Kaupmannahöfn, Copenhagen, 1943.

Doktor Valtýr segir frá. Úr bréfum Valtýs Guðmundssonar til móður sinnar og stjúpa 1878–1927. Edited by Finnur Sigmundsson. Íslenzk sendibréf V. Bókfellsútgáfan, Reykjavík, 1964.

Frásögur um fornaldarleifar 1817–1823. First volume. Edited by Sveinbjörn Rafnsson. Stofnun Árna Magnússonar, Reykjavík, 1983.

Frásögur um fornaldarleifar 1817–1823. Second volume. Edited by Sveinbjörn Rafnsson. Stofnun Árna Magnússonar, Reykjavík, 1983.

Geir biskup góði í Vínarbréfum 1790–1823. Edited by Finnur Sigmundsson. Íslenzk sendibréf VII. Bókfellsútgáfan, Reykjavík, 1966.

Gömul Reykjavíkurbréf 1835–1899. Edited by Finnur Sigmundsson. Íslenzk sendibréf VI. Bókfellsútgáfan, Reykjavík, 1965.

Hafnarstúdentar skrifa heim. Sendibréf 1825–1836 og 1878–1891. Edited by Finnur Sigmundsson. Íslenzk sendibréf IV. Bókfellsútgáfan, Reykjavík, 1963.

Konur skrifa bréf. Sendibréf 1797–1907. Edited by Finnur Sigmundsson. Íslenzk sendibréf III. Bókfellsútgáfan, Reykjavík, 1961.

Magnús Stephensen, Brjef. Edited by Hið íslenzka Fræðafélag í Kaupmannahöfn. Safn Fræðafélagsins um Ísland og Íslendinga IV. Copenhagen, 1924.

Sendibréf frá íslenzkum konum 1784–1900. Edited by Finnur Sigmundsson. Helgafell, Reykjavík, 1952.

Skrifarinn á Stapa. Sendibréf 1806 1877. Edited by Finnur Sigmundsson. Íslenzk sendibréf I. Bókfellsútgáfan, Reykjavík, 1957.

Þeir segja margt í sendibréfum. Edited by Finnur Sigmundsson. Bókaútgáfan Þjóðsaga, Reykjavík, 1970.

12

Information structure and unmarked word order in (Older) Germanic

ROLAND HINTERHÖLZL

12.1 Introduction

Basic OV and basic VO order have been accounted for by different settings of the head-complement parameter, raising the question of how languages with mixed OV/VO order should be treated. Apart from (modern) Yiddish, we find mixed word orders in older stages of the Germanic languages. While mixed word orders in Old English and Old Nordic have been subject to thorough investigation and heated debates concerning their correct analysis (cf. I. G. Roberts 1997, Pintzuk 1999), the discussion on older stages of German has been mute owing—as I will show—to the incorrect assumption that German already was an OV-language in its oldest accessible stage.

In this chapter, I will correct this picture by showing that Old High German should not be analyzed as a language with basic OV order plus a higher degree of extraposition. The core of the chapter is thus a careful empirical investigation of word-order variation in the OHG-Tatian translation, which reveals that word order in OHG was determined primarily by information-structural restrictions. The most important empirical results are that arguments and predicates, which are not subject to extraposition in modern German, occupy a postverbal position in embedded clauses when they are part of the focus domain of the clause.

The results are then embedded into a phase-based model of the interaction between syntactic, prosodic, and information-structural conditions that define the characteristic word-order properties of OV- and VO-languages, without assuming differences in the base structure or a directionality parameter in argument licensing (cf. Haider and Rosengren 2003). This approach crucially depends on dissociating unmarked word order and basic word order and having the unmarked word order in a language be determined by two types of mapping conditions at the interfaces with PF and LF.

12.1.1 *Word order change in Germanic*

One of the most intriguing developments in the history of the Germanic languages, next to the grammaticalization of Verb Second (V2), is the change in basic word order in English and Scandinavian. This development involved a change from the presumed Indo-European basic OV order to the basic VO order in these languages. In this scenario, German (and Dutch) retained the inherited base order (modulo some changes in the application of extraposition).

The traditional explanation of this phenomenon is to assume that the loss of case led to the positional marking of grammatical functions. However, this account faces serious difficulties if we consider the development of Dutch and Icelandic, since Dutch has also lost its case distinctions but retained OV order, while Icelandic has preserved its rich case morphology, but nevertheless changed to basic VO order (cf. Hróarsdóttir 2000, this volume, for additional discussion of this issue).

Recently an alternative approach for the change in word order in English was proposed that assumes that the change from OV to VO is due to language contact and grammar competition in Early Middle English (EME) (cf. Pintzuk 1999, Kroch and Taylor 2000). This approach is based on the so-called double base hypothesis (Pintzuk 1999), according to which word-order variation follows from the co-existence of competing grammars that differ with respect to the head parameter of VP and IP. Within this approach, it is often assumed that VO orders are an EME-innovation that was brought about by language contact between Anglo-Saxons and the Scandinavian settlers in the tenth century (cf. Kroch and Taylor 1997, Trips 2002).

12.1.2 *Word order variation in Germanic*

If we look at word order regularities in the older stages of the Germanic languages, we find that both OV- and VO-properties already existed in Old English (OE), rendering the contact scenario sketched above less plausible. Furthermore, if we look at Old High German (OHG) and Old Nordic (ON) (cf. Hróarsdóttir 2000, this volume) we find a similar kind of variation in word order, suggesting that these mixed word order properties should not be treated in terms of language contact, but may simply be part of the common Germanic inheritance. In the following, I will restrict myself to a discussion of word order variation in OE and OHG.

The examples in (1) illustrate typical OV-properties in OE. For example, in (1a) the direct object and the verb particle precede the finite verb in final position within an embedded clause and in (1b) the non-finite verb precedes the finite auxiliary in sentence-final position as is typical in OV-languages. The same state of affairs, maybe less surprisingly, also holds in OHG, as is illustrated in (2).

(1) a. þæt he his stefne up ahof (Pintzuk 1999: 71)
that he his voice up raised

b. forþon of Breotone nædran on (Pintzuk 1999: 117)
because from Britain adders on
scippe lædde wæron
ships brought were

(2) a. soso zi In gisprochan uuas (T 37, 5)
how to them spoken was

b. thaz then alton giqu&an uúas (T 64, 13a)
what to-the old said was

However, we also find ample evidence of properties in OE that one would associate with VO-languages, as is illustrated in (3). For instance in (3a) and in (3b), the direct object and a small-clause predicate follow the selecting verb in an embedded clause. In (3c), the verb particle which is stranded by movement of the finite verb, arguably an instance of V2, does not appear in a clause-final position, as is common in OV-languages, but precedes a (heavy) predicate in that position and in (3d), the manner adverb follows the verb it modifies. These properties are typical of VO-languages.

(3) a. þæt ænig mon atellan mæge ealne (Pintzuk 1999: 36)
that any man relate can all
one demm
the misery

b. forðam ðe he licettað (van Kemenade 1987: 35)
because that they pretended
hie unscyldige
themselves innocent

c. he ahof þæt cild up (van Kemenade 1987: 36)
he raised the child up
geedcucod and ansund
quickened and healthy

d. þæt he þæt unaliefede dod (van Kemenade 1987: 36)
that he the unlawful did
aliefedlice
lawfully

Similar examples can be found in the OHG Tatian text. For example in (4a), the subject follows the selecting verb in an embedded clause and in (4b), the participle follows the selecting auxiliary, an order that is ungrammatical in modern standard German (though there are dialects that allow for right-branching verb clusters), but typical for a VO-language.

(4) a. thaz gibrieuit uuvrdi al these umbiuuerft (T 35, 9)
that listed was all this mankind

b. thaz sie uuvrdin gitoufit (T 46, 25)
that they were baptized

While it has been argued for English that the variation illustrated in (3) should not be accounted for in terms of assuming a basic OV-grammar plus extraposition of heavy material and really calls for the assumption of an additional VO-grammar (cf. Pintzuk 1999), the type of evidence illustrated in (4) has been treated quite differently so far.

The standard treatment of OHG is that of an OV-language that allowed for a greater amount of extraposition than modern German (cf. Lenerz 1984). This view has been reinforced by the first large-scale empirical investigation of word order in OHG by Dittmer and Dittmer (1998). They investigated the syntax of the Tatian translation, the largest document of OHG dated before 850, by considering only data that differ in order from the Latin origin, called deviations here. The Tatian text is often viewed as a slavish translation of the Latin origin and VO properties, as the ones illustrated in (4), have been relegated to Latin influence. However, it has long been noted that the quite frequent deviations from the Latin word order can provide valuable evidence for genuine OHG syntactic structure. Dittmer and Dittmer have shown that these deviations display features familiar from New High German (NHG) and thereby strengthened the picture that OHG was basically a pure OV-language like NHG.

In this chapter, I will show that deviations from Latin word order going against NHG can also be found in significant numbers. In particular, I will provide convincing evidence that OHG had a right-peripheral focus position, explaining why and when constituents follow the selecting verb in embedded clauses in OHG. I will then show that the deviations in their totality allow the reconstruction of a genuine OHG system in which word order is largely determined by information structural parameters.

12.2 Word order variation and information structure

In this section, I will first argue against accounting for word order variation in terms of multiple bases and propose that variation in word order is better explained as resulting from the complex interaction between prosodic and information-structural requirements. Furthermore, I will argue that a change from basic OV order to basic VO order cannot be equated with a change in the head–complement parameter, but involves differences in a number of properties that are better accounted for by conditions working at the interface to PF and LF. As an alternative, I propose to dissociate base order and unmarked word order and show that information structure plays a crucial role in determining what counts as unmarked or marked word order in a language.

12.2.1 *Mixed word orders and the double-base hypothesis*

An important aspect of the data in (3) and (4) is that we find not only sentences with pure OV-properties and sentences with pure VO-properties in these languages, but very often we find sentences with mixed word orders. For instance, in (3a) it can be seen that while the direct object follows the selecting verb (and the modal)—a typical VO-property—the infinitive precedes the selecting modal, which is rather typical of an OV-language. In (3d) the direct object precedes the selecting verb—a typical property of an OV-language—while the manner adverb follows the verb that it modifies, which is typical of VO-languages and ungrammatical in OV-languages. Likewise in (4), while the subject follows the selecting verb (and the auxiliary), the participle precedes the auxiliary, as is typical in OV-languages.

Given this state of affairs, the question arises of how to account for mixed word orders in the older stages in these languages. My approach is to assume that OV/VO orders do not signal the presence of two grammars, contrary to the double-base hypothesis (cf. Pintzuk (1999), but that the variation illustrated in (1)–(4) is due to the expression of different information-structural (IS) categories within one grammar. The advantage of this approach is clear when it comes to the characterization of mixed word orders. Proponents of the double-base hypothesis not only have to assume that a speaker possesses two grammars or two settings of the head complement parameter, but also that he can switch between the settings of the head complement parameter within one sentence.

To give a concrete example, in order to derive the sequence *relate can all the misery* in (3a), proponents of the double-base hypothesis have to assume that the speaker can switch from an OV grammar—in order to derive the verbal complex Infinitive > finite Modal—to a VO grammar, to account for the object in postverbal position. Note specifically that the assumption of an OV grammar plus extraposition only leads to further difficulties in the double-base approach. First, there is the issue that OV languages like German and Dutch do not allow for extraposition of DP-arguments, raising questions about which properties in a grammar allow and constrain extraposition. Second, the question arises of which factors determine when a speaker uses extraposition and when he simply switches to a VO-grammar in such a scenario.[1] Summing up, it would be advantageous if mixed word orders can be accounted for as variation within one grammar.

12.2.2 *Differences between OV- and VO-languages*

I have argued in Hinterhölzl (2004) that differences in basic word order should not be explained in terms of the head complement parameter, since OV- and VO-languages differ in properties that cannot be subsumed under this parameter. First note that adjuncts that can occur between the subject and the vP in VO-languages are subject to restrictions that are absent in OV-languages, as is illustrated in (5). This has first been observed by Haider (2000) and is further discussed in Hinterhölzl (2001). The data in (5) show that the English middle field does not tolerate heavy constituents, while no problem arises in the German middle field.

(5) a. John (more) often (* than Peter) read the book

b. Hans hat öfter (als der Peter) das Buch gelesen
Hans has more often (than the Peter) the book read

Second, event-related adverbs, that is time, place, and manner adverbs, appear in the order T>P>M preverbally in OV-languages, but appear in the exact mirror order postverbally in VO languages (cf. Haider 2000, Hinterhölzl 2002), as is schematized in (6).

(6) a. C T P M–V OV-languages

b. C V–M P T VO-languages

[1] An interesting alternative proposal within the Kaynean approach is constituted by Biberauer and Roberts (2005). They account for mixed OV/VO orders by assuming a variable degree of pied-piping of a movement operation that involves checking of formal features on the verb. It remains to be seen though how this approach can cope with the adjunct data discussed in Section 12.2.2.

This raises the question of why the placement of exactly these adverbs and not of other, say, higher adverbs seems to be correlated with the head–complement parameter. In Hinterhölzl (2002, 2004), I have argued that the order found in German is basic and that the English order is to be derived from the German word order in terms of successive cyclic VP-intraposition that pied-pipes the adjuncts at each step.[2] Furthermore, I have argued that VP-intraposition came about due to a stylistic rule of light predicate raising that was operative in OE and affected typically event-related adjuncts since they were primarily realized as rather heavy NPs and PPs.

Since the head–complement order seems to be connected with properties in the grammar that are quite distinct, I argue that the head–complement parameter should be dispensed with and propose that word order properties should better be defined by globally operating interface constraints.

12.2.3 *Base order and unmarked word order*

When we dispense with the head–complement parameter and adopt the Universal Base Hypothesis (UBH) (see Kayne 1994), then it is necessary to distinguish between the base order, which is universally defined as Specifier–Head–Complement, and the unmarked word order of a language. For instance, the unmarked word order in German (OV) cannot be taken to be a basic property (to be identified with the base order) anymore, but has to be derived from other properties in the language.

Nespor, Guasti, and Christophe (1996) propose that the head–complement parameter is determined by the predominant, that is, unmarked, prosodic patterns in an early phase during language acquisition (the rhythmic activation principle). More specifically, they argue that the decisive information for the child is the placement of main prominence within the phonological phrase.[3] In Hinterhölzl (2004), I have adopted this approach and proposed that the unmarked word orders in the phrases of a language are determined by the predominant prosodic patterns in that language.

[2] In this approach a modern English sentence like (ia) is derived as illustrated in (ib–e).

(i) a. John visited them in Vienna on Friday.

b. [... [on Friday [in Vienna [John visited them]]]].

c. [... [on Friday [[John visited them] in Vienna t_{VP}]]].

d. [... [[John visited them] in Vienna] on Friday].

e. [$_{IP}$ John$_i$ [[[[$_{VP}$ t_i visited them]$_k$ in Vienna t_k]$_j$ on Friday t_j]]].

[3] To get prosodic bootstrapping off the ground, we have to assume that the child has access to the following universal principle that if a head and a non-head are combined in a single phrase it is the head (or the functional element) that is weak. So if the child encounters the pattern (weak strong), it knows that the head precedes the complement or that a functional element precedes a lexical category.

That a language can have several unmarked prosodic patterns is shown by German. While with DPs and PPs the unmarked prosodic pattern is (weak strong) ((w s)), the unmarked prosodic pattern with VPs is (s w). It is interesting to note that APs show both types of prosodic patterns with a strong preference for the verbal pattern (s w), as is illustrated in (7). Sentence (7a) shows the neutral order, while (7b) is rather marked and can only be used for specific communicative purposes.

(7)	a.	weil	Hans	[[auf die Maria] stolz]	ist	(unmarked)
	b.	weil	Hans	[stolz [auf die Maria]]	ist	(marked)
		since	Hans	(of the Maria) proud (of the Maria)	is	

In conclusion, I would like to propose that syntactic structures are not marked *per se* (say, in terms of complexity), but count as marked or unmarked if they realize marked or unmarked prosodic patterns. Since the unmarked word order in a language is defined by the predominant, that is to say, the most frequent, prosodic pattern in a language, the unmarked word order may differ from the universal base order. The frequency component in the definition of the unmarked word order is important for explaining word order change. If a marked prosodic pattern is more widely used and, by crossing a certain threshold, becomes unmarked, it may give rise to an additional or new unmarked word order. This is the approach to word order change that I will pursue in the following sections.

12.2.4 *Information structure and word order*

Since the expression of IS-categories influences the default mapping between syntactic structure and prosodic structure, as I will show below, IS plays an important role in defining marked and unmarked word orders.

Before we discuss the role of information structure in determining word order in OHG, let me point out which factors were thought in the traditional literature to determine word order in these languages. A much quoted point of view is constituted by the observation by Behaghel (1932) that pronouns and unmodified nouns tend to precede the verb, while modified nouns, PPs, and other heavy material tend to follow the verb that gave rise to the generalization in (8).

(8) Light elements precede heavy elements in OE, OI, and OHG.
(Behaghel 1932: *Das Gesetz der wachsenden Glieder*)

The statement in (8) raises a number of questions. First, the question arises which principle of grammar this tendency derives from. Second, note that (8) is a statement that forbids heavy elements in the middle field of OE (as well as in Old Icelandic (OI) and OHG), raising the question of whether there is a connection between (8) and the modern English facts, illustrated in (5) and (6) above. In Section 12.5, I will argue that (5), (6), and (8) can be derived from a prosodic condition on the mapping between syntactic structure and PF.

There is another generalization that emerges from an IS-analysis of the Tatian translation (to be revised below) that derives (8) as a mere corollary. Given that discourse-given elements are typically realized as light elements, while focused constituents may count as prosodically heavy elements since they receive stress, (9) derives the tendency expressed in (8).

(9) C background V (presentational) focus

The following notions are relevant for an IS-analysis of marked and unmarked word orders.

(a) The discourse status of an expression denoting a discourse referent: discourse-given elements are assumed to be part of the background, that is, part of the knowledge that is assumed to be shared by speaker and hearer, while discourse-new elements are assumed to be part of the presentational focus of the clause.
(b) Discourse-given elements can also be made prominent and be part of the focus domain of a clause. A typical case in question is a contrastively stressed pronoun.
(c) Taking into account the prosodic properties of focused elements, we further need to distinguish between wide and narrow focus. The three different notions of focus relevant here are illustrated in (10): broad and narrow presentational focus and contrastive focus. In (10), brackets mark the focused constituents and capitals mark a high pitch accent, which is typical for contrastively focused elements.

(10) a. What did John do? (broad presentational focus)
John [gave a book to Mary].

b. What did John give to Mary? (narrow presentational focus)
John gave [a book] to Mary.

c. John gave Mary [a BOOK], not a pen. (contrastive focus)

The distinction that prosody makes between narrow and broad focus is relevant for determining the unmarked prosodic pattern in the clause. Constituents that are broadly focused do not restructure with the verb. They are mapped into prosodic constituents according to the syntactic structure alone, while narrowly focused constituents integrate into the phonological phrase of the verb (cf. Nespor and Vogel 1986, Frascarelli 2000). This means that preverbal narrow focus gives rise to the prosodic pattern (s w), while postverbal narrow focus may strengthen the prosodic pattern (w s).

The direction of restructuring of the verb (induced by focus restructuring) is also important for licensing co-predicates of the verb in small clause constructions, verb particle constructions and idioms, determining whether a language counts as OV- or as VO-language on the surface.

The generalization in (9) allows us to account for word order variation within one grammar by taking into account the information-structural contribution of a constituent in the discourse: for instance, a direct object will precede the verb (in embedded clauses) if it is discourse-given, but will follow it if it is discourse-new.

According to the generalization in (8), a direct object will be placed preverbally if it is realized as a pronoun or single noun, but postverbally if it is made heavy by modification. As I have indicated above, the two conditions are not independent of each other, but it would be interesting to see which one of them is more basic (see Hróarsdóttir, this volume, for similar observations in ON).

12.3 Word order regularities in the OHG Tatian translation

In this section, I will report on the results of an empirical investigation of word order variation in the Tatian translation.[4] The results are augmented by a small-scale information-structural analysis of the deviations noted in the study by Dittmer and Dittmer (1998). I will concentrate on word order deviations in the right periphery of embedded clauses. Dittmer and Dittmer observe that the Tatian translation contains a lot of deviations from the original Latin word order. They note (a) that in main clauses Latin word order is regularly converted into V2-structures and (b) that in embedded clauses a middle field is typically formed by a general process of preposing of light elements that affects (single) nouns and pronouns, but also finite main verbs and auxiliaries.

These observations by Dittmer and Dittmer have generally been taken to show that the deviations always go towards the NHG grammar and that deviations from NHG should be relegated to adherence to the Latin origin. Dittmer and Dittmer do not address the question of whether the regularity with which light elements are preposed and heavy constituents remain in their postverbal position could not be due to an original OHG system.

[4] This empirical investigation was carried out by project B4 "The role of information structure in the development of word order regularities in Germanic" of the SFB 632 "Information Structure: the linguistic means for structuring sentences, utterances and texts" at the University of Potsdam and HU-Berlin (see www.sfb632.uni-potsdam.de/projects_b4eng.html).

The question becomes especially relevant as a close inspection of the data in the Tatian translation reveals that there are numerous deviations from the Latin original that are not compatible with the grammar of NHG, since they yield word orders which are highly marked or ungrammatical in NHG. We will have a careful look at these cases in the following subsection.

12.3.1 *Systematic deviations differing from NHG*

If the deviations from the Latin original went only towards the NHG-grammar, as is suggested by Dittmer and Dittmer, then one would not expect that verbs in final position in embedded clauses in the Latin original text are replaced by orders which place constituents after the verb. However, such cases are not rare at all and it can be shown that these deviations are systematic and reveal a genuine OHG system.

Considering adverbial clauses and relative clauses, it is noteworthy that the Tatian translation quite frequently deviates from the verb-final pattern grammaticalized in NHG. Here we are interested in cases in which the Tatian translation deviates from the Latin original but, instead of the verb-final pattern to be expected, displays an order in which the finite verb, in many cases an auxiliary, precedes a constituent that is more important from an informational point of view, as is illustrated in (11).

(11) [Lucerna corporis .est oculus./] si **fuerit** oculus tuus simplex./ totum corpus tuum lucidum erit.
[Liohtfaz thes lihhamen ist ouga/] oba thin ouga **uuirdit** lutter/ thanne ist al thin lihhamo liohter (T 69, 22ff)
[The light of the body is the eye.] when your eye becomes light then all your body is light

A similar strategy is evident in embedded clauses that correspond to a nominal expression or a participial construction in the Latin original. Since verb placement in these cases cannot be influenced by the Latin text, it is reasonable to assume that these (freely built) structures display genuine OHG word order. A number of these cases indicate that this genuine word order cannot simply be equated with the verb-final pattern, as is illustrated in (12) and (13). Also these cases can be subsumed under the rule that we noted to be relevant for (11): the light verb or auxiliary precedes the predicate noun, which is associated with the new information.

(12) [& ecce homo erat In hierusalem.]'/cui nomen simeon
senonu tho uuas man In hierusalem.'/
[behold there was a man in Jerusalem]
thes namo **uuas** gihezzan Simeon (T 37, 23ff.)
of name was named Simeon

(13) Beati misericordes
salige sint thiethar sint miltherze (T 60, 12)
blessed are those that are mercyful

The most convincing cases for displaying a genuine OHG word order are those in which the verb-final pattern in the Latin original is replaced with a non-verb-final pattern in the translation in embedded clauses. Note that this phenomenon can be found in all types of embedded clauses, in complement clauses as well as in adverbial clauses, although it appears most frequently in relative clauses. Examples of this type of word-order deviation are given in (14) for adverbial clauses, in (15) for relative clauses, and in (16) for complement clauses.

(14) ut in me pacem **habeatis**
thaz in mir **habet** sibba (T 290, 8)
so-that in me you-have peace

(15) [&obtulerunt ei/ omnes male habentes/ [...]]/ &qui demonia **habebant**
Inti bráhtun imo/ alle ubil habante/ [...]]/
[and brought to-him all evil bearers]
Inti thie thár **hab&un** diuual (T 59, 1)
and those that had devil

(16) non resistere malo
thaz ír **niuuidarstant&** ubile
that you not-withstand (the) evil

To sum up, we have seen that in all types of dependent clauses, including adverbial clauses, relative clauses, and complement clauses, we find a number of cases which display a word order in which the finite verb or auxiliary precedes a constituent, often a nominal, verbal or adjectival predicate which constitutes new information. As is evident in (11), (13), (14), (15), and (16), these constituents, while being "heavy" from an informational point of view, need not be heavy constituents from a prosodic point of view.

12.3.2 *Interpreting the pertinent variation in word order*

I would like to take a closer look at deviations from both the Latin original and from NHG in order to sketch out more precisely the regularities behind these deviations and to show that they are part of a genuine OHG grammatical system in which word order is predominantly determined by information-structural requirements. To this end, I will first argue that Behaghel's law is systematically outruled by information-structural requirements. I will then show that it can be derived from a (violable) interface condition that applies in the mapping between syntactic structure and prosodic structure.

First recall that (8) and (9) make slightly different predictions. In particular, (8) predicts (a) that light elements—that is, non-branching constituents—precede the verb independently of their information-structural role in the clause and (b) that heavy—that is, branching—constituents follow the verb irrespective of their IS-role in the clause. Both predictions turn out to be wrong.

In what follows, I will analyze the information-structural role of constituents that have been either preposed (into the middle field) or postposed into the postverbal domain in embedded clauses. Needless to say that preposed heavy constituents and postposed light constituents are of particular interest here. Since Dittmer and Dittmer (1998) have ordered and listed all cases of deviations, their study can be taken as a reliable basis for this small-scale empirical investigation.

According to Dittmer and Dittmer there are 142 cases in which a constituent is moved from the postverbal into the preverbal field in an embedded clause. The majority of these cases involve a pronominal subject or object that can be equally analyzed as prosodically light and as discourse anaphoric. Among the 30 cases with a nominal subject, there are 23 cases that involve a branching constituent that must be analyzed as heavy (the rest are biblical names such as *David*, *Herodes*, and *Johannes*.). These heavy subjects, however, are to be analyzed as belonging to the background and have been introduced in the previous discourse.

If, in turn, we investigate the deviations in which a constituent is postposed from the middle field, then it must be noted that among the few ten cases to be found (in very few cases a middle field can be identified in Latin), there are seven cases that involve a non-branching constituent that cannot possibly be analyzed as prosodically heavy. However, if we consider the information-structural role of these constituents, we note that they must be analyzed as being focused, as is illustrated in (17).

(17) Haec locutus sum vobis thisu sprahih iu
ut in me **pacem** habeatis thaz in mir habet **sibba** (T 290, 10)
in mundum **presuram** habebitis in therru weralti habet ir **thrucnessi**
'this I tell you that in me you-have peace in the world you-have unrest'

In example (17), which displays the context of example (14), the non-branching direct object is postposed, while the branching PP remains in the middle field. The information-structural analysis, however, reveals that this configuration is to be expected since we are dealing with contrastive statements in which the PPs represent contrastive topics and the direct objects provide the corresponding focused constituent.[5]

The example in (18) shows that word order in the Tatian translation cannot be explained with the simple rule that pronominals are preposed, while DPs remain in their base position in accordance with the Latin original. In (18), a weak pronoun has been placed in the postverbal domain against the Latin original. However, the context makes clear that the postposed pronoun is focused. The text passage requires that the pronoun be stressed. While modern German would use a focus particle (to strengthen the reflexive pronoun that cannot be stressed)—*wer sich selbst erniedrigt, der wird erhöht werden*—in OHG this focus is expressed syntactically by moving the pronoun into a right-peripheral focus position.

(18)	Quia omnis	bidiu uuanta iogiuuelih	(T 195, 16)
	qui se axaltat	thiedar sih arheuit	
	humiliabitur	uuirdit giotmotigot	
	qui **se** humiliat exaltabitur	inti therdar giotmotigot **sih** wirdit arhaban	
		and the one that humiliates himself will-be lifted	

'Therefore everyone that exalts himself will be humiliated but the one that humiliates himself will be uplifted.'

In conclusion, we have seen that Behaghel's law is fulfilled if the syntax including information structure does not impose more specific restrictions on its own. In this case, light elements are moved into the preverbal domain, while heavy elements move to or remain in the postverbal domain. This conclusion is further strengthened by the following observation. A careful analysis of the data in the Tatian translation also reveals that the generalization in (9) needs to be refined in as much as contrastive foci are concerned. While new information focus is typically realized in postverbal position, as we have seen above, contrastive focus is realized in preverbal position, as is illustrated in (19).

[5] An anonymous reviewer points out that the focused elements in (17) are contrastive and should be placed preverbally according to the generalization in (20). However, the contrast in (17) involves an opposition between the topical elements *in me* and *in the world*, while the postverbal constituents simply provide new information.

(19) tu autem cum ieiunas/ unge caput tuum/ & faciem tuam laua/ ne uideatis **hominibus**/ ieiunans. Sed patri tuo
thane thu fastes/ salbo thin houbit/ Inti thin annuzi thuah/ zithiu

thaz	thu	**mannon**	nisís	gisehan/	fastenti.
úzouh	**thimeno**	**fater**			
that	you	to-men	not-be	seen	fasting

'[when you fast, do not be like the hypocrites ...] when you fast, anoint your head and wash your face so that you do not appear to men to be fasting but to your Father.'

Note that in (19) the contrastive element *mannon* is placed preverbally against the order in the Latin original, signifying that we are dealing with an independent requirement of OHG. In this context it is interesting to note that PPs that are placed predominantly postverbally in accordance with Behaghel's law (due to their heaviness) appear preverbally when contrastively focused.

Therefore, the generalization in (9) that explains (in tandem with Behaghel's law) the variation in word order found in the Tatian translation needs to be revised as given in (20).

(20) C background contrastive focus V information focus

In this respect, OHG seems to pattern with Yiddish, which is a West Germanic language that has preserved mixed word orders. According to the description by Diesing (1997), constrastive foci pattern with background elements in occupying a preverbal position. Now, the question arises of which assumptions the pattern in (20) can be derived from. The pattern certainly calls for a syntactic representation of focus in OHG that distinguishes between contrastive focus and information focus. Given the UBH, arguments must be taken to move out of the VP to be licensed in (Case-) Agreement positions. If we then assume that a structural focus position is located above these licensing positions, the word order facts in (20) follow from the following requirements on the syntax of focus:

(a) The verb moves into the Focus head.
(b) A contrastively focused phrase moves into Spec,FocP.
(c) A constituent that represents new information focus just stays in the scope of the Focus head, while
(d) background elements move out of the scope domain of the Focus head (cf. Hyman and Polinsky, this volume, and Skopeteas and Fanselow, this volume). This is illustrated in (21).

(21) Assumptions about the syntax of focus (Hinterhölzl 2004)[6]
[C Background [$_{FocP}$ ContrastFocus V [$_{AgrP}$ InformationFocus [$_{VP}$]]]]

According to (21), word order in OHG obeys the same restrictions as word order in Yiddish. Yiddish is analyzed as a VO-language by Diesing (1997) since the preverbal position is preferred for strong DPs, while weak DPs tend to appear in postverbal position. Given that the VP is analyzed as the domain of existential closure (Diesing 1992), the base position of arguments must be taken to be postverbal. Haider and Rosengren (2003), applying the tests for OV/VO-properties of Vikner (2001), analyze Yiddish as an OV-language. In their approach, the OV–VO-distinction signals a difference in the base structure: arguments in VO-languages are projected in vP-shells, while arguments in OV-languages are projected in stacked adjunction structures, as is illustrated in (22a, b).

(22) a. [SU v-V [IO [t DO]]] VO-languages
b. [SU [IO [DO V]]] OV-languages

To account for the postverbal occurrence of arguments in Yiddish, they assume that Yiddish possesses an additional vP-shell (normally only available in VO-languages in their account) and optional verb movement into this additional head position. To highlight the particulars of the present approach, I will compare it with Haider and Rosengren's (2003) account. While Haider and Rosengren assume a directionality parameter that is operative in the base, namely, the directionality of argument-licensing within the vP (which forces the projection of vP-shells in VO-languages), I assume a universal VO-base order in which the unmarked word order is determined by information-structural and prosodic requirements. In particular, I propose that directionality is a prosodic property and thus only comes into play at the syntax–prosody interface.

The distribution of DP-types in Yiddish (and OHG) can be explained by assuming VO base order and the information structural restrictions given in (21) above. This account is superior to Haider and Rosengren's account, since they provide no obvious way of connecting verb placement and focusing. The OV-properties of Yiddish in the present account are treated as superficial properties derived in the syntax–prosody interface. As we will see in the following section, an important restriction on word order has to do with the

[6] See also Hyman and Polinsky (this volume) for arguments against a unique focus position in Aghem. The present proposal makes OHG (and Yiddish) look a lot like focus-licensing in Hungarian (cf. É. Kiss this volume).

headedness of phonological phrases. The preverbal placement of (separable) verb particles in Yiddish, which is treated as one of the main arguments by Vikner (2001) for the OV-character of Yiddish, follows from the direction of phonological restructuring of the verb in the present account. Assuming that constituents that build a complex predicate with the verb must form a prosodic unit with the verb, the preverbal occurrence of predicates follows from the directionality of phonological restructuring of the verb (towards left): a property that I propose is induced by the predominant direction of focus-restructuring (cf. Section 12.2.4, above).

In the following section, I will discuss what restrictions may lie behind Behaghel's law. The approach that I will pursue is that Behaghel's law is due to a violable interface condition that restricts the mapping between syntactic structure and prosodic structure and is phase-bound in its application.

12.4 Interface conditions and unmarked word order

The first question that arises with Behaghel's law, namely, how properly to define heaviness, I have already tacitly answered by equating heaviness with a branching constituent. In the following, I will propose that Behaghel's law should be properly understood as flowing out of a condition that defines the best match between syntactic structure and prosodic structure. The account that I am going to develop should also explain the data in (5) and (6), above. The goal of this approach is not to develop special conditions for VO-languages, as Haider (2004) does in proposing a specific *edge effect* in adjunct placement in English, but to propose a universal constraint that can also be taken to hold in OV-languages.

12.4.1 *A phase-based prosodic condition*

In this subsection, I will outline a constraint that applies in the V-domain in German. Then I argue that the same constraint can explain the data in (5) and (6) in English. The mechanism that I will employ is that prosodic mapping applies phase-wise such that one and the same constraint can be argued to apply in the V-domain in a language, but may fail to apply in another domain in this language, for instance in its I-domain, or middle field.

Note that there is a peculiar restriction that applies in the V-domain in German, that is to say, in verb clusters. German verb clusters are predominantly left-branching, but right-branching verb clusters are possible as long as the most deeply embedded cluster is left-branching (cf. Hinterhölzl 2006). A case in question is given in (23a). However, once a right-branching verb cluster is introduced, the verb cluster must also be right-branching at the next level up, as is illustrated by the contrast in (23b) and (23c).[7]

(23) a. weil er den Text [muß [lesen können]]
since he the text must read can
'since he must be able to read the text'

b. ??weil er den Text [[müssen [lesen können]] wird]
since he the text must read can will
'since he will have to be able to read the text'

c. weil er den Text [wird [müssen [lesen können]]]

The interesting question, now, is what the generalization illustrated in (23b, c) results from. In any event, the generalization cannot be derived from a hard syntactic condition in West Germanic, since we can find many instances that violate it, as is the case with IPP-infinitives in West Flemish. Example (24) involves a right-branching verb cluster headed by *willen* which itself sits on a left-branch with respect to the selecting auxiliary *een*.

(24) dan ze toch kosten [[willen [dienen boek (WF)
that they but could want-IPP that book
kuopen]] een]
buy have
'if they only could have wanted to buy that book'

A possible solution is to relegate the contrast in (23b, c) to a violable interface condition that determines the best match between a given syntactic structure and a prosodic output structure. Given that left- and right-branching verb clusters are mapped onto left- and right-headed phonological phrases, a possible candidate for such an interface condition is (25).

(25) *Phonological Transparency* (see Hinterhölzl to appear)
A right-headed phonological phrase (in a verb cluster) must sit on a right branch with respect to the syntactic head that is to become its prosodic sister

Condition (25) can probably be formulated in a more elegant way. But it is meant to account for the patterns in (26), which indicate the deeper reason

[7] Verb-cluster formation is argued in Hinterhölzl (2006) to involve XP-movement of the dependent infinitives into two different functional Specifiers in the V-domain of the selecting verb for licensing purposes. Given that the selecting verb moves into the highest head position in the V-domain, left- and right-branching verb clusters are derived by spelling out the dependent infinitives in the higher or lower Specifier in this account.

which probably lies behind the condition in (25): once fixed, stress tends to stay in a peripheral position within a certain domain.

Tree (26b) is okay since stress remains left-peripheral and (26c) is okay since stress remains right-peripheral, while stress in (26a) is neither left- nor right-peripheral unless H itself is stressed. (23b) violates the condition in (25). The violation can be circumvented if the (already) right-branching verb cluster is spelled out in the lower Specifier in the V-domain of the selecting verb (cf. n. 7), as is the case in (23c). In other words, (25) ensures a transparent, that is, monotonous, mapping between syntactic structure and prosodic structure.

(26)

a.* b. OK c. OK

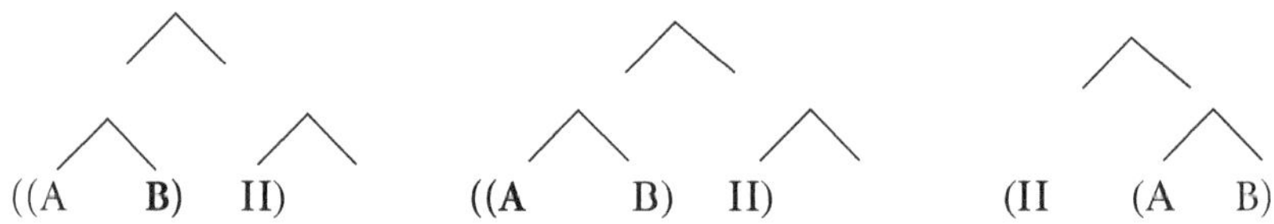

However, this condition must be taken to be restricted to the V-domain in German. The crucial evidence comes from restrictions on VP-topicalization in German. Right-branching verb clusters, with the exception of those involving an IPP-infinitive, cannot be topicalized in German. This restriction cannot be explained with a condition that applies in the C-domain, since the C-domain in German clearly tolerates right-branching prosodic constituents. In Hinterhölzl (2006), it is argued that the restrictions on VP-topicalization follow from the Phase Impenetrability Condition (see Chomsky 2001) and phase-based restrictions on spell-out which require that the verb cluster to be topicalized is spelled out in the higher specifier in the V-domain of the selecting verb, thereby accruing a violation of the mapping condition in (25).

This solution, however, is feasible only in a framework in which spell-out applies cyclically, including the cyclic, that is, phase-wise, application of conditions that map syntactic structure onto prosodic structure (see Féry and Ishihara, this volume). More specifically, we have seen above that interface conditions can apply phasewise in the sense that they may hold in one phase (the V-domain in German) without holding in another phase (the I-domain, or middle field, in German).

12.4.2 *Interface conditions and the OV–VO parameter*

Note that condition (25) is a condition that can be taken to apply in the V-domain and in the I-domain (excluding the subject) in VO-languages. In fact, (25) can be used to rule out (5a), since *more often than Peter* is a right-headed prosodic constituent that sits on a left branch with respect to the verb phrase that is to become its prosodic sister. Along the same lines, we may assume that event-related adjuncts are placed postverbally in English (see (6b) above), since their preverbal occurrence would induce violations of the condition in (25). This is why VP-intraposition is obligatory with heavy event-related adjuncts in modern English. The only exceptions are light, that is, non-(right-)branching adverbs, as is illustrated in (27).

(27) a. John carefully read the book.
b. *John with care read the book.

The fact that German tolerates heavy adjuncts (and arguments) in the middle field (see (6a) above) indicates that this condition cannot be taken to apply in the German middle field, but—as we have seen above—is still active in the German V-domain. Furthermore, note that the condition in (25) can be taken to be the basis of Behaghel's law if we assume that it applied in the middle field in OHG, OE, and ON: while pronouns and bare nouns do not induce violations of the condition in (25), the same condition would require modified nouns and PPs to be placed postverbally.

How can we explain the differential development of the three languages? In Section 12.3 we saw that the prosodic condition in OHG was regularly outruled by information-structural requirements: discourse-given heavy DPs are regularly preposed, as are heavy constituents that receive a contrastive-focus interpretation, while light elements were postposed when they were discourse-new or non-contrastively narrowly focused. This means that information-structural requirements led to the production of a high number of marked prosodic structures with the result that the prosodic constraint was restricted to the V-domain in German. Word order in the German middle field is subject to another interface constraint that guarantees scope transparency, as we will see below.

OV word order in an approach subscribing to the UBH must be taken to be derived by obligatory movement of vP-internal material into licensing positions in the middle field (cf. Hinterhölzl 2006). The reasons for motivating licensing movement out of the vP must also be taken to be valid in VO-languages. In the present framework we have two options to account for VO orders on the surface: (a) spell-out in the lower base position (called silent scrambling in Hinterhölzl 2002) or (b) spell-out in the licensing position with vP-movement around it.

The second option can be excluded since there is ample empirical evidence that VO orders in English cannot be derived by object movement that spells out the higher copy plus vP-movement around it. First, note that the vP cannot be topicalized excluding the direct object, as is illustrated in (28). Furthermore, note that the object cannot be separated from the verb and appear in its *scope position* between adverbs, as is illustrated by the contrast between German and English in (29a, b). The intended reading of (29a) is possible in the order given in (29c), where the direct object arguably occupies a vP-internal position.

(28) a. John wanted to buy something yesterday...
... * and buy John did a book

b. Hans wollte gestern etwas kaufen...
... und gekauft hat er heute ein Buch.
... and bought has he today a book

(29) a. *John met every day two girls in their classrooms.
(Temp > DO > Loc)

b. Hans traf jeden Tag zwei Mädchen in ihren
Hans met every day two girls in their
Klassenzimmern.
classrooms

c. John met two girls in their classrooms every day.

The only option that remains is to assume that DPs that undergo licensing movement are spelled out in the vP in English. This option can be derived from the condition in (25), if we assume that (25) requires the spell-out of arguments in the lower position. Haider and Rosengren (2003) use the availability of scrambling as evidence for motivating a distinction in the base structure between OV- and VO-languages: in their account, scrambling is possible only if arguments are projected in stacked adjunction structures, as illustrated in (22b), above. In the present account, there is no distinction in the base between OV- and VO-languages. Both language types are assumed to allow for scrambling, as is also evidenced by the parallel readings in (29b, c), but scrambling in VO-languages is argued to be silent due to an interface condition operating on the PF-side of the computation.

The contrast in (29) shows that word order in German displays scope in a transparent manner and can thus be taken to obey another interface constraint that restricts the mapping between syntax and LF.

(30) *Scope transparency* (see Hinterhölzl to appear)
If *a* has scope over *b*, the spell-out copy of *a* should c-command the spell-out copy of *b*.

Information structure interacts with syntactic structure through the imposition of scope requirements, as outlined in (21). Obeying the information-structural requirements in (21) has weakened the application of the prosodic mapping condition in (25), but strengthened the scopal mapping condition in (30) in the history of German, leading to a language in which word order is fully scope transparent.

In the history of English, on the other hand, the prosodic condition was retained or even reinforced in the middle field, leading to the development of a pure VO-language, that is to say, of a language that shows a large degree of PF-transparency as defined in (25).

In conclusion, word order variation can be accounted for by the differential expression of information structural categories within one grammar and word order preferences (Behaghel's law) are due to violable interface conditions that define the ideal mapping between syntactic structure and prosodic structure in the course of the derivation. Furthermore, I have argued that the distinction between OV- and VO-languages should not be accounted for by a distinction in the base but can be reduced to the workings of two parallel interface constraints that define the optimal mapping from syntax to PF and from syntax to LF, respectively. Finally, I have laid out the historical conditions that led German into developing into a scope transparent OV-language (see (21)).

Part III

Experimental and Psycholinguistic Approaches

13

Effects of givenness and constraints on free word order

STAVROS SKOPETEAS AND GISBERT FANSELOW

13.1 Word order and givenness

Clark and Clark (1977) and Clark and Haviland (1977) have proposed that constituent order is influenced by a principle requiring that given information precede new information in a sentence (the G ≻ N-principle) (see also Hörnig and Weskott, this volume). The importance of this principle for linearization in the world's languages has been established in quite a number of studies; see Siewierska (1993) for instance. Given elements come to precede new material by various mechanisms. Some languages use a simple reordering rule (scrambling) for having linear order conform with the G ≻ N-principle, while others take recourse to passivization (A-movement in the sense of Chomsky 1981), apparently because of a general lack of constituent order flexibility (see Mathesius 1975: 156ff., Tomlin 1995: 538, Prat Sala 1997: 99). Other givenness effects come about by the choice of different constructions that are not syntactically related to each other. For example, the G ≻ N-principle affects the choice among the double-object and the NP PP construction for English verbs with two objects (Collins 1995). Following Oehrle (1976), many have argued for a purely lexical relation between the two constructional options of *give* and similar verbs. Likewise, givenness influences the order of structural adjuncts in German, but since adjuncts do not scramble (Haider and Rosengren 2003), given-before-new order is base-generated in this domain.

This chapter evolved within the project D2 "Typology of Information Structure" in the SFB 632 "Information Structure" (University of Potsdam/Humboldt University Berlin, financed by the Deutsche Forschungsgemeinschaft). We want to thank Caroline Féry, Hubert Haider, Sam Hellmuth, Robin Hörnig, Ivona Kučerová, and Malte Zimmermann for helpful comments. Special thanks are due to Rusudan Asatiani, Shorena Bartaia, Tamar Khizanishvili, and Tamar Kvaskhvadze for consulting us about the syntactic and information structural properties of Georgian.

The present chapter is concerned with such syntactic mechanisms that allow given material to precede new information. We present the results of an elicitation experiment investigating the syntactic reflexes of givenness in twelve languages. Particular attention is paid to German and Georgian, but our perspective is broadened by the results for American English, Czech, Dutch, Canadian French, Greek, Hungarian, Konkani, Yucatec Maya, Prinmi, and Teribe. Deviations from normal word order turn out to be licensed in all but one (Greek) of these languages in the interest of respecting the G ≻ N-principle.[1] Given objects are fronted by A-movement (Chomsky 1981). Languages differ as to whether they employ simple A-scrambling (Déprez 1989; Mahajan 1990) or change grammatical functions in order to move a given XP leftwards. The former option seems preferred when available, but German fails to make use of its structural potential here.

A′-movement in the sense of Chomsky (1981) is normally not used as a syntactic reflex of givenness. That A′-movement is unavailable for this purpose is not unexpected. After all, A′-movement displaces operators (such as *wh*-phrases), and "givenness" is not a concept corresponding to a semantic operator.[2] The absence of a syntactic reflex of givenness in Greek may be due to the fact that the (relevant) syntactic inventory of Greek consists of A′-movement operations only, which are useless for the expression of givenness. When we observe that Hungarian nevertheless places given phrases into A′-positions, this suggests that they are not connected to operator status here, as assumed by É. Kiss (2003).

13.2 Givenness and word order: An elicitation experiment

The experiment presented here is part of QUIS, a tool for linguistic fieldwork created by the SFB 632 (University of Potsdam/Humboldt University Berlin, see Skopeteas *et al.* 2006). In the present study, the participants were asked to describe a sequence of two scenes presented in pictures. In the first scene, an individual, "a boy" for example, was introduced. In the second scene, this individual was involved in an event which is likely to be encoded by a transitive verb with two arguments, as in "the boy is kicking a man on the shoulders". One of the verb's arguments is thus given information, the other, new information. The critical description concerns the second scene (henceforth, "target scene"). The information structure of this description depends on the first

[1] See Krifka (2007) for an overview of the devices used for indicating givenness.

[2] See Kučerová (2007) for a different view. A discussion of her proposal is beyond the scope of the present chapter.

scene (henceforth, "context scene") introducing either the agent or the patient, as shown in the experimental conditions in (1).

(1) AG/GIV: (=agent/given & patient/new)
At the time of the target description, the agent (and not the patient) is part of the given information.
PAT/GIV: (=agent/new & patient/given)
At the time of the target description, the patient (and not the agent) is part of the given information

In natural discourses, the constellation AG/GIV & PAT/NEW is the most frequent contextual situation, while AG/NEW & PAT/GIV occurs less frequently (Du Bois 2003: 34ff.). A comparison of this pair of contexts should reveal whether there is a preference for given information to precede new information. For languages in which agents precede patients in canonical word order, canonical word order and G ≻ N order are harmonically aligned in the condition AG/GIV & PAT/NEW, while they are in conflict in the condition AG/NEW & PAT/GIV.

The two conditions concerning givenness were implemented as follows. A total of eight sets of two-picture sequences were constructed, involving eight different events with an agent and a patient. Each informant has been confronted with the conditions in (1) twice (the further items related to further experimental conditions, see Skopeteas *et al.* 2006 for details). Since we are interested in the effect of givenness on the linear arrangement of subjects and objects, only those reactions of the participants entered our analysis which consisted of clauses with an agent and a patient. Reactions to the stimuli that did not meet this requirement were excluded from further analysis. Furthermore, in languages with a determiner system such as German, the choice of an (in-)definite determiner in an NP indicates whether the participant's response conformed with the intended distribution of givenness among agent and patient. When both arguments of the target scene were referred to by indefinite NPs, this suggests that the participant did not recognize the intended referential identity of one of the arguments of the target scene with the referent introduced in the context scene. Similarly, when both arguments were expressed by definite NPs, this suggests that the participant may have assumed that the referents of both arguments are included in the situational common ground (perhaps taking into account that the instructor knows the pictures). In both cases, the response does not meet critical assumptions about the common ground. For this reason, such data has also not entered our analysis.

13.3 The object-fronting strategy

Six languages in our sample simply front the object in contexts in which the patient is given and the agent new: Georgian, Czech, Hungarian, Konkani, Prinmi, and Teribe.

13.3.1 *Georgian*

Georgian speakers produced four different word orders in our data set:[3] SOV (2a), SVO (2b), OSV (2c), and OVS (2d).

(2) (a) [sc-1] {A man is walking ...}
[sc-2] k'ac-i kal-s e-kač-eb-a
man-NOM woman-DAT (IO.3)OV-move.up-THM-PRS.S.3.SG
'... the man is lifting a woman.'
(AG=SBJ/first; condition AG/GIV)

(b) [sc-1] {A man is standing ...}
[sc-2] ... k'ac-i u-rt'q-am-s kal-s.
man-NOM (IO.3)OV-hit-THM-PRS.S.3.SG woman-DAT
'... the man is hitting the woman.'
(AG=SBJ/first; condition AG/GIV)

(c) [sc-1] {There is a box on the table ...}
[sc-2] ... qut-s k'ac-i a-gd-eb-s
box-DAT man-NOM NV-(IO.3)throw-THM-PRS.S.3.SG
'... a man is throwing the box.'
(AG=SBJ/non-first; condition PAT/GIV)

(d) [sc-1] {The boy is standing on the stairs ...}
[sc-2] ... bitʃ'-s a-gd-eb-s gogo
boy-DAT NV-(IO.3)throw-THM-PRS.S.3.SG girl(NOM)
k'ib-i-dan.
stair-INS-from
'... a girl is throwing the boy from the stairs.'
(AG=SBJ/non-first; condition PAT/GIV)

All descriptions obtained in the condition AG/GIV are canonical active clauses (AG=SBJ/first) with the agent preceding the patient (see Table 13.1). This result

[3] Thirty-two native speakers of Georgian, students at the University of Tbilisi, were interviewed in Tbilisi (Georgia) in October 2005 and January 2006. Experimental sessions were conducted by Stavros Skopeteas (October 2005) and Rusudan Asatiani (January 2006) and transcribed by Rusudan Asatiani, Shorena Bartaia, and Nutsa Tsetereli.

TABLE 13.1 Georgian data

	AG/GIV		PAT/GIV	
	n	%	*n*	%
TOTAL	64		64	
Other	22		13	
ag=sbj/first	42	100	31	60.8
ag=sbj/non-first	–	–	20	39.2
ag=non-sbj/non-first	–	–	–	–
ag=non-sbj/first	–	–	–	–

was expected, since the preference for given information to precede new information is in accordance with the canonical word order in this condition. The condition PAT/GIV involves a conflict between the preference for agents to precede patients and the G ≻ N-principle. As an effect of this conflict, we find active clauses with non-canonical word order (AG=SBJ/non-first), in addition to sentences with canonical order.[4] The effect of givenness on word order thus manifests itself in a significant difference in the percentage of non-canonical active sentences in which patients precede agents between the condition AG/GIV (0%) and the condition PAT/GIV (39.2%) ($t_{23} = 3.99$, $p < .05$).[5]

A brief consideration of the grammatical system of Georgian suggests that the simple reordering observed in our experiment has been effected by A-movement. There is ample evidence for an asymmetrical relation between the verbal arguments, with the subject c-commanding the object in base order (cf. A.C. Harris 1981, S.R. Anderson 1984, Boeder 1989, Joppen-Hellwig 2001, Amiridze 2006). The behaviour of the target position of movement with respect to argumental binding belongs to the clear diagnostics for the movement type (see Déprez 1989, Mahajan 1990). When the landing site

[4] The word-order change in the given-patient condition is optional in all the languages we have investigated. This could mean that there is no grammatically unique solution to the conflict between the ordering principles agent > patient and given > new. The gradient nature of the givenness effect might also indicate that givenness triggers word-order changes only in combination with further factors not controlled in the experiment, so that the experimental design was not strong enough to always yield a givenness effect.

[5] In the condition AG/GIV & PAT/NEW, 3 of 42 agent expressions (7.1%) are pronominal. In the condition AG/NEW & PAT/GIV, two of the clauses with a given patient (with OVS and SVO order, respectively) involved a pronominal object (2 of 51 clauses, i.e. 3.9%). In the condition AG/GIV & PAT/NEW, 33 out of 42 clauses are V-medial (78.5%), the others V-final. In the condition AG/NEW & PAT/GIV, 25 out of 31 clauses with an initial agent (80.6%) and 9 out of 20 clauses with an initial patient (45%) are V-medial. We do not assume an effect of givenness on V position. See below for a brief discussion of the related structural facts.

is an A-position, new binding relations can be entered after movement. A′-movement, however, does not create new binding options. For Georgian, the creation of new binding possibilities in the context of object preposing has been documented by McGinnis (2004).

Next to the binding properties, the availability of movement of a constituent across (embedded) clause boundaries is often assumed to be a diagnostics for A′-movement. The absence of such long-distance dependencies suggests that the movement operation is of the A-type (but such a conclusion is not inevitable, see Müller and Sternefeld 1994). The following examples from Georgian show that long-distance scrambling is not possible. Example (3a) illustrates the canonical construction containing a matrix and a subordinated clause. Long-distance dependency is grammatical in (3c) only, that is, only in the context of *wh*-movement, which is a case of A′-movement. O ≻ S order in simple reordering is thus created by A-movement in Georgian.

(3) (a) soso a-mb-ob-s rom bavšv-s
Soso(NOM) NV-(IO.3.SG)say-THM-PRS.S.3.SG that child-DAT
kotan-s a-čven-eb-s.
pot-DAT NV-(B.3.SG)show-THM-PRS.S.3.SG
'Soso says that he shows the pot to the child.'

(b) *kotan-s a-mb-ob-s soso rom
pot-DAT NV-(IO.3.SG)say-THM-PRS.S.3.SG Soso(NOM) that
bavšv-s a-čven-eb-s.
child-DAT NV-(B.3.SG)show-THM-PRS.S.3.SG
(intended) 'It is a pot that Soso says that he shows to the child.'

(c) ra-s a-mb-ob-s soso rom
what-DAT NV-(IO.3.SG)say-THM-PRS.S.3.SG Soso(NOM) that
bavšv-s a-čven-eb-s?
child-DAT NV-(IO.3.SG)show-THM-PRS.S.3.SG
'What does Soso say that he is showing (it) to the child.'

Georgian speakers never used passive voice in sentences with given objects. Georgian verbs form passive voice by affixation and show all irregularities of morphological passives such as defective verbs and deponent verbs (see A.C. Harris 1981, Aronson 1994, Hewitt 1995). Passive diathesis, which implies promotion of the patient to subject and demotion of the agent to adjunct status, is possible in Georgian, but it occurs only in written styles and is rare. Ivanishvili and Soselia (1999) extracted more than a thousand sentences including a passive verb from Georgian short stories and found only about 20 sentences which can be analyzed as involving verbal diathesis. Passive morphology, rather, shows up with deponent verbs and is used to express potentiality, deliberation, inchoation, etc. Native speakers report that though they could productively form passives, also in the contexts presented in our production study, they would never do it in this context because passive belongs to a formal register.

Object preposing may, but need not be, accompanied by a change in verb position, as observed above (see Skopeteas and Fanselow 2009 for detailed discussion). Skopeteas and Fanselow (2009) argue that the base order is SOV and that V medial orders result through an optional V-fronting operation.[6] The medial placement of the verb then allows an analysis similar to the one for German, in which the verb moves to a higher functional head position. If we want to maintain that object preposing involves A-movement in Georgian in the case of given objects, the following analysis comes to mind: the object is first scrambled across the subject (= A-movement), and then it undergoes a purely formal, non-operator movement to the position preceding the verb, as sketched in (4) (see similar proposals for German in Frey 2004 and Fanselow 2003).

(4) $[_{FP}$ Object $[_{F}$ Verb $[t_{object}$ subject $[t_{object}\ t_{verb}]]]]$

This analysis is supported by the observation that the ability of scrambled phrases to establish new possibilities for binding holds independently of the position of the V, in other words, the observations made above for SOV and OSV order also hold for SVO/OVS order as shown in the following examples:

(5) (a) qvela avtor-i$_i$ a-gd-eb-s
every(NOM/DAT) author-NOM NV-(IO.3)throw-THM-PRS.S.3.SG
tav-i-s$_i$ c'ign-s.
RFL-GEN-DAT book-DAT
'Every author$_i$ is throwing his$_i$ book.'

(b) *tav-is-i$_i$ avtor-i a-gd-eb-s
RFL-GEN-NOM author-NOM NV-(IO.3)throw-THM-PRS.S.3.SG
qvela c'ign-s$_i$.
every(NOM/DAT) book-DAT
(intended) 'His$_i$ author is throwing every book$_i$.'

[6] See also Skopeteas, Féry, and Asatiani (2009) for evidence from a rating experiment.

TABLE 13.2 Czech data

	AG/GIV		PAT/GIV	
	n	%	*n*	%
TOTAL	64		64	
Other	19		38	
ag=sbj/first	45	100	15	57.7
ag=sbj/non-first	—	—	9	34.6
ag=non-sbj/non-first	—	—	2	7.7
ag=non-sbj/first	—	—	—	—

Note: The data was collected in Leipzig (Germany, Jan. 2008) and transcribed by Denisa Lenertová. Sixteen native speakers of Czech participated and provided four data points per condition.

(c) qvela c'ign-s$_i$ a-gd-eb-s
every(NOM/DAT) book-DAT NV-(IO.3)throw-THM-PRS.S.3.SG
tav-is-i$_i$ avtor-i.
RFL-GEN-NOM author-NOM
'His$_i$ author is throwing every book$_i$.'

13.3.2 *Further languages*[7]

Czech is known for its word-order freedom and uses simple reordering of the clausal constituents in cases in which English uses passive voice (see Mathesius 1975: 156ff.). The absence of weak crossover effects suggests that OS orders are derived by A-movement. OS orders in Czech create new binding relations (see Kučerová 2007: 139). Example (6a) exemplifies the canonical SVO word order, elicited in the condition AG/GIV. The further examples illustrate two responses to the condition PAT/GIV: a sentence involving OVS order in (6b) and a sentence involving passivization in (6c) (see proportions in Table 13.2). The dominant strategy in the condition PAT/GIV is reordering, though a few passive sentences were produced as well.[8] The percentage of patient-first sentences in this condition differs significantly from the non-occurrence of

[7] If not otherwise indicated, the further data sets have been collected with eight native speakers (two descriptions per condition).

[8] Benz (2008) reports a similar result for Russian. In a data set obtained by the same experimental procedure and material, Russian speakers ($n = 8$) produced only canonical active sentences in the condition AG/GIV & PAT/NEW (30 valid sentences), and three active sentences involving OVS order (out of 29 valid sentences) in the condition AG/NEW & PAT/GIV.

patient first sentences in the condition AG/GIV (paired samples t-test: $t_{11} = 4.06$, $p < .05$).[9]

(6) (a) [SC-1] {A woman stands in the room.}

[SC-2] Žena uhodila muže, který k
woman:NOM hit:PAST:F man.ACC which.M.NOM to

ní stojí zády.
her:DAT stand:3.SG backwards

'The woman hit the man that was standing with his back side towards her.' (AG/GIV)

(b) [SC-1] {A chain is lying on the ground.}

[SC-2] a ten řetízek bere pes do
and the:ACC chain:ACC take:3.SG dog:NOM in

tlamy.
mouth:GEN

'... and a dog takes the chain in his mouth.' (PAT/GIV)

(c) [SC-1] {There is a woman that is starting to walk on the street.}

[SC-2] a je napadena cizím mužem...
and be:3.SG attack:PASS.PTCP:F strange:INS man:INS

'and is attacked by a strange man.' (PAT/GIV)

The choice of a marked OS order for given objects is thus not mandatory in Czech (in contrast to claims made in Kučerová 2007). Passivization competes with scrambling in the expression of givenness.

Konkani (Goa, India; Southern Indo-Aryan) is an SOV language. There are no syntactic analyses of word-order alternation in Konkani, but scrambling is a widespread property of Indo-Aryan languages (see Mahajan 1990, Kidwai 2000 for Hindi–Urdu). Konkani lacks an active–passive distinction and morphological means of encoding topics. Our data shows a pattern similar to other V-final languages: when the subject is given, speakers produce only canonical sentences (see 7a); when the object is given, it optionally scrambles over the subject resulting in OSV word order (see illustration in 7b and proportions in Table 13.3). The difference of the means of proportions of OS orders in the two conditions is significant (paired samples t-test: $t_5 = 3.4$, $p < .05$).

[9] The obtained frequencies were normalized through the *arcsine-square root* transformation in order to meet the normality requirements of parametric tests. All probabilities reported in this paper are calculated through one-tailed paired samples *t-tests*. We adopt $p < .05$ as significance level for all data sets.

TABLE 13.3 Konkani data

	AG/GIV		PAT/GIV	
	n	%	*n*	%
TOTAL	16		16	
Other	5		6	
ag=sbj/first	11	100	5	50
ag=sbj/non-first	–	–	5	50
ag=non-sbj/non-first	–	–	–	–
ag=non-sbj/first	–	–	–	–

Note: The Konkani data presented here was collected (in Panaji, India, May 2005) and transcribed by Caroline Menezes.

(7) (a) [SC-1] {Here a man is standing near a slide.}
[SC-2] Hanga to daadlo slide-acher ek rath
here that man slide-OBL.LOC a tire
dukhal-ta.
push-PRS.3.SG
'Here that man is pushing a tire on the slide.' (AG/GIV)

(b) [SC-1] {There is a snake here.}
[SC-2] Atan te sarop-a-che shempden sunen ghaans
now that snake-OBL-POS tail dog bite
mara-ta.
hit-PRS.3.SG
'Now a dog is biting the tail of that snake.' (PAT/GIV)

The Konkani data are reminiscent of Hindi, Sinhala, and Tamil (see Herring and Polillo 1995, cited after Morimoto 2000) in that given objects can, but need not be, placed in front of new subjects. Hindi and Sinhala have (obligatory) indefinite determiners and Hindi can mark definiteness by a definite determiner; the same seems to be true of Konkani.

Teribe (Panama/Costa Rica; Chibchan) is another V-final language which, however, differs from Georgian and Konkani in not allowing OSV word order. Objects always occur left-adjacent to the verb; adverbs and non-argument NPs are realized postverbally. Teribe has a direct–inverse voice alternation which does not affect the grammatical status of the arguments (see Quesada 2000, Quesada and Skopeteas 2008), as is suggested by the morphological case properties of pronominal elements and the fact that preverbal agents in direct voice and postverbal subjects in inverse voice backwards-control the subjects

TABLE 13.4 Teribe data

	AG/GIV		PAT/GIV	
	n	%	n	%
TOTAL	64		64	
Other	11		13	
ag=sbj/first	53	100	15	29.4
ag=sbj/non-first	–	–	36	70.5
ag=non-sbj/non-first	–	–	–	–
ag=non-sbj/first	–	–	–	–

Note: The Teribe data was collected by Diego Quesada and Stavros Skopeteas (Térraba reservation, Costa Rica, Apr. 2007). Four native speakers of Teribe participated and provided 16 data points per condition (in four different field sessions).

of matrix control-predicates. Inverse and direct voices are associated with different word orders, however. In the SOV word order, the verb occurs in the direct voice form. In the OVS word order, the verb is marked for inverse voice and the subject constituent is accompanied by an obviative marker. Binding properties show that subject constituents may bind into objects in the SOV order, but not vice versa. The OVS order creates new binding possibilities: the subject may bind into the antecedent object.

(8) (a) bapinga$_i$ klara klara ba$_i$ opinga e woydë.
teacher stand stand 3.SG pupil that like
'Every teacher$_i$ likes his$_i$ pupil.'

(b) *ba$_i$ bapinga opinga$_i$ klara klara e woydë.
3.SG teacher pupil stand stand that like
(intended) 'His$_i$ teacher likes every pupil$_i$.'

(c) ba$_i$ opinga e woydë bapinga$_i$ klara klara dë.
3.SG pupil that like teacher stand stand OBV
'Every teacher$_i$ likes his$_i$ pupil.'

The Teribe data set is presented in Table 13.4 (see further discussion in Quesada and Skopeteas 2008). As in the previous languages, the condition AG/GIV only triggers sentences in canonical order (9a), while the condition PAT/GIV licenses inverse clauses in which the object constituent precedes the subject. The difference of the means is significant (paired samples t-test: $t_3 = 60.77$, $p < .05$).

(9) (a) [SC-1] {There is a man standing.}
[SC-2] domer walë poskak.
man woman push
'The man pushes a woman.' (AG/GIV)

(b) [SC-1] {There is a man standing.}
[SC-2] domer poska-ya walë dë
man push-IPFV.INV woman OBV
'A woman pushes the man.' (PAT/GIV)

Prinmi (Yunnan, China; Tibeto-Burman) is a V-final language (see Ding 1998). There are no voice operations which could affect the alignment of θ-roles to syntactic functions. The canonical order of the arguments is agent≻beneficiary≻theme, but this order is not rigid. Crucially, two suffixes occur in this language that relate to different concepts of topicalization:[10] the suffix *-bbo* ("external topic" after Ding 1998) occurs at the right edge of sentence initial constituents only. Material in this position has the properties of left dislocation: it must not have a syntactic relation to an element in the clause and it can be resumed in the main clause by a co-indexed pronoun. The available evidence does not allow for conclusions as to whether movement or *in situ* generation is involved (which is a general problem in East Asian languages; see Xu 2006), but it shows clearly that the sentence initial position which is optionally delimited by this suffix is not an A-position. The suffix *-ggi* ("internal topic"), on the other hand, occurs at the right edge of constituents in all possible positions of a sentence, optionally marking constituents which are part of the common ground. In sum, we assume that Prinmi OSV sentences come in two versions: (a) the object scrambles over the subject landing in a clause internal position, (b) the object occupies a sentence initial extra-clausal position. We assume that (b) is the case when the initial object is accompanied by the external topic marker *-bbo*.

In the condition AG/GIV & PT/NEW, all valid responses exhibit SOV order, as exemplified in (10a). The given agent is always marked with the "internal topic" marker. No instance of the external topic marker *-bbo* appears in our data set. In the condition AG/NEW & PT/GIV, we obtained either canonical clauses or OSV clauses as exemplified in (10b). In both cases, only the marker *-ggi* occurs in the data. The proportion of AG/non-first clauses is significantly higher in the condition PAT/GIV (paired samples t-test, $t_4 = 3.98$, $p < .05$). The Prinmi data are set out in Table 13.5.

[10] Our account is based on Ding (1998) and on our interpretation of the experimental dataset of project D2.

TABLE 13.5 Prinmi data

	AG/GIV		PAT/GIV	
	n	%	*n*	%
TOTAL	16		16	
Other	4		8	
ag=sbj/first	12	100	3	37.5
ag=sbj/non-first	–	–	5	62.5
ag=non-sbj/non-first	–	–	–	–
ag=non-sbj/first	–	–	–	–

Note: The Prinmi data was collected (in Yunnan, China, Jan. 2005) and transcribed by Shizhi Ding.

(10) (a) [SC-1] {A woman stood there.}
[SC-2] Deanón'a hmezhá=ggi zzîn-dìn xxié=rao.
just.now=REL woman=TOP sit-INS lift=N.INV
'The woman just mentioned was lifting chair.' (AG/GIV)

(b) [SC-1] {A flag is here.}
[SC-2] Diebbòn=ggi dê peatèmi=ggon kéarao.
flag_CLF=TOP this guy=TOP.AGT push=N.INV
'This guy is pushing the flag.' (PAT/GIV)

If our interpretation of *-ggi* is correct, Prinmi is a further instance of a language that expresses givenness by means of A-movement.

Hungarian constitutes a problem for the view that givenness-related changes in word order come about by A-movement only. The canonical constituent order in Hungarian is VSO. Preverbal arguments occupy designated A′-positions for topic and focus (see É. Kiss 1998a: 256). The grammar allows passive formation, but passives seem excluded in the context we examined for stylistic reasons (they belong to the formal register and occur mostly in written styles). All valid descriptions in the condition AG/GIV involve orders in which the subject precedes the object, as in (11a). The condition PAT/GIV triggers patient fronting in a subset of the descriptions as exemplified in (11b) (see proportions in Table 13.6). The difference in the proportion of fronted objects in this condition is statistically significant (paired samples *t*-test: $t_6 = 5.05$, $p < .05$).

(11) (a) [SC-1] {A man is walking.}
[SC-2] A férfi rángat egy nőt.
DEF man pull:3.SG.PRS INDEF woman-ACC
'The man is pulling a woman.' (AG/GIV)

TABLE 13.6 Hungarian data

	AG/GIV		PAT/GIV	
	n	%	*n*	%
TOTAL	16		16	
Other	5		13	
ag=sbj/first	11	100	5	45.5
ag=sbj/non-first	–	–	6	54.5
ag=non-sbj/non-first	–	–	–	–
ag=non-sbj/first	–	–	–	–

Note: The data was collected (in Piliscsaba, Hungary, 2006–7) and transcribed by Krisztián Tronka.

(b) [SC-1] {There is a barrel on the bottom of the picture.}
[SC-2] A hordót most felemeli egy nő.
DEF barrel-ACC now up-lift-3.SG INDEF woman
'A woman is lifting up the barrel.' (PAT/GIV)

Given non-subjects are therefore able to appear in an A′-position. However, É. Kiss (2003) argues that this A′-position is not filled by operator movement. If she is correct, there is no reason why purely given XPs should not be able to appear in an A′-position in Hungarian.

13.4 The passivization strategy

13.4.1 *German*

Thirty-two native speakers of German participated in the experiment, all students of the University of Potsdam.[11] The following types of response were elicited: descriptions with an active verb (see 12a, c) and descriptions with a passive verb (12b); in both cases, the descriptions instantiate the canonical word order. Active descriptions (12a) were elicited in both conditions, while passive sentences (12b) were only elicited in the condition PAT/GIV. There was a single description with non-canonical word order in the condition PAT/GIV (12c). It involves a weak object pronoun which is obligatorily placed at the highest position of the midfield in German for syntactic rather than information-structural reasons.

[11] German experimental sessions were conducted and transcribed by Katharina Moczko and Andreas Pankau (University of Potsdam, January–March 2006).

TABLE 13.7 German data

	AG/GIV		PAT/GIV	
	n	%	*n*	%
TOTAL	64		64	
Other	19		17	
ag=sbj/first	45	100	36	76.6
ag=sbj/non-first	—	—	1	2.1
ag=non-sbj/non-first	—	—	10	21.3
ag=non-sbj/first	—	—	—	—

(12) (a) [SC-1] {A boy stands on a carpet...}
[SC-2] ... dieser Junge schubst eine grüne Sektflasche um...
'... this boy pushes a green champagne-bottle.'
(decoded as AG=SBJ/first; condition AG/GIV)

(b) [SC-1] {A girl is running...}
[SC-2] ... das Mädchen wird von einem Mann gegriffen und umgeschmissen...
'... the girl is grasped and knocked down by a man.'
(decoded as AG=NON-SBJ/non-first; condition PAT/GIV)

(c) [SC-1] {A boy stands annoyed on a staircase...}
[SC-2] ... und plötzlich schubst ihn ein Mädchen von hinten. (PAT/GIV)
'... and suddenly a girl pushes him from the backside.'
(decoded as AG=SBJ/non-first; condition PAT/GIV)

The effect of givenness on word order manifests itself in the significant difference in the percentage of passive clauses between the condition AG/GIV (0%) and the condition PAT/GIV (21.3%) (paired samples *t*-test: $t_{23} = 3.61$, $p < .05$) (see proportions in Table 13.7).

In the condition AG/GIV & PAT/NEW, 14 out of the 45 agent expressions (23.7%) are pronominal. In the condition AG/NEW & PAT/GIV, six out of the 47 patient expressions in German (11.3%) are pronominal. Three out of six clauses with a pronominal patient contain a passive verb, which suggests that passivization is more frequent with pronominal patients (50% passive clauses with pronominal patients vs. seven out of 41, i.e. 14.5% passive clauses with lexical patients), though the small numbers do not allow us to prove the statistical significance of this difference.

In contexts with a given patient and a new agent, participants of the German experiment used passivization rather than simple reordering for producing G ≻ N order. This outcome is surprising for two reasons. First, the result does not fit comparative and theoretical expectations. While previous cross-linguistic studies (Mathesius 1975; Prat Sala 1997) have shown that some languages use passives in contexts in which others use simple reordering, these studies explain this difference in terms of overall word-order flexibility. Syntactically, German *is*, however, a free-word-order language allowing scrambling and movement of the object to Spec,CP, so one does not expect the preference for grammatical function change found in our data. Second, our result seems incompatible with corpus data. Weber and Müller (2004) found more OVS (144) than SVO (88) sentences in contexts with given patients and new agents. In natural (newspaper) text, the givenness of an object seemingly licenses preposing in main clauses without passivization. Kempen and Harbusch (2005) observed an unexpected rigidity of word order in embedded clauses: they found only four instances of O ≻ S order with two lexical arguments against 1,476 instances of S ≻ O order (0.2%) in a corpus of written texts (NEGRA), and one instance of O ≻ S against 132 instances of S ≻ O (0.7%) in a corpus of oral texts (VERBMOBIL). In general, structures in which an object is fronted across an underlying subject but not placed into the preverbal position of a main clause are very rare: there are between one and two occurrences of such structures per 1,000 sentences.

The fact that our experiment yielded marked word order in Georgian and further languages excludes the possibility that some overall suboptimal design property is responsible for the absence of a simple word-order variation in German.

Let us first deal with the contrast between corpus studies and our experimental findings. The movement of a given object to first position observed in corpora is not necessarily triggered by the givenness of the object itself. The trigger for such displacements may be another pragmatic function compatible with givenness. Only half of the objects in OVS sentences are given and precede new subjects in Weber and Müller (2004), a figure that is reminiscent of Speyer's (2007) observation that only half of all XPs in Spec,CP represent "backward looking centers", that is, given topics. It is certain that the appearance of XPs in first position can fulfill other functions than expressing givenness.

Givenness was directly controlled in our experiment (so that its effects should be visible in the results), but further pragmatic factors are not likely to show an effect in our production study. Natural texts differ from our experimental situation in that the fronting of given XPs may be called for because of subsequent text (enhancement of anaphoric options, the preparation or execution of topic shifts, etc.). All such potential continuing functions are not expected for the description of the second picture in the mini-texts elicited from our informants. Contrast plays no role in our experimental setting either, and marked O ≻ S order in our simple picture description design cannot be motivated by further communicative needs such as directing the hearer's (=instructor's) attention. We thus believe that our results show the *pure effects of givenness* that are blurred by a multitude of other factors in natural texts (animacy, weight, contrast to other referents, etc.).

Speyer (2007) shows that what goes to first position (Spec,CP) is primarily determined by contrast, the introduction of brand-new information, and the introduction of elements standing in a POSET relation (see E. F. Prince 1997) to given material. These factors override "topicality" in the choice of the element occupying Spec,CP. Frey (2004) shows that it is not "familiarity" (strongly related to givenness) but "aboutness" that determines whether an element moves leftwards to a topic position. This is in line with our findings: the givenness of an object is not a sufficient reason for movement across the subject.

We can now turn to the first question: why is givenness not able to trigger movement in German? Half of the answer is easy. The left peripheral position Spec,CP is an A′-position. Movement to A′-positions creates operator–variable dependencies. Categories such as contrast, focus, and topicality can be interpreted along these lines, but pure givenness is normally not amenable to a treatment as an operator-variable relation. Given material therefore cannot reach A′-positions directly (though they may be attracted there if they additionally bear operator properties).

However, the left periphery of German main clauses also hosts non-operator material: the leftmost category of the narrow clause (TP) can go to Spec,CP, too (see Fanselow 2003, Müller 2004, Frey 2005). Typically, this leftmost position of TP is occupied by subjects and sentence and temporal adverbs, but objects can reach this position by scrambling, too. The difficult half of finding an answer to the question of why given objects cannot be fronted thus is: why is givenness not sufficient for the application of object scrambling in German? Why is scrambling of accusative objects so rare in natural texts, and non-existent in our data?

13.4.2 *Further Languages*

Four further languages will be presented in this section that—though structurally different from German to different degrees—show the same data pattern in this experimental condition: American English, Québec French, Dutch, and Yucatec Maya.

All four languages have productive passivization, but they differ from German in that A-scrambling is not grammatically available. Nevertheless, non-canonical orders are possible in these languages as well (see inversion, preposing, and dislocation in Birner and Ward 2004 for English), but they involve operator movement, and are thus not expected to be invoked by a simple asymmetry in givenness. Dutch is similar to German in terms of SOV base order and the verb-second effect. Objects may scramble over adverbs (see Neeleman and Reinhart 1998) but not over subjects (at least not in non-contrastive contexts).

The examples in (a) illustrate canonical sentences produced in the condition AG/GIV and the examples in (b) illustrate expressions obtained in the condition PAT/GIV, showing that givenness asymmetries can trigger passivization.

(13) (a) [SC-1] {A woman is standing with her hands on her hips.}
[SC-2] La femme frappe un homme dans l'dos.
'The woman is hitting the man in the back.' (AG/GIV)
(b) [SC-1] {There is a young boy running.}
[SC-2] Le jeune garçon est attrapé par un homme.
'The young boy is caught by a man.' (PAT/GIV)

(14) (a) [SC-1] {There's a man walking.}
[SC-2] Now the man is attacking a woman. (AG/GIV)
(b) [SC-1] There's a woman who's walking.}
[SC-2] Now she's attacked by a man from behind. (PAT/GIV)

(15) (a) [SC-1] {A man is standing on the end of the glide.}
[SC-2] De man duwt een band de glijbaan op.
'The man pushes a tire up the glide.' (AG/GIV)
(b) [SC-1] {There is a small table on a staircase.}
[SC-2] Eh, de tafel wordt door een meisje van de trap geduwd.
'The table is pushed from the staircase by a girl.' (PAT/GIV)

The results for these three languages are summarized in Table 13.8. The differences in the two conditions are in accordance with our hypothesis and are significant in all languages (French data: $t_7 = 2.39$, $p < .05$; English data: $t_{13} = 2.48$, $p < .05$; Dutch data: $t_6 = 2.42$, $p < .05$).

In Yucatec Maya (Yucatán/Quitana Roo, Mexico; Maya), objects precede subjects in canonical word order. Like most Mayan languages, Yucatec Maya is V-initial, the canonical position of postverbal arguments being VOS (Durbin

TABLE 13.8 Canadian French, American English, and Dutch data

	Canadian French				American English				Dutch			
	AG/GIV		PAT/GIV		AG/GIV		PAT/GIV		AG/GIV		PAT/GIV	
	n	%	*n*	%	*n*	%	*n*	%	*n*	%	*n*	%
TOTAL	16		16		48		48		16		16	
Other	1		5		14		24		5		2	
ag=sbj/first	15	100	8	72.7	34	100	20	83.7	11	100	10	71.4
ag=sbj/non-first	—	—	—	—	—	—	—	—	—	—	—	—
ag=non-sbj/non-first	—	—	3	27.3	—	—	4	16.3	—	—	4	28.6
ag=non-sbj/first	—	—	—	—	—	—	—	—	—	—	—	—

Note: The data from Canadian French was collected and transcribed by Alain Thériauld (Québec, Dec. 2005–Feb. 2006), and the data from Dutch by Hanneke Van Hoof (Rijswijk, the Netherlands 2006). The data from American English was collected and transcribed by Elizabeth Medvedovsky (Chicago, Dec. 2005) and involves a larger sample (24 native speakers).

and Ojeda 1978). However, this order occurs only very rarely in corpora and is difficult to process (see Skopeteas and Verhoeven 2005, Gutiérrez Bravo and Monforte y Madera 2007). In our experiment, we encountered a single VOS clause, given in (17c), but since this is a single instance of this order, we cannot draw conclusions about its interaction with the examined discourse conditions. As already observed in corpora (see Skopeteas and Verhoeven 2005, Gutiérrez Bravo and Monforte y Madera 2007), the order that most frequently occurs in corpora is SVO, involving a left-dislocated subject constituent (constituents in this position may be resumed *in situ* by a co-referent pronoun or noun and cannot occur in some subtypes of subordinate clauses). Hence, in the condition AG/GIV we obtained 20 descriptions with SVO order as illustrated in (16a). Given that the basic VOS order hardly has a chance of being realized, two structural configurations allow for patients preceding agent constituents. The first structural configuration which results in a fronted patient is left dislocation of the object involving A′-movement of this constituent. Active clauses with left-dislocated patients do not occur in our data set at all. The second configuration is passive voice, exemplified in (16b).

(16) (a) [SC-1] {There is a man standing.}
[SC-2] te'el-a' le máak-o' táan u ch'ik le
this-D.1 DEF man-D.2 PROG A.3 take(B.3.SG) DEF
sáarten-o'.
pan-D.2
'In this (scene), the man is taking the pan.' (AG/GIV)

TABLE 13.9 Yucatec Maya data

	AG/GIV		PAT/GIV	
	n	%	*n*	%
TOTAL	32		32	
Other	11		8	
ag=sbj/first	20	95.2	10	41.6
ag=sbj/non-first	—	—	1	4.2
ag=non-sbj/non-first	1	4.8	13	54.2
ag=non-sbj/first	—	—	—	—

Note: The data presented in this table was produced by two speakers that gave 16 descriptions per condition each (in eight field sessions). The data was collected and transcribed by Stavros Skopeteas and Elisabeth Verhoeven (Aug. 2006). A previous data-set based on a pilot version of the same experiment was collected with 12 speakers in Dec. 2004 and showed the same data pattern. All informants are native speakers and inhabitants of Quintana Roo, Mexico.

(b) [SC-1] {The pan is on the little table...}
[SC-2] le sáarten-o' tuún liík's-a'l yo'l le
DEF pan-D.2 PROG:A.3 lift-PASS.INCMPL on DEF
meèsa-o'.
table-D.2
'The pan is being lifted from the table.' (PAT/GIV)

(c) [SC-1] {Here is a chair...}
[SC-2] lela' muka'h u koh le
it:D1 IMM.FUT A.3 push(SUBJ)(B.3.SG) DEF
k'áanche'-o' máak-o' be'òraáh le chàan
chair-D.2 person-D2 now DEF little
boòláah yéetel hun-p'éel máartillóoh.
ball with one-CL.INAN hammer
'[on] this [picture], a man is going to push the chair.'
(PAT/GIV)

Table 13.9 shows that Yucatec Maya chooses the passivization option (13 clauses, i.e. 56.5%). Hence, Yucatec Maya displays the same pattern as German or English. The proportion of passive clauses is significantly higher in the condition PAT/GIV ($t_7 = 5.11$, $p < .05$, measured on items due to the low number of subjects, $n = 2$).

13.5 Canonical word-order strategy

The previous sections presented two strategies for placing given patients in front to new agents. In a third language type, patients are not fronted at all when they outrank agents in givenness. This data pattern is attested in

TABLE 13.10 Greek data

	AG/GIV		PAT/GIV	
	n	%	*n*	%
TOTAL	32		32	
Other	15		12	
ag=sbj/first	17	100	20	100
ag=sbj/non-first	–	–	–	–
ag=non-sbj/non-first	–	–	–	–
ag=non-sbj/first	–	–	–	–

Note: The data was collected and transcribed by Thanasis Georgakopoulos and Yannis Kostopoulos in collaboration with George Markopoulos (Athens, Jan.–Mar. 2006).

Greek. Greek is a VSO language with preverbal operator positions. There is morphological medio-passive formation, which is strongly defective and lexically varies in the availability of passive or middle voice. Crucially, for most verbs occurring in our experiment, the medio-passive form has reflexive and reciprocal functions blocking its use in passive voice.

The alternative to passive formation in Greek would be an OVS order which is derived by movement of the object to an operator position. Preverbal object topics trigger a co-referent proclitic pronoun resulting in a construction that is known as clitic left dislocation, which under standard assumptions involves A′-movement (see Alexiadou and Anagnostopoulou 2000). It has been already noted in the literature that clitic left dislocation in Greek is functionally equivalent to passivization in English (see Warburton 1975). However, accounts on the discourse function of clitic left dislocation do not assume that this construction is triggered by givenness asymmetries but refer to discourse concepts such as *contrastive topicalization* (Iatridou 1995) or *linkhood* of the object constituent (see Alexopoulou and Kolliakou 2001). Hence, we do not expect to obtain instances of clitic left dislocation in a context that involves a simple givenness asymmetry of the arguments, and this prediction is borne out (see Table 13.10).

(17) (a) [SC-1] {A man feels like dancing.}
[SC-2] aftós o ádras
this:M.SG.NOM DEF:M.SG.NOM man:M.SG.NOM
pjáni mja jinéka apó
catch:3.SG INDEF:F.SG.ACC woman:F.SG.ACC from
tus ómus.
DEF:M.PL.ACC shoulder:M.PL.ACC
'This man catches a woman on the shoulders.' (AG/GIV)

(b) [SC-1] {It is a man who sits on the floor.}
[SC-2] ci éna aGoráki xtipái
and INDEF:NOM.SG.N boy:NOM.SG.N hit:3.SG
ton ádra.
DEF:ACC.SG.M man:ACC.SG.M
'... and a boy hits the man.' (PAT/GIV)

13.6 Discussion

Our cross-linguistic results show that the condition AG/GIV induces canonical sentences across languages, while the condition PAT/GIV licenses deviations from the canonical pattern (see summary in the last column of Table 13.11). The second column in Table 13.11 shows that a subset of the sample languages have (in-)definite markers, including Prinmi, which has a suffix marking the discourse status of the NP. The next column shows that a subset of the languages displays an overt case opposition for the distinction between agent and patient constituents. Basic word order is given in the next column. All grammars have some structural configuration that involves an operator position and may serve as the landing site of A′-movement. The grammars differ with respect to the availability of operations that allow movement to an argument position. Some grammars have scrambling, and some grammars allow passivization. Note that for some languages these structural possibilities

TABLE 13.11 Summary

	def.	case	w.o.	Operator Position	Argument Position: Reordering	Argument Position: Passive	Difference in PAT/GIV
Georgian	–	+	SOV	+	+	*restricted*	reordering
Konkani	+	+	SOV	+	+	–	reordering
Prinmi	+	+	SOV	+	+	–	reordering
Teribe	–	–	SOV	+	+	–	reordering
Am. English	+	–	SVO	+	–	+	passivization
Can. French	+	–	SVO	+	–	+	passivization
Dutch	+	–	SOV	+	(+)	+	passivization
Yucatec Maya	+	–	VOS	+	–	+	passivization
German	+	+	SOV	+	+	+	passivization
Czech	–	+	SVO	+	+	+	reordering
Greek	+	+	VSO	+	–	*restricted*	(no difference)
Hungarian	+	+	SVO	+	+	*restricted*	reordering

are subject to constraints of a stylistic or structural nature, so that they do not constitute real options in the production process we examined. For example, passive voice is stylistically excluded in Georgian and Hungarian and does not occur due to defectivity and semantic blocking in our Greek sample.

It should be noted that the typological classification in Table 13.11 relates to the expression of givenness asymmetries and not to word-order freedom in general, that is, the same languages may show different data patterns in other discourse conditions.

Definite and indefinite articles contrast for the distinction of given and new specific referents in the examined context. Hence, the availability of definiteness in a language may be assumed to have a compensatory effect on the use of syntactic means for the encoding of givenness. However, our data shows that this is not the case. The use of marked word order and the choice between scrambling and passivization are independent of the presence/absence of definiteness markers.

General word-order freedom is well correlated with the existence of morphological case distinctions and with the VO–OV distinction. As mentioned above, the general tendency in our small language sample is that OV languages can apply their scrambling operation to given arguments. Hindi and Turkish (Temürcü 2001) support this tendency. Dutch and German are the only exceptions in this context, with German being a scrambling language but not applying the process for givenness marking. The idea suggests itself that it is the verb-second property of these languages that comes in the way of fully exploiting the potential of OV languages, but our sample is too small to really warrant such a conclusion, and it is unclear how such a negative correlation between verb-second movement and the use of scrambling for givenness marking could be derived from syntactic models.

We argued that givenness-induced word order variation depends on the availability of argument positions in syntax. Our data suggests the following generalizations:[12]

(18) a. Givenness asymmetries do not induce A′-movement.
b. Givenness asymmetries may induce A-movement (either of the reordering or of the passive type).

These generalizations are supported by rich evidence from our data set. All languages in our sample have the structural option of moving an XP to an operator position, but with one possible exception: speakers never choose this option for expressing givenness. Strong support for the view that givenness

[12] See very similar results and generalizations in Neeleman and van de Koot (2008).

is incompatible with operator/A′-movement is found in the Greek data. Passivization in Greek does not apply in the examined context and clitic left dislocation would be the only way to achieve a linearization which satisfies the pragmatic preference for given information to precede new. This construction involves A′-movement, however, which is not induced by givenness asymmetries, as shown by our experimental results and in line with the previous literature on the information-structural properties of clitic left dislocation.

However, generalization (19) fails to explain the Hungarian data. Though object fronting involves A′-movement in this language, this construction is in effect induced by givenness asymmetries. This finding suggests that A′-movement in Hungarian has different syntactic or information structural properties. If É. Kiss (2003) is correct, the relevant exceptional property of Hungarian would lie in the fact that A′-movement to topic positions is not operator movement in Hungarian. The discussion in Baker (2000) shows that Hungarian is not exceptional in failing to align certain A′-positions with operator status.

Turning now to the types of A-movement that occur across languages, we observe a preference for A-scrambling in our language sample. As mentioned above, German is an exception to this in that it does not apply its scrambling operation for the expression of givenness contrasts. In a sense, Czech is an exception, too, because it is the only language in which we find variation between scrambling and passive in the picture descriptions. All other languages adopt a single strategy for expressing givenness only.

A large part of the data set is explained through the available options in the grammar. Speakers of Konkani, Prinmi, and Teribe do not have a passivization option and they (optionally) select reordering in order to linearize the propositional content according to the pragmatic preferences. Speakers of American English, Canadian French, Dutch, and Yucatec Maya do not have the scrambling option; hence, they select passivization. German and Czech illustrate the critical case in which both options are available, but their reactions are not uniform: German speakers opt for passivization, while Czech speakers prefer reordering in the same context.

We found that given material is only optionally fronted in the languages of our sample. The optionality of givenness fronting in our data is compatible with those grammatical models that view scrambling as a (syntactically) "non-triggered" operation (Haider and Rosengren 2003) or assume that scrambled structures are base-generated (Fanselow 2001). Likewise, the choice of passive morphology is syntactically untriggered in many grammatical models (though it may trigger a number of syntactic consequences such as A-movement). In other words, our results corroborate the view that scrambling and passivization are not only optional from a purely syntactic point of view but also in terms of the expression of information structure. Our findings therefore disfavor grammatical models in which there is no untriggered movement and models such as classical OT in which conflicts such as those between ordering principles are (nearly) always resolved in a unique way. Our findings favor syntactic models in which a "gradient" conflict resolution is not exceptional or models in which the actual choice between syntactic constructions is not part of the theory of syntax.

We have not considered the impact of prosody in the present chapter. Given phrases are obligatorily deaccented in German (see Féry and Herbst 2004, Féry 2006, Féry and Ishihara, this volume). Where deaccentuation is the formal counterpart of givenness, the word-order changes induced by givenness can be conceived of as being caused in an indirect way. The object position is typically the most prominent one prosodically, and the fronting of given objects has the effect of "optimizing" prosodic structure in the sense that deaccented material leaves that prominent position. We believe that the optionality of givenness fronting can be understood in terms of this interaction between syntax and prosody.

14

Investigating effects of structural and information-structural factors on pronoun resolution

ELSI KAISER

14.1 Introduction

The ability to interpret informationally impoverished referential forms (e.g. *s/he, it*) is crucial for successful language comprehension. However, this task appears rather daunting, given that a form such as *she* can, at least in principle, refer to any female human referent—in other words, any member of a rather sizeable set. The "search space" is commonly assumed to be constrained by the nature of the mapping between referential forms and referent salience. Specifically, it is widely agreed that there exists a connection between the degree of reduction of referential forms and the level of salience of their antecedents, such that the most reduced (and semantically least informative) forms refer to the most salient referents—in other words, to the most prominent, most accessible referents. A range of factors has been argued to influence salience and to make referents good antecedents for subsequent pronouns, but questions remain regarding the nature of the interaction between the different factors.

This chapter explores two questions related to reference resolution: (i) How do information-structural factors and syntactic factors (in particular focusing, pronominalization, and subjecthood) interact to guide the process of

I would like to thank Rebekka Puderbaugh, Sasha Eloi, and Joyce McDonough for assistance with the experiment described here. I would also like to thank Christine Gunlogson, Jeffrey Runner, and Michael Tanenhaus for useful feedback and comments on earlier versions of some aspects of this chapter. Thanks also to audiences at *Sinn und Bedeutung 10* (Berlin 2005) and the 81st Annual Meeting of the LSA (Anaheim 2007) where some parts of this research were presented. Many thanks are also due to Caroline Féry and Thomas Weskott for their insightful comments. Needless to say, all mistakes and misunderstandings are mine.

reference resolution? (ii) Is the interpretation of a sentence-initial anaphoric form connected to the presence or absence of other referential forms later in the same sentence? The first question looks backwards from an anaphoric form with the goal of clarifying what aspects of earlier discourse influence pronoun interpretation. In contrast, the second question looks forward to investigate whether upcoming discourse has an impact on reference resolution.

The results of the sentence-completion experiment discussed in this chapter are compatible with the view that multiple, differently weighted factors influence reference resolution (see Kaiser and Trueswell 2008 and others). The results indicate that a subject-position pronoun is most likely to be interpreted as referring to the agentive subject of the preceding sentence, regardless of whether the subject is discourse-old and pronominalized (e.g. *He criticized the businessman. He*...) or discourse-new and contrastively focused (e.g. *It was the secretary who scolded her. She*...). Assuming that pronouns refer to salient antecedents, this indicates that (agentive) entities realized in the grammatical role of subject can be highly salient even when they are not discourse-old or pronominalized (what some might call "topical"). The results also provide some indication that although clefting a focused object makes it more likely to be the antecedent of a subsequent pronoun, focusing may contribute less to a referent's salience than subjecthood and pronominalization.

As for the effects of subsequent discourse, the results support the view that a purely backward-looking approach to reference resolution is insufficient (see also Centering Theory, Grosz, Joshi, and Weinstein 1995). More specifically, the data show that whether an ambiguous pronoun in subject position refers to the preceding subject or the preceding object is connected to the referential properties of the remainder of the clause—specifically, whether or not the other argument from the preceding sentence is mentioned in the pronoun-containing clause. A pronoun that refers to the *preceding subject* is more likely to be followed by a subsequent mention of the preceding object than a pronoun that refers to the *preceding object* is to be followed by a mention of the preceding subject. This pattern suggests that if a less salient referent (object) is "upgraded" by being interpreted as the antecedent of a subject-position pronoun, then subsequent mention of the higher-salience referent (subject) in the same clause is avoided.

As a whole, the analyses presented in this chapter show that subjecthood plays an important role both when looking back at preceding discourse and when looking forward to subsequent discourse. The structure of the chapter is as follows. Section 14.2 provides an overview of the notion of salience and discusses some factors argued to influence referent salience. In Section 14.3, I explore the idea—discussed in Centering Theory (e.g. Grosz, Joshi, and

Weinstein 1995)—that the interpretation of a pronoun in subject position depends on or has consequences for what comes later in the same sentence (see also Winograd 1972, Hobbs 1979, Kehler 2002 for relevant discussion). Section 14.4 presents the results of the sentence completion experiment, and conclusions and directions for future work are discussed in Section 14.5.

14.2 Salience

It is commonly assumed in the tradition of attention-based approaches to reference resolution that use of different kinds of referential expressions (pronouns, demonstratives, full NPs, etc.) is correlated with the level of salience/accessibility of their antecedents (e.g. Givón 1983, Ariel 1990, Gundel, Hedberg, and Zacharski 1993). Generally, researchers assume that the most reduced anaphoric expressions refer to the most salient antecedents, where "salient" is used to mean entities that are currently most prominent, at the center of attention. If we accept this view, we can use pronouns as a tool to investigate salience: we can probe what factors make an entity likely to be the antecedent of a pronoun in subsequent discourse, and thus learn more about what factors influence salience.[1] Previous research has argued that a range of factors increase the likelihood of subsequent pronominalization (see Arnold 1998 for a summary), including (i) occupying the grammatical position of subject, (ii) being given/old information, and (iii) being realized as a pronoun.

Many of the factors claimed to affect subsequent pronominalization seem to be linked with the notion of topicality. A precise definition of topicality, however, has proved hard to pin down, and the term is used differently by different researchers. According to Reinhart's aboutness-based definition, the topic is "the expression whose referent the sentence is about" (Reinhart 1982: 5). Gundel (1985) characterizes topics in terms of shared knowledge: "the topic of a speech act will normally be some entity that is already familiar to both speaker and addressee" (Gundel 1985: 92). Strawson defines the topic of an utterance as "what is of current interest or concern" (Strawson 1964: 104).

[1] Assuming that referential forms can be mapped directly onto a salience scale is not entirely unproblematic (see Kaiser 2003, Kaiser and Trueswell 2008). Our research suggests that not all referential forms are sensitive to the same factors to the same degree, and thus the mapping between referential forms and degree of salience of the antecedent may well be more complex than is normally assumed. However, this chapter focuses only on one referential form, namely, personal pronouns in English, and our conclusions should be interpreted as restricted to this form. For present purposes, we assume that English personal pronouns can be used as a tool to probe referent salience. However, even if one regards this salience-related assumption as problematic, the results are nevertheless informative, as they shed light on the factors influencing pronoun interpretation.

In recent Centering-theoretic work, E. F. Prince (2003) and Beaver (2004) use the term "topic" for the backward-looking center in Centering Theory (Grosz *et al.* 1995)—a use which links topicality with pronominalization and givenness. Although it is not the aim of this chapter to provide a precise definition of topicality, it is worth noting that many of the factors that have been claimed to influence salience have also been linked to the notion of topicality.

However, not all researchers agree that topicality-related factors are the primary determinants of referent salience, and some have suggested that focusing is what makes referents cognitively prominent. In the remainder of this section, we examine the claims regarding subjecthood, givenness, and pronominalization, which could be regarded as being related to topicality, and then turn to research on focusing.

14.2.1 *Effects of subjecthood, givenness, and pronominalization*

One of the most widely discussed salience-influencing factors is grammatical role. A sizeable body of research suggests that entities realized in subject position are more salient than those in non-subject positions (e.g. Chafe 1976, Brennan, Friedman, and Pollard 1987, Crawley and Stevenson 1990; see also Gordon, Grosz, and Gilliom 1993, *inter alia*), a claim that has received support from corpus work and psycholinguistic research (see also Fiedler *et al.*, in this volume, for typological discussion concerning the special status of subjects). (We focus here on pronouns in subject position. The interpretation of object-position pronouns has also been investigated in previous work, e.g. Smyth 1994, Chambers and Smyth 1998.)

Another factor that has been linked to salience is the distinction between given and new information. On the basis of a detailed corpus analysis of a range of German texts, Strube and Hahn (1996, see also Strube and Hahn 1999 for further details) claim that old/given referents are more salient than new referents. According to their view, an anaphoric expression that follows a sentence containing a discourse-old referent and a discourse-new referent tends to refer to the discourse-old one (see also Ariel 1990 and Givón 1983 on topicality effects). Further evidence regarding givenness comes from corpus work on the Oceanic language Yapanese by Ballantyne (2004).

A further factor that has been argued to influence salience is the referential form with which an entity is realized. According to Kameyama (1996), the salience of a referent in non-subject position increases if that referent is pronominalized, to the extent that the referent can "compete" with a non-pronominalized entity in subject position (see also Grosz, Joshi, and Weinstein 1986). In Beaver's (2004) Optimality-theoretic approach to anaphora resolution, the constraint SALIENT FORM similarly "implies that being pronominalized makes a referent salient in the future" (Beaver 2004: 31 n. 30; see also Kehler 2002: 169).

Given that factors such as subjecthood, givenness, and pronominalization have been claimed to increase salience, and that these factors are often regarded as related to topicality, one might be tempted to conclude that a salient referent is a topical referent. However, not all researchers agree that salience is connected to topicality, as the next section makes clear.

14.2.2 *Effects of focusing*

This section examines claims that focusing increases referent salience—more specifically, that contrastive focus increases salience. The existing psycholinguistic work on this topic has, for the most part, used structures such as it-clefts (*It was Mary who called Lisa*), which express contrastive/identificational focus.[2] In keeping with the existing emphasis on contrastive focus, the experiment described in Section 14.4 uses *it*-clefts and *in-situ* focus constructions[3] in contrastive contexts, and thus the conclusions drawn on the basis of this experiment apply only to contrastive focus.

A large body of cognitive psychology research indicates that focused entities pattern differently from non-focused entities. In an early study, Hornby (1974) presented participants with pictures and sentences and asked them whether the sentence matches the picture. When presented with cleft sentences (e.g. *It is the girl who is riding the bicycle*), participants were better at detecting mismatches when the mismatching information was focused than when it was presupposed (see also Cutler and Fodor 1979). Related work suggests that people attend more to the focused/non-presupposed part of a sentence (Zimmer and Engelkamp 1981, but see Delin 1990), and remember focused information better than non-focused information (e.g. Singer 1976, Birch and Garnsey 1995; see also Birch and Rayner 1997).

Extending the investigation of focusing effects to reference resolution, Almor (1999) showed that definite NPs referring to focused antecedents are read faster than definite NPs whose antecedents are not in focus, regardless of whether the focused entity is the subject or object. His results suggest that reference to focused antecedents is processed more easily than reference

[2] Corpus studies show that regarding all clefts as structures where the clefted constituent is contrastively focused and the rest of the sentence is presupposed is an oversimplification (see e.g. Delin 1990). Nevertheless, the *it*-clefts used in the experiment described here were straightforward in that the focused constituent was contrastively focused new information and the rest of the sentence was old/presupposed information.

[3] Green and Jaggar (2003) argue that *in-situ* focus in English can be interpreted contrastively, contrary to É. Kiss (1998).

to non-focused antecedents. Further evidence for effects of focusing comes from Birch, Albrecht and Myers (2000), who argue, based on sentence completion results, that focused concepts are more salient and have a "stronger memory trace" (Birch *et al.* 2000: 302) than non-focused concepts. More recently, Foraker and McElree (2007) claimed that clefting makes referents more distinctive in memory but does not cause them to be actively maintained in the focus of attention. Their results fit with the other studies insofar as they suggest that focused entities are represented differently from non-focused entities.

In sum, the existing literature leaves us at an intriguing crossroads. A number of factors—including subjecthood, givenness, and pronominalization—are claimed to increase the discourse salience of an entity and thus make that entity a good antecedent for a subsequent pronoun, and many of these factors seem to be related to topicality. However, a number of other studies have found that contrastive focus affects the discourse representation of an entity, rendering it more prominent in people's mental models of the discourse.

14.2.3 *Comparing focus to other factors*

Taking important steps to reconcile the claims discussed in the preceding sections, Arnold (1999), Cowles (2003), and Cowles, Walenski, and Kluender (2007) investigated how focus, subjecthood, and givenness influence salience. Because these experiments provide the starting point for much of the research presented in this chapter, this section summarizes their key findings.

Given that not all researchers agree on how the term "topic" should be used and that this chapter does not aim to provide a definition of topichood, I will adapt a neutral approach and describe the referents in Arnold's and Cowles's experiments in terms of (i) their grammatical role, (ii) their referential form (full NP vs. pronoun), and (iii) their discourse status (given/old vs. new). In adopting this approach, I do not intend to suggest that topicality is unimportant. Whether an entity is realized in subject or object position, realized with a pronoun or a full NP, and so on, depends on and has consequences for higher-level issues connected to aboutness and topicality that merit further research (see also Fiedler *et al.*, this volume, for related discussion concerning the subject position and topicality).

In a series of experiments, Arnold (1999, see also Arnold 1998) used pronouns as a tool for probing which referent in the preceding discourse is the most salient, and how salience is influenced by topicality and contrastive focus. On the basis of a rating study comparing focused (clefted) objects and matrix subjects (which she analyzes as sentence topics, following a large body of research on the connection between subjecthood and topicality) to other referents, Arnold concluded that topics and foci are more salient than other referents. In two further experiments, she pitted pronominalized, discourse-old subjects against focused objects, and discourse-new subjects against focused objects. She concluded that (i) pronominalized, discourse-old subjects—which she views as discourse-topics—are more salient than focused objects (both *in-situ* and clefted), but that (ii) discourse-new subjects are relatively less salient than discourse-old subjects. She suggests that a focused object is better able to compete with a discourse-new full-NP subject than with a discourse-old pronominalized subject, which she regards as an established discourse-topic. She views salience as a competitive phenomenon, "where the representations of different referents in a particular discourse compete for activation" (Arnold 1999: 28).

In related research, Cowles (2003, see also Cowles *et al.* 2007) reached a somewhat different conclusion concerning the effects of topicality and focus. Cowles tested three types of antecedent: (i) focused subjects, (ii) discourse-old full NP subjects, and (iii) discourse-new full NP subjects. On the basis of a cross-modal naming study, she concluded that "[a]ll three information statuses [discourse-topic, sentence-topic and contrastive focus, EK] appear to make their referent more likely to be interpreted as the antecedent of a subsequent pronoun" (Cowles 2003: 93). Thus, in contrast to Arnold, who found that established discourse-topics (pronominalized subjects) are more salient than focused objects, Cowles's results lead her to conclude that "two information-structure types [topic and focus, EK] that are considered distinct... appear to have the same psychological effect" (Cowles 2003: 94). However, as Cowles *et al.* (2007) note, detailed analysis of the results hints that focusing may not have quite as powerful an effect as subjecthood does, although they note that this is not entirely clear.

The differences in Arnold's and Cowles's findings—in particular Arnold's conclusion that (discourse) topics are more salient than foci and Cowles's conclusion that topics and foci have comparable effects on salience—may stem from methodological differences. However, the disparity could also stem from differences in grammatical role and pronominalization. Arnold's finding that discourse-old pronominalized subjects (which she analyzes as discourse-topics) are more prominent than focused objects could be related to their respective grammatical roles (and/or NP form) in addition to their topic/focus status. In Cowles's experiments, both topics and foci were full names in subject position, which may have meant that (i) Cowles's foci (in subject position) were more prominent than Arnold's foci (in object position), due to the difference in grammatical role and/or that (ii) Cowles's topics (full NPs in subject position) were less prominent than Arnold's pronominalized subjects, due to effects of NP form.

Thus, although existing experimental work suggests that focused entities and discourse-old subjects (which some analyze as topics) are more salient than other referents (see also Navarretta 2002), questions remain as to how the different factors involved (focus, subjecthood, pronominalization/givenness) relate to each other. The experiment reported here builds on Arnold's and Cowles's work by investigating both focused subjects and objects, as well as pronominalized subjects and objects.

14.3 Connection between pronoun interpretation and mention of other referents

The work on reference resolution discussed in the preceding sections focused primarily on what aspects of the preceding discourse influence the interpretation of pronouns in subject position. In this section I consider the role of subsequent discourse, namely, whether the interpretation of a subject pronoun is connected to the presence of other referential forms later in the same sentence. It is clear that interpretation of a subject-position pronoun can be influenced by the semantics of the verb that follows the pronoun (see e.g. Winograd 1972), but here we focus on the potential effects of another kind of post-pronominal information, namely the referential properties of the rest of the clause.

The idea that the interpretation of the initial (subject) pronoun is connected to the likelihood that another referent will or will not be mentioned is addressed in Centering Theory, an influential computationally based model of the local-level component of attentional state in discourse. In this section, I first review some basic tenets of Centering Theory, and then in Section 14.3.1 I outline how, according to Centering, the presence or absence of other referents in the pronoun-containing clause is connected to the interpretation of the subject-position pronoun.

Centering Theory, formulated on the basis of extensive corpus work, makes predictions regarding how focus of attention, choice of referring expression and local coherence are connected (Grosz, Joshi, and Weinstein 1995). According to Centering, entities mentioned in an utterance (called "centers") are ranked in terms of how salient (how "central") they are. It is commonly assumed that in English, this ranking is done by grammatical role, with subjects being more salient than objects. Centering Theory provides an implementable way of analyzing the coherence relations between pairs of utterances—in other words, of assessing how much of a load they impose on the human language processing system. The coherence of the transition from one utterance to the next utterance depends (i) on whether, (ii) in what position, and (iii) with what form the most central entity from one sentence is mentioned in the next sentence. For example, using a subject pronoun in Utterance 2 to refer to the most central entity mentioned in Utterance 1 results in a more coherent transition than use of an object pronoun in Utterance 2 to refer to the most central entity in Utterance 1. Centering Theory posits four transition types, of differing levels of coherence (i.e. involving different amounts of processing load), summarized in the Appendix.[4]

Although the original Centering-theoretic work by Grosz *et al.* (1995 and earlier versions) is not a pronoun-resolution algorithm, an algorithm for pronoun interpretation based on Centering principles was proposed by Brennan, Friedman, and Pollard (1987, BFP). The BFP algorithm assumes that more coherent transitions are preferred over less coherent transitions (i.e. there is a bias to minimize processing load), and the algorithm assigns pronominal reference accordingly (see also Poesio et al. 2004). Thus, pronouns are resolved in such a way that the resulting transition is as coherent as possible.

14.3.1 *Effects of subsequent referents on pronoun interpretation*

To illustrate how, in Centering Theory, the interpretation of the initial pronoun is connected to the likelihood of subsequent mentions of other entities, let us consider (1). I start by considering a situation where the subject pronoun at the start of the third sentence is interpreted as referring to the preceding *object*:

(1) a. John was sitting by the window. Peter tickled him$_{\text{John}}$. He$_{\text{John}}$ laughed.

b. John was sitting by the window. Peter tickled him$_{\text{John}}$. He$_{\text{John}}$ tickled {him$_{\text{Peter}}$/Peter} back.

According to Centering Theory, in a context such as (1), if the subject pronoun *he* in the third sentence is interpreted as referring to the preceding object (*John*), then continuation (1a)—in which the preceding subject Peter is not mentioned—is more coherent than continuation (1b), in which the preceding subject Peter is mentioned in a non-subject position later in the sentence (see Kehler 1997 for discussion of similar examples).

Centering Theory allows for the possibility that not only grammatical role but also different syntactic constructions (e.g. clefts) and semantic/pragmatic

[4] The reader is referred to Grosz *et al.* (1995) and Walker, Joshi, and Prince (1998) and the references cited therein for further details (see also Poesio, Stevenson, di Eugenio, and Hitzeman 2004 for recent and extensive Centering Theoretic analyses of large corpora).

factors influence which entities are most central. However, for the purposes for this chapter, I make the simplifying assumption that the predictions regarding the consequences of the initial pronoun's interpretation are the same for clefts and SVO sentences. Centering Theory also makes claims regarding the referential form used to refer to the preceding subject (Rule 1, see e.g. Walker *et al.* 1998: 4), which I do not discuss here. In the subsequent discussion, pronominal and full NP mentions are grouped together (as in (1b)), but I acknowledge the importance of distinguishing them in future work.

Let us now consider a situation where the subject pronoun *he* in the third sentence refers to the preceding full NP subject (Peter), as in (1c, d). In this situation, according to Centering Theory, a continuation that mentions the preceding object (1d) is as coherent as one which does not (1c). However, neither is as coherent as continuation (1a), in which the subject pronoun refers to the preceding pronominalized object and no other referents are mentioned.

(1) c. John was sitting by the window. Peter tickled him_{John}. He_{Peter} laughed.

d. John was sitting by the window. Peter tickled him_{John}. He_{Peter} pushed {him_{John}/John} over.

If we follow Brennan *et al.* and assume that maximally coherent transitions are easiest to process and to be preferred, we expect that in sequences like (1), (i) a subject pronoun referring to the preceding (pronominalized) object will probably not be followed by a mention of the preceding subject later in the sentence (i.e. (1a) is more coherent than (1b)), whereas (ii) a subject pronoun that refers to the preceding subject may or may not be followed by a mention of the preceding object ((1c) is as coherent as (1d)).

As Kehler (1997) notes, the idea that the maximally coherent interpretation of "he" depends on post-pronominal information seems to be at odds with the idea that pronouns in subject position are interpreted quickly and incrementally (e.g. Arnold, Eisenband, Brown-Schmidt, and Trueswell 2000; see Kehler 1997). However, a sizeable body of psycholinguistic research suggests that the human language processing system can consider different structures/interpretations in parallel (e.g. Tanenhaus and Trueswell 1995 for an overview). This raises the possibility that upon encountering a pronoun, the language comprehension process does not pick a single antecedent but keeps track of ranked alternatives. We could have a system that creates an initial ranking of possible referents incrementally, but is capable of changing the ranking of the alternatives later. If this is on the right track, the claim that interpretation of the sentence-initial pronoun is connected to, or has consequences for, what gets mentioned later seems less implausible.

14.4 Experiment

The experiment discussed in this chapter has two main aims: (i) to investigate how information-structural and syntactic factors (focus, pronominalization, and subjecthood) influence subsequent pronoun use, and (ii) to shed light on whether the interpretation of one referential form (the pronoun in subject position) is connected to the likelihood of other referents being mentioned in subsequent parts of the utterance.

To investigate these questions, I used a sentence-completion task based on dialogues between two hypothetical speakers: speaker A and speaker B. On critical trials, speaker B's response contained the critical target sentence, and ended with a prompt pronoun. Participants played the role of speaker B, and their task was to provide a natural-sounding continuation beginning with the prompt pronoun.

In the critical sentence, syntactic form (cleft vs. SVO) and the grammatical role of the focused constituent[5] were manipulated, as illustrated in (2). This resulted in four conditions: [Cleft.Object = focus], [Cleft.Subject = focus], [SVO.Object = focus] and [SVO.Subject = focus]. In the clefted conditions, the preceding context and the syntactic construction mark one of the arguments as being in contrastive focus (see Rooth 1992); the rest of the sentence is presupposed. In the SVO conditions, the focus–presupposition division is not encoded by a particular syntactic construction but follows from the preceding context. The argument in the presupposed part of the sentence is discourse-old and pronominalized (subject in (2a, c), object in (2b, d)).

(2) a. [Cleft.Object = focus]
A: The maid scolded the bride.
B: No, that's wrong! It was the secretary$_{\text{focus}}$ that she scolded. She . . .

b. [Cleft.Subject = focus]
A: The maid scolded the bride.
B: No, that's wrong! It was the secretary$_{\text{focus}}$ who scolded her. She . . .

c. [SVO.Object = focus]
A: The maid scolded the bride.
B: No, that's wrong! She scolded the secretary$_{\text{focus}}$. She . . . .

d. [SVO.Subject = focus]
A: The maid scolded the bride.
B: No, that's wrong! The secretary$_{\text{focus}}$ scolded her. She . . .

[5] Not all factors are fully crossed in this experiment, partly due to the nature of the phenomena under investigation. In this design, pronominalization and focus are in complementary distribution: a particular referent is either pronominalized or focused.

If one follows Reinhart (1982), Givón (1983), Arnold (1999), Cowles (2003) and others in connecting subjecthood with topicality, then it seems to follow that subjects in the [Object = focus] conditions are topics, and subjects in the [Subject = focus] conditions are presumably contrastive topics. However, it seems that some other researchers would disagree with this classification: E.F. Prince (2003) and Beaver (2004) would presumably regard the discourse-old pronominalized object as the the topic in the [Subject = focus] conditions, not the focused discourse-new subject. Since a particular definition of topicality is not central for my claims, I will not commit to one view over the other; rather, I will discuss the arguments in terms of their grammatical role, NP form, and focus status.

The verbs used in the 16 target items were agent–patient verbs (as defined by Stevenson, Crawley, and Kleinman 1994), and the three human referents in each target had the same gender; (stereotypically) male or female. The experiment also contained 16 fillers. Participants ($n = 24$) were asked to provide a natural-sounding continuation sentence using the pronoun prompt. They were instructed to imagine that someone had just made the claim in part A, and they were now responding to this person's statement by saying part B out loud and providing a continuation. Parts A and B, as well as the prompt pronoun, were presented to the participants in writing. The responses of the participants were recorded and transcribed. The referent of the prompt pronoun was coded by two coders working independently. If it was not clear who the pronoun referred to, the item was coded as "unclear". The continuations were also coded for whether the other argument of the immediately preceding sentence (i.e. whichever argument the pronoun did not refer to) was mentioned. In this chapter I only discuss a simple level of "other argument" coding, one that does not take into account referential form or specific grammatical position of the other argument.

14.4.1 *Predictions*

14.4.1.1 *Predictions regarding the interpretation of subject pronouns* Researchers differ as to whether they view anaphor resolution as being a process that is determined by a single factor or guided by multiple constraints (e.g. Ariel 1990, Arnold 1998, Strube and Hahn 1999, Kaiser and Trueswell 2008.). If a single factor determines reference resolution, the predictions are straightforward. If subjecthood is what matters, the prediction is that the prompt pronoun will consistently refer to the immediately preceding subject, regardless of NP form or focus status. (Note that the subjects in the experiment are agentive so the design cannot distinguish subjecthood from agentivity.) If pronominalization

TABLE 14.1 Predicted effects of subject-pronoun interpretation on subsequent mentions

	Antecedent of pronoun	Is other referent mentioned later in the sentence?
[Subject = focus]	Preceding subject	Yes or no
	Preceding object	No
[Object = focus]	Preceding subject	Yes or no
	Preceding object	(Yes or no)

is crucial, pronouns should always refer to the pronominalized argument in the preceding sentence. If focusing is what matters, prompt pronouns are predicted to refer to the focused referent, regardless of grammatical role.

Turning now to the predictions of a multiple-factor view, if we assume that all three factors contribute equally, we predict an equally strong preference for the subject of the immediately preceding clause in all four conditions. Let us first consider the [Subject = focus] conditions. In both the [SVO.Subject = focus] and the [Cleft.Subject = focus] conditions, pronominalization pushes towards the object, but subjecthood and focusing both point to the subject. Thus, if the three factors contribute equally, the subject is predicted to be the preferred antecedent. Turning now to the [Object = focus] conditions, in both the [SVO.Object = focus] and the [Cleft.Object = focus] conditions, subjecthood and pronominalization point towards the subject, whereas focusing points to the object. Again, if all factors matter equally, the subject is predicted to be the preferred antecedent.

However, it may well turn out to be the case that the three factors are not all weighted equally. In fact, constraint-based models of language processing (e.g. MacDonald, Pearlmutter, and Seidenberg 1994, Trueswell, Tanenhaus, and Garnsey 1994, Tanenhaus and Trueswell 1995; see also Bates and MacWhinney 1989) claim that language processing is guided by constraints of differing weights (see also Arnold 1998). I return to this issue in the results section.

14.4.1.2 *Predictions regarding initial pronoun interpretation and subsequent mentions* The Centering-Theory-derived predictions for how the interpretation of the initial pronoun is connected to the likelihood of the other referent being mentioned in the remainder of the sentence are shown in Table 14.1. These predictions are largely derived from the BFP algorithm, which rests on the assumptions articulated in Centering Theory such as the ranking of the different types of transitions that hold between utterances (see Section 14.3). As mentioned above, for ease of exposition I assume that the predictions are the same for clefted and non-clefted sentences. In addition, I only discuss a simple level of "other referent" coding that does not take into account the other referent's referential form or grammatical position. In this regard, the data analysis presented in this chapter is best regarded as a simplified adaptation of Centering Theory.

In conditions where the subject is focused and the object is pronominalized (e.g. *The secretary*$_{\text{focus}}$ *scolded her. She*...), the BFP algorithm predicts a strong asymmetry between subject interpretations and object interpretations of the sentence-initial pronoun prompt. In subject-focus conditions, if the prompt pronoun refers to the preceding object, the prediction is that the preceding object may or may not be mentioned later in the prompt-pronoun sentence ("yes or no" in Table 14.1). This is because the resulting transitions[6] are equally coherent in Centering terms (both are "smooth-shift" transitions). However, if the prompt pronoun refers to the preceding object, the prediction is that the preceding subject is not mentioned. The transition resulting from not mentioning the subject ("continue", a very coherent transition) and the transition resulting from mentioning the subject ("rough shift", a very incoherent transition) are two extremes of the Centering transition hierarchy, and thus if we follow BFP, we predict a preference for the most coherent transition over the least coherent transition and thus a preference for the subject to not be mentioned.

In conditions where the object is focused and the subject is pronominalized (e.g. *She scolded the secretary*$_{\text{focus}}$. *She*...), we do not expect such a clear asymmetry between subject interpretations and object interpretations. If the prompt pronoun is used to refer to the preceding subject (e.g. *She scolded the secretary. She was really angry.*), the prediction is that the preceding object (*secretary*) may or may not be mentioned. This is because both situations lead to equally coherent transitions. (Both involve a continue transition.) However, if in these conditions the pronoun refers to the preceding (focused) object (e.g. *She (maid) scolded the secretary. She really didn't like being scolded by the maid.*), then according to the traditional view of transition rankings and the BFP algorithm, the expectation is that there will be a preference for the preceding subject to be mentioned in the continuation sentence (which results in a retain transition), as compared to being left unsaid (which results in a less coherent smooth-shift transition). However, a number of Centering researchers have expressed doubts regarding the traditional view that smooth-shift transitions really are less coherent than retain transitions (see e.g. Poesio

[6] It is commonly assumed that four transition types exist, listed here ranked from most coherent to least coherent: Continue > Retain > Smooth-shift > Rough-shift. An overview of some central ideas in Centering Theory is provided in the Appendix, and further details are available in Grosz *et al.* (1995) and Walker *et al.* (1998).

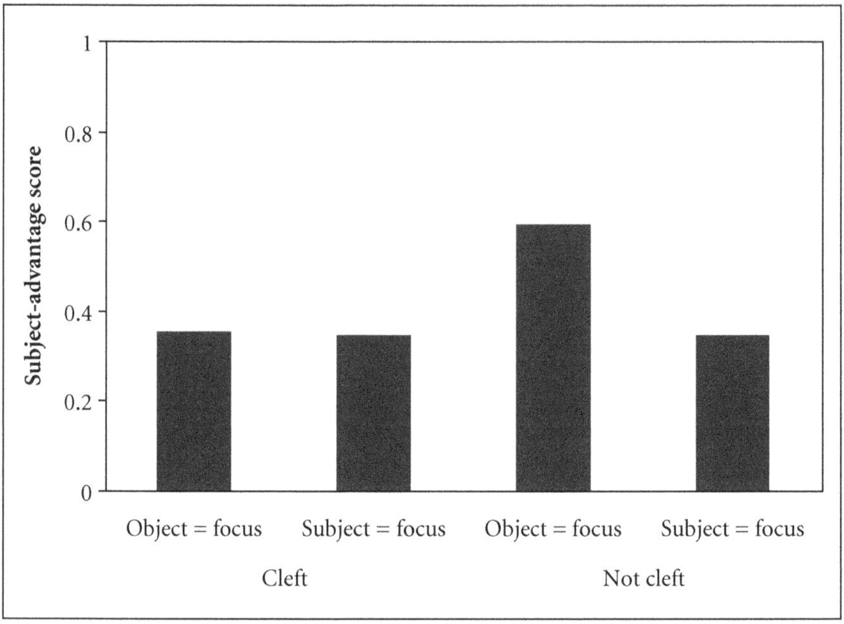

FIGURE 14.1 Subject-advantage scores (proportion of subject continuations minus proportion of object continuations)

et al. 2004), and thus a more reasonable prediction might be that if the initial pronoun refers to the preceding object, the subject may or may not be mentioned.

14.4.2 *Results and discussion*

14.4.2.1 *Interpretation of subject pronouns* The continuations of the participants show a strong overall bias to interpret a prompt pronoun as referring to the agentive subject of the immediately preceding clause, as illustrated in Figure 14.1. The Figure shows the average subject advantage scores obtained by subtracting the proportion of object continuations from the proportion of subject continuations. A positive number means that there were more subject continuations than object continuations, a negative number, that there were more object continuations than subject continuations. There were significantly more subject interpretations than object interpretations in all four conditions.

The overall subject preference shows that (i) grammatical role makes an important contribution to the salience of focused and pronominalized entities and that (ii) neither focus nor pronominalization outweighs subjecthood as there is no overarching preference for pronouns to refer to either focused or pronominalized antecedents. Rather, what we see is that pronouns are most likely to be interpreted as referring to (agentive) subjects, regardless of whether the subjects are pronominalized or focused (either in a cleft or *in situ*). This finding highlights the importance of taking grammatical role into account when investigating effects of topicality and focusing, and could be taken as an indication that subjects do not need to be discourse-old or pronominalized in order to be salient.

Although these results attest to the importance of grammatical role, should we conclude that subjecthood is the lone factor that determines pronoun interpretation? A number of reasons argue against such a conclusion. First, a large body of existing work has found reliable effects of other factors in addition to subjecthood (see Section 14.3). Second, as discussed in the predictions, given the set-up of this particular experiment, a subject bias could arise even if multiple factors are underlyingly contributing to referent salience and pronoun interpretation. Thus, an overall subject bias does not provide evidence that only subjecthood matters.

Thirdly, as can be seen in Figure 14.1, the subject advantage does not seem to be equally strong in all four conditions; [SVO.Object = focus] shows a slightly stronger subject advantage than the other three conditions. However, analyses of variance by subjects and by items indicate that the differences between [SVO.Object = focus] and the other conditions are not overwhelming, in some cases reaching only marginal significance or less. This means that we should be careful not to attribute too much importance to the asymmetry visible in Figure 14.1 without further empirical evidence. Nevertheless, the results provide preliminary support for the view that multiple factors guide pronoun interpretation.

In fact, the results for most of the conditions seem to fit well with the view that multiple factors matter (i.e. pronominalization, focusing, and subjecthood all play a role) but also provide some hints that focusing may matter less than subjecthood and pronominalization. In such a situation, the expectation is that pronominalized objects should compete with subjects more than focused objects. In other words, conditions where subjecthood and pronominalization coincide (i.e. the [Object = focus] conditions, [SVO.Object = focus] and [Cleft.Object = focus]) should exhibit a stronger subject preference than conditions where subjecthood and focus coincide (i.e. the [Subject = focus] conditions, [SVO.Subject = focus] and [Cleft.Subject = focus]). This prediction fits with the finding that [Subject = focus] conditions exhibit a numerically weaker subject preference than the [SVO.Object = focus] condition, but raises the question of why the [Cleft.Object = focus] condition does not exhibit as strong a subject preference as [SVO.Object = focus]. A possible reason for the different behavior of the two [Object = focus] conditions has to do with the fact that the [Cleft.Object = focus] condition is the only condition in which the object linearly precedes the subject (and the subject is in a non-prominent syntactic position). In light of existing claims regarding effects of linear order on salience (e.g. Gernsbacher and Hargreaves 1988, 1992), one might expect the object to receive a boost in this condition and thus to weaken the subject preference (see Arnold 1998, 1999 on salience as a competitive phenomenon). If this interpretation is on the right track, then—although there is no clear main effect of cleft vs. SVO—it suggests that clefting a focused object increases its chances of being referred to by a subsequent pronoun.[7]

In sum, if we attribute a greater influence to pronominalization and grammatical role than to focusing and also take into account the potential object-boosting effects of clefting and/or linear order in the [Cleft.Object = focus] condition, we correctly predict the observed data. However, further research should be conducted to confirm whether focusing is indeed weighted less heavily. Crucially, even on the basis of the current results, it is clear that grammatical role needs to be taken into account when investigating effects of notions such as "topic" and "focus" on salience and pronoun interpretation.

14.4.2.2 *Initial pronoun interpretation and subsequent mentions* Now we turn to the question of whether the interpretation of subject pronouns is influenced by (or influences) what is mentioned later in the same clause. Figure 14.2 shows what percentage of subject and object continuations (i.e. continuations in which the prompt pronoun was interpreted as referring to the preceding subject vs. object) mention the other referent from the immediately preceding clause (collapsing across clefts and SVO sentences).

As can be seen in the figure, the interpretation of the initial pronoun is indeed connected to the likelihood of the other referent being mentioned later in the sentence. Overall, there were fewer mentions of the other referent in those continuations in which the sentence-initial prompt pronoun was used to refer to the preceding object than in those continuations in which the prompt pronoun was used to refer to the subject. Furthermore, more detailed analyses

[7] Parallelism may be an additional factor that one might expect to play a role. According to Smyth (1994), pronouns prefer antecedents that are in the same syntactic position as the pronoun. However, according to Smyth, this parallel-position bias holds only when both sentences have the same global structure and matching thematic roles (e.g. *Peter pushed John. Alex pinched him.*) However, an examination of participants' continuations suggests that the required degree of matching across sentences does not appear to be consistently present, which casts doubt on the idea that parallelism is at work here.

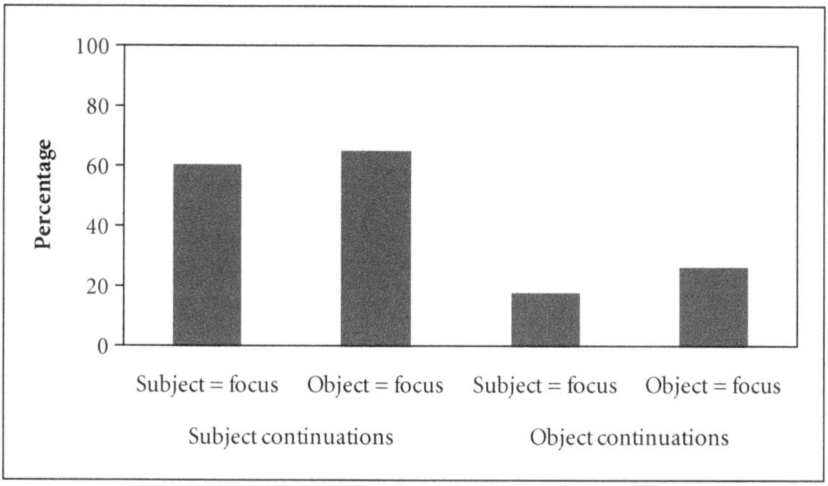

FIGURE 14.2 Percentage of subject continuations and object continuations that mention the other argument from the preceding sentence.

suggest that there do not appear to be overarching effects of topicality, focusing or syntactic form on the likelihood of the other referent being mentioned.

Let us now take a closer look at the [Subject = focus] and [Object = focus] conditions, to see how the data fit with the predictions sketched out in Section 14.4.1.2. In the [Subject = focus] conditions, there is an asymmetry between subject continuations and object continuations, as subject continuations mention the other referent more frequently (> 60%) than object continuations do (< 20%). This fits with the Centering-based predictions in Table 14.1.

In the [Object = focus] conditions, there is an asymmetry in the same direction as with the [Subject = focus] conditions—the other referent seems more likely to be mentioned in subject continuations than in object continuations (approx. 65% vs. < 30%). These data do not fit with the predictions outlined in Section 14.4.1.2. that subject and object continuations would show similar rates of mention, nor do they support the more specific prediction that object continuations would be more likely to involve mention of the other argument than subject continuations.

In sum, the data support the predictions for the [Subject = focus] conditions, but not the predictions for the [Object = focus] conditions. However, although those specific [Object = focus] predictions are not supported, the striking asymmetry between subject and object continuations nevertheless shows that the interpretation of subject-position pronouns is indeed connected to the likelihood of other referents being mentioned: (i) if the initial pronoun refers to the preceding subject, the continuation often mentions the preceding object as well, but (ii) if the initial pronoun refers to the preceding object, the preceding subject is less likely to be mentioned. Thus, it seems that grammatical role is what matters; there are no clear across-the-board effects of pronominalization, focusing, or syntactic packaging on the likelihood of the other referent being mentioned.

If we assume that subjecthood plays a central role in determining salience, one way of describing the results is as follows: if the sentence-initial pronoun is interpreted as referring to a highly salient referent (the preceding subject), the sentence may also include mention of a lower-salience referent later on. However, if the sentence-initial pronoun is interpreted as referring to a previously lower-salience referent (preceding object), the remainder of the sentence is less likely to make reference to the higher-salience referent of the preceding sentence. In other words, if a less-salient referent is promoted (becomes more salient) by being interpreted as the antecedent for a subject-position pronoun, language users avoid subsequent mention of the preceding higher-salience referent (preceding subject) in a lower-salience position in the same clause. Furthermore, these patterns could also be related to the nature of the semantic relation between the pronoun-initial clause and the preceding clause (e.g. a cause-effect relation or a resemblance relation, see Kehler 2002, Kehler, Kertz, Rohde, and Elman 2008), a question worthy of future investigation.

These findings have implications for our view of reference resolution because they provide further support for the view that pronoun resolution depends on both preceding (pre-pronominal) and subsequent (post-pronominal) information. The data are compatible with an approach which assumes that even though a comprehender can process a sentence incrementally and constructs a weighted set of possible antecedents for a pronoun upon first encountering it, information later on in the sentence can have an impact on the weighting of these alternatives (see also Winograd 1972 on the effects of semantic information).

14.5 Conclusions

This chapter investigated two questions related to reference resolution: (i) How do information-structural factors and syntactic factors (in particular focusing, pronominalization, and subjecthood) interact to guide the process of reference resolution? (ii) Is the interpretation of a sentence-initial anaphoric form connected to the presence or absence of other referential forms later in the same sentence? The first question looks backwards from an anaphoric form, towards preceding discourse, whereas the second question looks forward, towards subsequent discourse.

When it comes to the first question, the results of the experiment discussed here are compatible with the idea that reference resolution is guided by multiple interacting factors (see also Kaiser and Trueswell 2008 and others). The results indicate that pronouns are most likely to be interpreted as referring to (agentive) subjects regardless of whether they are pronominalized or focused. Participants' continuations indicate that—at least in the contexts investigated here—a contrastively focused, discourse-new, subject is preferred over a discourse-old object. If pronouns are used to refer to highly salient antecedents, our results show that subjects can be highly salient without being discourse-old or pronominalized. In addition, the results provide some hints of (i) object-clefts boosting the salience of the focused object (and/or resulting in a loss of salience of the subject) and (ii) a strength asymmetry between pronominalization, subjecthood and focusing, as it seems that pronominalization and subjecthood may contribute more than focusing. These findings, however, are only tentative and need to be investigated more specifically in further work.

The second question sheds light on the question of whether a purely "backward-looking" approach is sufficient for pronoun interpretation. The results support the idea that the interpretation of a pronoun is influenced not only by properties of preceding discourse, but also by what else is mentioned later in the clause containing the pronoun (see Grosz *et al.* 1995, see also Winograd 1972 on effects of semantics). The results show that the interpretation of the subject pronoun (whether it refers to the preceding subject or object) is connected to the likelihood of the other argument from the preceding sentence being mentioned in the same clause: subject continuations were more likely to include a mention of the other argument than object continuations. This finding could be interpreted as follows: If a less salient referent (object) is upgraded by being interpreted the antecedent for a subject-position pronoun, subsequent mention of the preceding high-salience referent (preceding subject) in a lower-salience position in the same clause is dispreferred.

Of course, many questions remain open, and further research should be conducted to investigate the validity of the claims discussed here. For example, given data suggesting that different factors may be weighed differently, we would like to know more about the extent of cross-linguistic variation. In addition, the relation between agentivity and subjecthood merits further study. The experiment discussed here investigated agentive subjects, and thus agentivity and subjecthood are fully correlated. To find out whether the structural notion of subjecthood or the semantics of agentivity is behind the strong effects of subjecthood on subsequent pronominalization, agentive and non-agentive subjects (e.g. experiencers) need to be compared. To learn more about the expectations that participants build up over the course of a sentence after having encountered an anaphoric expression, future work should also test directly the on-line prediction that comprehenders may need to revise their initial preferred interpretation of a pronoun based on subsequent information (see also Kehler 1997). Furthermore, better to understand the nature of post-pronominal effects, it would be useful to investigate the semantic relations between sentences, to see whether different completion types correlate with different inter-sentential relations (e.g. cause-effect vs. similarity, see Kehler 2002, Kehler *et al.* 2008 for details).

In conclusion, the findings presented here fit with the idea that reference resolution is influenced both by preceding discourse and subsequent discourse. Taken as a whole, the results of the two analyses indicate that subjecthood, in particular, plays an important role in guiding both the backward- and forward-looking aspects of pronoun resolution.

Appendix: Centering Theory

Some of the core ideas, constraints, and rules underlying Centering Theory (from Walker *et al.* 1998: 3–6) are provided here.

- "The set of forward-looking centers, Cf(U_i, D), represents discourse entities evoked by an utterance U_i in a discourse segment D" (Walker *et al.* 1998: 3).
- "The highest-ranked member of the set of forward-looking centers is referred to as the preferred center, Cp." (Walker *et al.* 1998: 3). The ranking of the Cf list can depend on various factors. Centering-based algorithms implemented for English often use a grammatical-role based ranking, in which subject is ranked above object.

"CONSTRAINTS: For each utterance U_i in a discourse segment D consisting of utterances $U_1, \ldots, U_m$:

1. There is at most one backward-looking center, Cb(U_i, D).
2. Every element of the forward-looking centers list, Cf(U_i, D), must be realized in U_i.
3. The center, Cb(U_i, D) is the highest-ranked element of Cf(U_{i-1}, D) that is realized in U_i." (Walker *et al.* 1998: 3)

"RULES: For each utterance U_i in a discourse segment D consisting of utterances $U_1, \ldots, U_m$:

1. If some element of $Cf(U_{i-1}, D)$ is realized as a pronoun in U_i, then so is $Cb(U_i, D)$.
2. Transition states are ordered. The Continue transition is preferred to the Retain transition, which is preferred to the Smooth-Shift transition which is preferred to the Rough-Shift transition." (Walker *et al.* 1998: 4)

Centering transitions:

	$Cb(U_i) = Cb(U_{i-1})$ (or $Cb(U_{i-1})$ undefined)	$Cb(U_i) \neq Cb(U_{i-1})$
$Cb(U_i) = Cp(U_i)$	CONTINUE	SMOOTH-SHIFT
$Cb(U_i) \neq Cp(U_i)$	RETAIN	ROUGH-SHIFT

15

Given and new information in spatial statements

ROBIN HÖRNIG AND THOMAS WESKOTT

15.1 Introduction

Under what circumstances does a speaker tell her addressee *The box is under the table*? This sounds like a trivial question. She will do so if the box is under the table. Yet, how do circumstances differ prompting a speaker to instead utter *Under the table is the box*? Maybe an English speaker would avoid this latter utterance anyway, but the corresponding marked structure *Unter dem Tisch ist die Schachtel* is frequent in German. In this chapter we examine from a psycholinguistic point of view under what circumstances unmarked and marked (German) variants of spatial relational sentences are easier to comprehend and, to a certain extent, under which circumstances a speaker will utter which variant. We argue that information-structural properties determine the appropriateness of the utterances in different contexts, the givenness of the entities and their places involved in spatial relational sentences. Section 15.2 introduces two candidates competing for the default partitioning of given and new information in spatial-relational sentences with unmarked word order. Section 15.3 extends the competition to marked sentences and a third candidate. Section 15.4 investigates the mixed discourse status of the winner's preverbal constituent by comparing two approaches to contextual licensing. Section 15.5 substantiates the information-structural basis of the experimental evidence and, presenting a fourth candidate, reconsiders the default partitioning of unmarked sentences. The conclusion is presented in Section 15.6.

This research was funded by the Deutsche Forschungsgemeinschaft. We thank Gisbert Fanselow, Caroline Féry, Reinhold Kliegl, Klaus Oberauer, and Andrea Weidenfeld, with whom we work or have been working on the issues addressed in this chapter.

15.2 The default given–new partitioning of unmarked spatial-relational sentences

15.2.1 *Candidate 1, the original*

An early proponent of the important role of the given–new distinction in language was Chafe:

> As a starting point for our tentative discussion here, we can recall that language permits the transfer of information from the mind of the speaker to the mind of the hearer. [...] Now, typically it is the case that the speaker assumes that some of the information he is communicating is new; it is information he is introducing into the hearer's mind for the first time. But typically it is also the case that some of the information in the sentence is not new. Some of it is information which the speaker and hearer share at the time the sentence is spoken. This shared information constitutes a kind of starting point based on concepts already "in the air," to which the new information can be related. The old information may be shared from the common environment in which both the speaker and the hearer are interacting. Frequently it is also shared on the basis of the sentences already uttered. (Chafe 1970: 211)

Some information in utterances is given or old. Given information is part of the common ground, the set of beliefs shared by speaker and addressee. Information can be contextually given in the perceptually available extralinguistic context shared by speaker and addressee. A second option is that information can be discourse-given through previous mentioning. Information can also be given in both respects. Chafe's basic idea is that (English) declarative sentences can be decomposed into a default partitioning of given and new information, which he calls *least marked*. New information, or part of it, receives the highest pitch accent. Sentence (1) presents Chafe's example (5a), a spatial relational sentence:

(1) The box is under the table.

Chafe stipulates a context in which someone has asked *Where is the box?* rendering the box discourse-given. Chafe assigns (1) the default partitioning (P1) ("P" stands for partitioning):

(P1)

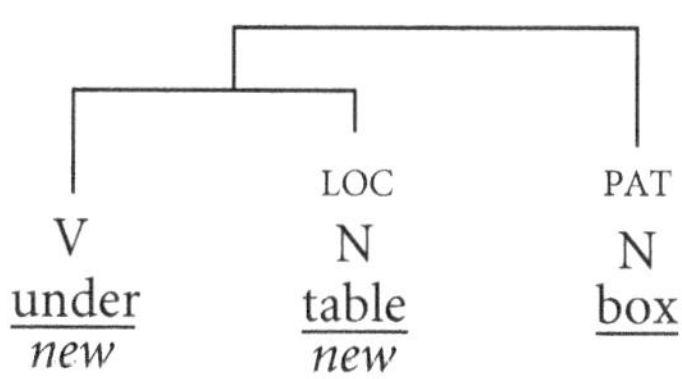

The noun *table* in (1) receives the highest pitch accent and counts as new; the box is classified as given (not tagged by Chafe). Though not preserved in (P1), the proposed default partitioning coincides with a given–new ordering of the denotata of the two noun phrases *the box* and *the table*.

From a processing perspective we can pick up the notion of a given–new default partitioning and hypothesize that a default partitioning is easier to process than a marked partitioning. Clark and Haviland (1977) directly translate Chafe's claim that shared information constitutes a starting point to which new information can be related into processing assumptions. They argue that the integration of new information proceeds in three successive steps: (1) determine the given information in the sentence; (2) access the antecedent in memory; and (3) attach the new information to the antecedent in memory. Sentence processing is generally considered to be easier if given information precedes new information for two reasons. First, providing given information early in the sentence speeds up antecedent access. Secondly, early encoded new information increases memory load because it must be held in abeyance until integration becomes possible. Integrating new information is therefore faster and more accurate if given information is encoded first.

Yekovich, Walker, and Blackman (1979) confirmed a given–new advantage in sentence processing. They manipulated the order of given and new information in SVO target sentences via preceding context sentences and showed that identical target sentences were read faster with a given–new than with a new–given order. If the given–new advantage is indeed a general principle of language processing, the spatial relational sentence in (1) should be easier to comprehend if the box is given and the table is new rather than the other way around. Such a finding would corroborate the default partitioning of given and new information proposed by Chafe. The partitioning in (P1) is chosen as candidate one.

15.2.2 *Nominating candidate 2*

Chafe claims that, though only the noun *table* is prosodically tagged as new, the whole prepositional phrase is new. This is an interesting point. The prepositional phrase *under the table* denotes the place of the box and the interlocutor does usually not know the place of the box when she asks, *Where is the box?* The answer usually adds the information about the place of the box to the common ground. However, if someone answers the question with (1) and if this reply is intended to help the addressee, the speaker must assume that the table and its place are part of the common ground. This precondition is immanent in the idea of linguistic localization, to which we now turn.

We assume that spatial relational statements like *The box is under the table* are generally uttered to localize an entity linguistically. Such statements are used to inform the addressee about the as yet unknown place of the *locatum* (or *located object* or *figure*), the entity which is linguistically localized. The locatum usually figures as the grammatical subject of the statement—*the box*, for instance. The prepositional phrase *under the table* is the *locative expression*. The object noun phrase within the locative expression denotes the *relatum* (or *reference object* or *ground*), the entity relative to which the locatum is localized, for instance, *the table*. Miller and Johnson-Laird (1976) postulate that there are two nested search domains involved in a spatial relational localization. The first search domain surrounds the place of the relatum. The second search domain, containing the place of the locatum, is a subdomain of the first one which is determined dependent on the spatial preposition applied to the relatum.[1] This conception induces a distinct asymmetry on the arguments of a spatial preposition, the relatum and the locatum, which is commonly acknowledged in the spatial language literature (Clark 1973; Talmy 1983; Levelt 1984; Herskovits 1986; Landau and Jackendoff 1993; Levinson 1996). Importantly, Miller and Johnson-Laird claim that the first search domain, the place of the relatum, should be contextually given, so it is not a domain that has to be searched for. That the place of the relatum is part of the common ground implies that the relatum is part of the common ground, too. The place of the locatum is not part of the common ground. The locatum may be part of the common ground, like the box in Chafe's example, or not.

Placement tasks in different extralinguistic contexts have corroborated the role asymmetry in spatial relational localizations. Huttenlocher and Strauss (1968) employed a context which itself established an asymmetry in the givenness of places of two entities. They showed that young children (4½–5 and 6½–7 years old) are much faster and more often correct in placing a mobile block relative to a fixed block if instructed by a spatial relational sentence assigning the relatum role to the fixed block and the locatum role to the mobile block than with roles assigned the other way around. Hence, instructions were much easier to comprehend when the place of the relatum was given and the place of the locatum was new. The same effect has been obtained with adults (Clark 1972; L. J. Harris 1975). Huttenlocher and Weiner (1971) implemented a displacement task, a variant of the placement task which does not establish a contextual asymmetry. Children displaced one of two misarranged

[1] Miller and Johnson-Laird (1976) distinguish between the *intrinsic* and the *deictic* reading of spatial prepositions, a distinction which we consider to be irrelevant for the issues addressed in this chapter.

blocks as instructed by spatial relational sentences. They much more often displaced the locatum relative to the relatum than the reverse, showing that the utterance itself induced a powerful asymmetry on the extralinguistic context. This outcome is expected on the linguistic localization account. Displacing a locatum updates the common ground such that the place of the locatum is new and the place of the relatum is given.

Huttenlocher suggested that "perceived agentivity" constitutes a congruency between a mobile entity and the sentence subject. To dissociate grammatical and logical subject, Huttenlocher, Eisenberg, and Strauss (1968) implemented the placement task instructing children (4th grade) by means of the transitive verbs *push* and *pull* in the active or passive voice. The children were faster and more often correct in placing a mobile truck relative to a fixed truck if the former figured as agent, in other words, as the grammatical subject of active sentences and the non-subject of passive sentences. Huttenlocher and Weiner (1971) reported a corresponding result for the displacement task. Children (4th grade) strongly tended to displace the truck in the agent role with active and passive instructions alike. How can we explain these findings on the linguistic localization account? The *push/pull* sentences assign neither a locatum nor a relatum role. We argue that agentivity affects the mapping of the arguments of *push/pull* sentences, if apprehended as placement instructions, onto the arguments of spatial prepositions. Pulling something implies being in front of something; pushing something implies being behind something. Thus, the agent is preferably mapped onto the locatum and the non-agent is correspondingly mapped onto the relatum. Placement behavior is then explained in terms of the role asymmetry of linguistic localizations.

One piece of evidence provided by Huttenlocher and Weiner (1971) seems to contradict the linguistic-localization account. Children (4th grade) placed two mobile blocks to match a spatial relational instruction. The linguistic-localization account predicts that the relatum is placed first to establish the first search domain, but the children most often placed the locatum first. Additional evidence shows a peculiarity of this task. The authors implemented the same task with *push/pull* instructions. In contrast to the previous results, the children preferred placing the truck denoted by the grammatical subject first, with active and passive instructions alike. Order of mention determined the placement order in this task. Without going into the details in the present chapter, we attribute this effect to the topic status of the entity mentioned first. In contrast to the displacement task, in which both entities are already part of the common ground, both entities are added to the common ground in the present task. What we observe is that participants prefer adding the topic to the common ground first.

To summarize, the experimental evidence corroborates the linguistic-localization account if one agrees that pushers and pullers are preferably assigned the locatum role and that the preference for placing the topic first when placing two mobile entities does not disprove the linguistic-localization account. In the face of this evidence, we nominate candidate 2, which attests that the locative expression conveys new information although the relatum is given. Candidate 2 is underspecified as regards the discourse status of the locatum. As we will see, the rules of the competition, a spatial reasoning task, fix the discourse status as being new.

15.2.3 *The competition*

In cognitive psychology, there is a well-established experimental paradigm of (spatial) relational reasoning. In the simplest variant of such an experiment, participants solve linear three-term problems. They receive a pair of relational premises and their task is, for example, to derive a conclusion from them. We consider only determinate pairs. Each premise of a determinate pair relates two adjacent entities of a linear array of three entities, allowing the reasoner to infer the relation between the two remote entities. Examples (2a) and (2b) present premise pairs describing the arrangement of three pictures of animals hanging next to each another on a wall. One can validly conclude from these premise pairs, for example, that the deer is to the left of the donkey.

(2) a. The snake is to the left of the donkey. The deer is to the left of the snake. [BA–CB]

b. The donkey is to the right of the snake. The deer is to the left of the snake. [AB–CB]

The first premise of a pair introduces two new entities. In second premises, one of the two noun phrases anaphorically refers back to a given entity; the other noun phrase introduces a third new entity.[2] Examples (2a) and (2b) differ with respect to the role of the antecedent *the snake* in the first premise. The important variation for our present concerns is the role assigned to the anaphoric expression *the snake* in the second premise. Examples (2a) and (2b)

[2] It is customary in such premises to equip all nouns with a definite determiner, irrespective of the discourse status of the denoted entity. We are not aware of any reasoning study that has tested whether it facilitates comprehension of a second premise if the discourse status is marked by definite and indefinite determiners.

share the same second premise with the relatum being given and the locatum being new. Sentences (3a) and (3b) show the converse second premise with the locatum being given and the relatum being new.

(3) a. The donkey is to the right of the snake. The snake is to the right of the deer. [AB–BC]

b. The snake is to the left of the donkey. The snake is to the right of the deer. [BA–BC]

Traditionally, the four different premise pairs are characterized by their term order, provided in squared brackets to the right of the examples. B figures as antecedent in the first premise and as anaphoric expression in the second premise; A is the non-antecedent in the first premise; C designates the new entity in the second premise.

The question is how the discourse status of places comes into play. Huttenlocher's idea was that people solving a three-term problem construct an integrated representation of the described array. She argued that the effects obtained in placement tasks should likewise be reflected in integration processes in solving three-term reasoning tasks. We adopt her view within Johnson-Laird's (1983) mental-model theory. On this theory, people construct a mental model from a description, a representation structurally equivalent to the described situation. Entities occupy places in spatial mental models and people mentally place entities in constructing such models. In reading a first premise of (2) or (3), reasoners construct an initial mental model with a snake on the left and a donkey on the right. The snake and the donkey as well as their places are given in the initial mental model. The second premise is interpreted in the context of the initial model, the extralinguistic context, which happens to be acquired through the discourse context. Reasoners try to integrate the third entity into the initial model. When they succeed in doing so, they can read off the conclusion from the fully integrated model. On this account, the reasoner apprehends the second premise as a placement instruction as to where to put the new entity in the model. If this assumption is correct, the second premise should be easier to comprehend if the place of the relatum is given in the initial model.

Partitioning (P2) presents the given–new partitioning of candidate 2, the second premises in (2); (P3) corresponds to (P1) and presents the given–new partitioning of candidate 1, the second premises in (3). We assign no discourse status to spatial prepositions. The second premise repeats the preposition of the first premise in (2a) and (3a) but not in (2b) and (3b), and the prepositions could be classified as given and new, respectively. Since preposition repetition did not influence the comprehensibility of second premises in our studies, we ignore their discourse status. Studies reporting a difference are mentioned below.

(P2)

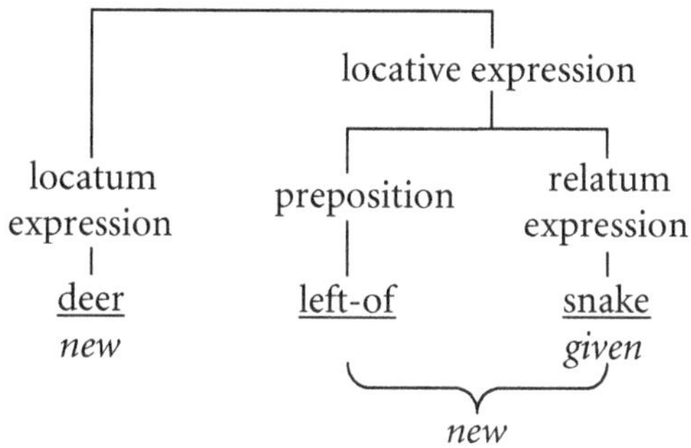

(P3)

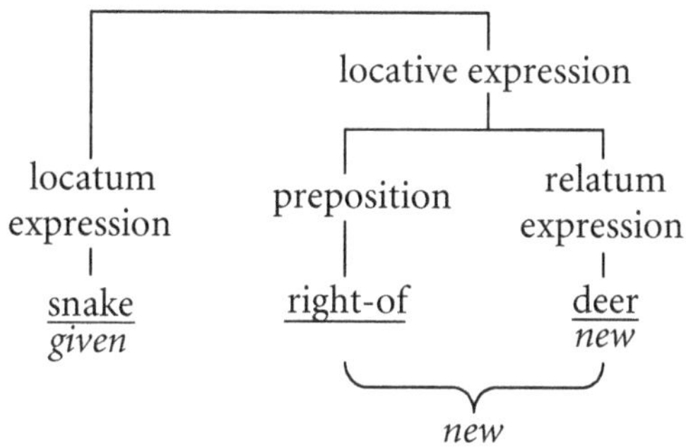

In (P2), the locatum is new and the relatum is given; the locative expression denotes the new place of the locatum. In (P3), the locatum is given and the relatum is new. The discourse status of the locative expression is less obvious. The locative expression *to the right of the deer* denotes the place of the given locatum, the snake. The snake is already placed in the initial mental model and its place is therefore given. In its relation to the place of the deer, however, the place of the snake is not part of the initial mental model because the deer is new and not yet placed in the initial model. The second premises in (3) introduce the information conveyed by the locative expression into the reasoner's mind for the first time. We therefore classify the locative expression in (P3) as new.

We now ask whether candidate 2 (P2) or candidate 1 (P3) more adequately captures the default given–new partitioning of spatial relational sentences with unmarked word order as revealed by processing a second premise in a spatial relational reasoning task. We reviewed fifteen papers reporting on a comparison between BC and CB premises. Studies on comparatives (e.g. *taller/shorter*) are included, as it has been shown that people mentally order the involved entities into a spatial array. Six papers reported a consistent CB advantage and eight papers partly reported a CB advantage. Some of these latter papers report a CB advantage in comparing AB–CB against BA–CB, but no difference between BA–CB and AB–BC; that is, the effect depended on the role of the antecedent or, confounded with it, on whether the second premise repeated the preposition of the first premise. Two papers report an advantage for AB–BC over BA–CB. Though our overview is not necessarily exhaustive, we think it is representative. A more detailed overview is found in Evans, Newstead, and Byrne (1993). Importantly, no single study reported a consistent given–new advantage. We recognize a trustworthy trend supporting the linguistic localization account. The preliminary winner, then, is candidate 2. We could deny a given–new advantage in processing a second (spatial) relational premise, but we think that previous experiments failed to demonstrate a given–new advantage because the benefit of an accelerated antecedent access was often insufficient to compensate for the difficulties in integrating a new relatum.

15.3 Extending the competition to marked sentences

We maintain a given–new advantage for processing a second premise in spatial relational reasoning although such an advantage has not been consistently observed in previous experiments. We argue that the given–new advantage in these studies was counteracted by the difficulty in integrating a new relatum predicted on the linguistic-localization account. The problem is that previous studies used (spatial) relational premise pairs like those presented in (2) and (3) in which the locatum figures as the grammatical subject and precedes the relatum. An exception is a study by Foos (1980), who used a construction with the relatum figuring as grammatical subject, for example, *The church has the school on its north side.* These constructions produced a BC advantage, but the finding does not decide whether aside from the advantage of a given relatum there was an additional given–new benefit. Huttenlocher (1968) applied the active–passive manipulation of her placement tasks to three-term reasoning problems by using the transitive verbs *to lead* and *to trail* in describing the order of runners in a race. She found a general disadvantage for problems formulated in the passive voice in both accuracy and response latencies. Moreover, problems in the active voice were more often solved correctly with a CB second premise while problems in the passive voice were more often solved correctly with a BC second premise. Response latencies revealed that CB second premises in the passive voice were more difficult than the other three forms of problems, which did not differ from one another. We interpret this latter result as showing that the general problem with the passive voice could be compensated for if the relatum was given and the order was, accordingly, given–new.

In our own experiment, we combined a given relatum with a given–new order by reversing the word order of the spatial relational premises. Examples (2′) and (3′) present the German second premises together with their English translations known from (2) and (3). The word order of these premises is considered unmarked since the grammatical subject is the preverbal constituent (abbreviated as *unmarked premises*). The relatum of the unmarked second premise in (2′) is given, thus the order is new–given. The relatum of the unmarked second premise in (3′) is new, thus the order is given–new.

(2′) Der $\text{Hirsch}_{\text{new}}$ ist links von der $\text{Schlange}_{\text{given}}$.
'The deer_{new} is to the left of the $\text{snake}_{\text{given}}$.'

(3′) Die $\text{Schlange}_{\text{given}}$ ist rechts vom $\text{Hirsch}_{\text{new}}$.
'The $\text{snake}_{\text{given}}$ is to the right of the deer_{new}.'

(4) Links von der $\text{Schlange}_{\text{given}}$ ist der $\text{Hirsch}_{\text{new}}$.
'To the left of the $\text{snake}_{\text{given}}$ is the deer_{new}.'

(5) Rechts vom $\text{Hirsch}_{\text{new}}$ ist die $\text{Schlange}_{\text{given}}$.
'To the right of the deer_{new} is the $\text{snake}_{\text{given}}$.'

In German, the positions of the locatum expression and the locative expression are easily swapped, as in (4) and (5), corresponding to (2′) and (3′), respectively. The word order of the premises in (4) and (5) is considered marked since the prepositional phrase occupies the preverbal position and the grammatical subject follows the verb (*marked premises*). As a result of the change in constituent order, the ordering of given and new information is reversed. (P4) and (P5) present the given–new partitionings of the marked second premises. From a processing perspective, the marked word order optimizes the distribution of given and new information in (P4) since it reconciles the linguistic localization account with the alleged given–new advantage. Compared to (P2), integration should be even faster since the antecedent can be accessed earlier in the initial model, so we elect (P4) as candidate 3. In contrast, the distribution of given and new information in (P5) should be the worst. Integrating a new relatum relative to a given locatum, expected to be difficult anyhow, is rendered even more difficult by the delayed antecedent access.

(P4)

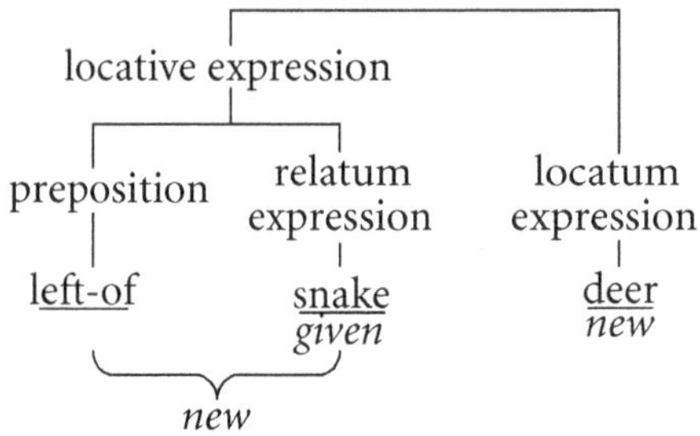

(P5)

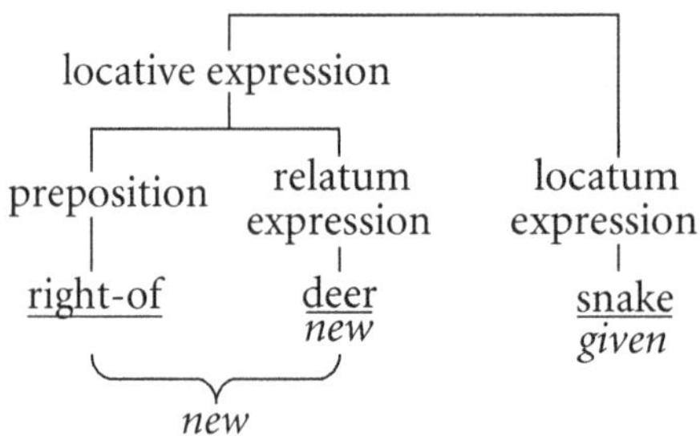

Hörnig, Oberauer, and Weidenfeld (2005) tested unmarked and marked German premise pairs, measuring self-paced sentence-comprehension times for each premise presented one after another. They predicted that integrating a second premise is easier if the relatum is given and the locatum is new and, in addition, there is a given–new advantage. Both predictions were clearly confirmed by the results. For unmarked second premises, comprehension times were slightly shorter with a given relatum and a new–given order than with a new relatum and a given–new order, though not significantly. Importantly, comprehension times were strongly influenced by word order. With a given relatum, marked word order facilitated integration considerably compared to unmarked word order, but marked word order rendered integration especially difficult with a new relatum. Hence, comprehensibility of marked second premises was strongly affected by the discourse status of the involved entities. The comparatively weak influence on unmarked second premises indicates that the given–new advantage almost compensated for the difficulty integrating a new relatum in this study on German.

The results show that a given–new partitioning of a marked sentence as in (P4) is strongly preferred over the partitioning in (P5). A much weaker corresponding preference holds for unmarked sentences, in line with the evidence reported in the previous section. Since the partitioning in (P4) was also much easier to process than that in (P2), the outstanding winner is candidate 3. The curious conclusion is that the least marked partitioning comes along with a marked word order.

15.4 On the mixed discourse status of the winner's preverbal locative expression

The finding reported in the previous section agrees with the idea that a marked structure restricts the set of appropriate contexts of utterance compared to the corresponding unmarked structure. In fact, we observed that the context much more strongly affected the comprehensibility of marked than of unmarked sentences and, furthermore, that an adequate context rendered marked sentences remarkably easy to comprehend. We ascribe to such a context the ability to license the marked word order. As we discussed the fact that the context established a given–new ordering with respect to the denotata of the two noun phrases in (4), we characterize the contextual licensing as *given–new licensing*. The corresponding ordering in (5) is new–given and we observe that (5) is not contextually licensed. Comprehension times measured for first premises support a licensing approach. Without any prior context, marked first premises were not contextually licensed and, indeed, marked first premises took longer to comprehend than unmarked ones. However, the difficulties in processing a marked first premise were much less serious than in interpreting a non-licensed marked second premise. Obviously, providing no context at all disrupts the processing of marked structures much less strongly than providing an inappropriate context.

So far, we have concluded that given–new licensing explains the good comprehensibility of marked spatial-relational sentences. Given–new licensing, however, is ignorant of the fact that the full preverbal constituent denotes a new entity, the place of the locatum. In this section, we will confront given–new licensing with the concept of *poset licensing*, which has been proposed to explain contextual licensing of marked word order. After introducing poset licensing in its original form, we will argue that, and how, it can be extended to the spatial domain and how it accounts for marked word order constructions in German spatial relational sentences. Finally, we will experimentally evaluate the two licensing accounts, given–new and poset.

15.4.1 *Poset licensing: The original proposal*

In its original form, poset licensing was introduced into linguistics as a condition on the discourse referent of the preverbal constituent of a marked word-order structure (see Hirschberg 1985, Ward 1988, E. F. Prince 1999). The condition states that the referent of the preverbal non-subject must stand in a poset relation to another previously established referent in order for the marked structure to be pragmatically appropriate: "[T]he entity represented by the preposed constituent must be related, via a partially ordered set relation, to one or more entities already evoked in the discourse model" (Hirschberg 1985). A partially ordered set (= poset) relation is a relation R that orders a set according to the following criteria: it must be reflexive (aRa), antisymmetric (if aRb and bRa, then $a = b$), and transitive (if aRb and bRc, then aRc). Strict partial orderings are transitive and irreflexive ($\sim aRa$) and therefore asymmetric (if $a \neq b$ and aRb, then $\sim bRa$).

Poset licensing has been shown to account for the pragmatic appropriateness of marked structures like English left-dislocation and topicalization (see Shaer *et al.* 2008). This claim is exemplified in (6), taken from E. F. Prince (1999: 8), where both types of marked structures are licensed, the left dislocations in (6c, d) and the topicalization in (6e):

(6) a. She had an idea for a project.

b. She's going to use three groups of mice$_{i,j,k}$.

c. One$_i$, she'll feed them$_i$ mouse chow, just the regular stuff they make for mice.

d. Another$_j$, she'll feed them$_j$ veggies.

e. And the third$_k$ she'll feed e$_k$ junk food.

The formal criteria formulated above identify the relation in (6) as a case of a strict partially ordered set relation between the set $B = \{A_i, A_j, A_k\}$ consisting of the three groups (sets) of mice mentioned in (6b), and the respective proper subsets of B, the groups of mice A_i, A_j, and A_k, being mentioned in the sentences (6c–e) successively. Formally, the poset relation in the example can be represented by the ordered pair $\langle B, \subset \rangle$; the ordering relation *subset-of* fulfills the criteria for strict partial orderings: it is transitive, irreflexive, and asymmetric.[3] Note that the antecedent referent is not merely *evoked* by the context of utterance, but is explicitly mentioned in (6b) and thus appropriately characterized as being salient. The availability of the antecedent entities licenses the marked word order in the subsequent sentences, that is, the preposing of the non-subjects denoting the three groups of mice in the sentences (6c–e).

[3] In the strict set-theoretic sense, the *subset-of* relation does not hold between B and A_i, A_j, A_k, but rather between the power set of B, $\wp(B)$, and the (disjunct) subsets of $\wp(B)$, $A_i = \{mouse_1, \ldots, mouse_i\}$, $A_j = \{mouse_j, \ldots, mouse_k\}$ and $A_k = \{mouse_l, \ldots, mouse_m\}$; since nothing in our argument hinges on this distinction, we will gloss over it in what follows.

15.4.2 *Extending poset licensing to the spatial domain*

In order to check whether poset licensing can be extended to cases of spatial statements, we have to transfer the criteria for poset relations from "proper" discourse referents, individuals and sets built from them, to the perhaps less obvious domain of spatial discourse referents: regions independent of the individuals that they host. Consider the following example:

(7) a. In front of me, I see three animals standing next to each other.
 b. On the left, there's a donkey.
 c. On the right, there's a deer.
 d. In the middle, there's a snake.

To check whether the marked word orders in (7b–d) are poset licensed, we have to answer two questions: (i) is there an antecedent referent evoked in the discourse model, and, if so, (ii) do the referents of the preposed elements stand in a salient poset relation to that antecedent?

The answer to (i) is in the positive: the antecedent referent is the configuration denoted by the phrase *three animals standing next to each other*. We may conceive of the denotation of this antecedent as the structured mereological object C consisting of the three regions at which the three animals are located. In this case, the antecedent entity is not explicitly mentioned, but is appropriately characterized as being "evoked" by (7a).

To answer (ii), we first have to determine the denotation of the preposed phrase of (7b), *on the left*. We assume that it denotes a region l_1 for which it holds that it occupies the left part (the leftmost subregion) of the configuration C. The second step in answering (ii) is to determine whether the antecedent referent, C, and the referent of the preposed phrase of (7b), l_1, stand in a poset relation. The answer is yes, the poset being $\langle C, \sqsubset \rangle$, that is: the structured object C and the strict partial ordering induced on it by the mereological relation *part-of*. The discourse referents denoted by the two other preposed phrases in (7c) and (d) also fulfill the poset criterion: the *part-of* relation is transitive, irreflexive, and asymmetric. Also, the relation between the overall configuration and its parts is sufficiently salient. Thus, the marked word order of the sentences is poset licensed. This goes to show that poset licensing can indeed be extended so as to account for spatial assertions such as the ones in (7).

Returning to example (7), we have to point out that the preverbal constituents do not express spatial relations. This raises the question as to what happens in terms of poset licensing if we substitute the positional expression, the adverb *in the middle* in (7d), with a relational expression, the adverb *in between*:

(7) d′. In between (the donkey and the deer), there's a snake.

The denotation of the preverbal phrase in (7d′) is the same as in (7d). No matter how we conceive of this referent, positionally or relationally, it stands in a poset relation to the overall configuration and both ways of construing the relation are sufficiently salient. Hence, poset licensing also holds in the case of the preposed spatial relational adverb in (7d′).

Now, let us consider the case of spatial relational sentences with prepositions. We have seen that the preposing of a prepositional phrase is given–new licensed. A marked second premise was easier to comprehend if the preverbal relatum expression denoted a given entity than when it denoted a new entity. But can such a marked order also be said to be poset licensed? Consider the second premise in (4) again, repeated here:

(4) Links von der Schlange$_{\text{given}}$ ist der Hirsch$_{\text{new}}$.
'To the left of the snake$_{\text{given}}$ is the deer$_{\text{new}}$.'

In order for the marked word order of the second premise to be poset licensed, the denotation of the preverbal PP has to stand in a salient poset relation to an entity already evoked in the context. We again must first determine the denotation of the preverbal constituent, and then check its poset relation to the context. The preverbal PP *links von der Schlange* in (4) denotes a region l_3 for which it holds that it is to the left of the region l_4 occupied by the snake. We assume that the interpretation of the first premise has supplied us with the information that there are two regions, l_1 and l_2 such that l_1 is to the left of the donkey and is occupied by the snake, and that the donkey, accordingly, is located at l_2. Thus, l_1 and l_4 can be identified, since they both fall under the description *is occupied by the snake*. The obvious candidate for the antecedent entity to which l_3 stands in a salient poset relation thus is l_1. This means that the poset relation is explicitly expressed by the semantic content carried by the PP. The preverbal PP asserts the poset relation *left-of*, thereby making it salient. In a manner of speaking, it wears its licensing poset relation on its sleeve. We conclude that the marked word order in a spatial relational statement with prepositions is not only licensed by given–new, but also by poset. This raises the question as to which kind of licensing explains the results of Hörnig *et al.* (2005).

15.4.3 *Poset licensing versus given–new licensing: Experimental evidence*

Spatial relational sentences with prepositions cannot keep given–new licensing apart from poset licensing. Whenever the relatum expression denotes a given entity, the preverbal locative expression denotes a place which is poset-related to the given place of the relatum. Disentangling the two types of licensing becomes possible by splitting up the relatum expression from the relational expression. In German, a spatial relational assertion can be expressed with an adverbial locative instead of the prepositional phrase. This construction, consisting of a spatial adverb adjoined to the PP containing the relatum expression, has the desired property of independent variability of the position of the adverb and the PP. All four word orders of the second premises in (8a–d) are grammatical.

(8) a. Der Hirsch$_{\text{new}}$ ist von der Schlange$_{\text{given}}$ aus links. (SOA)
the deer$_{\text{new}}$ is as from the snake$_{\text{given}}$ left

b. Von der Schlange$_{\text{given}}$ aus ist der Hirsch$_{\text{new}}$ links. (OSA)
as from the snake$_{\text{given}}$ is the deer$_{\text{new}}$ left

c. Von der Schlange$_{\text{given}}$ aus links ist der Hirsch$_{\text{new}}$. (OAS)
as from the snake$_{\text{given}}$ left is the deer$_{\text{new}}$

d. Links von der Schlange$_{\text{given}}$ aus ist der Hirsch$_{\text{new}}$. (AOS)
left as from the snake$_{\text{given}}$ is the deer$_{\text{new}}$

Sentence (8a) exhibits the unmarked word order, whereas (8b–d), with the PP in preverbal position, are all marked. The three marked orders vary the relative position of the PP and the locative adverb *links* "left". In (8b), the PP is preposed but the locative adverb carrying the relational information is stranded in sentence-final position. In (8c), both the PP and the adverb have undergone preposing. In (8d), the adverb has been adjoined to the left of the preposed PP, yielding a construction most similar to the one with a preposition presented above in (4).

In discussing the licensing conditions of (8b–d) we take the relatum *the snake* as given and the locatum *the deer* as new. The preverbal relatum expression ensures given–new licensing for all three structures. However, the precondition for poset licensing is only met by (8c) and (8d). Only the preverbal constituent of these two structures, the relatum expression together with the spatial adverb, denotes a new place that is poset-related to the given place of the relatum. The critical structure is therefore (8b). If in a comprehension study (8b) behaves the same as (8c, d), there is no evidence in favor of poset licensing. However, if (8c, d) turn out to be licensed while (8b) is not, poset licensing accounts for the results of Hörnig *et al.* (2005) with spatial prepositions.

Hörnig, Weskott, Kliegl, and Fanselow (2006) reran the study of Hörnig *et al.* (2005) using the material illustrated in (8a–d) with given and new relata to assess contextual licensing. Contextual licensing was strong for (8c, d) as previously established for (4). Sentence (8b) also exhibited a significant licensing effect, though reliably less strong than for (8c, d). Contextual licensing is therefore a matter of degree, with given–new licensing marked word order not as strongly as poset. In concluding this section, we note that the extension of the notion of poset licensing from "ordinary" discourse referents to referents in the domain of space (regions) has provided us with a plausible explanation for the good comprehensibility of statements with marked word order.

15.5 Information structure beyond integration requirements

Based on the results from the last section, we can now elaborate on the integration process. Starting from the dissociation of objects and places immanent in poset licensing, Hörnig *et al.* (2006) specified how to integrate a new locatum: (1) affirm that the relatum is given; (2) access the antecedent together with its place in the initial model; (3a) extend the initial model by a new place according to the spatial expression; and (3b) place the new locatum at the new place. On this view, integration benefits from the early encoding of a preverbal constituent fully specifying the new place, because step 3a, adding the new place, can be initiated early. This assumption adds a second aspect to the explanation of the remarkably good comprehensibility of marked spatial relational statements involving a given relatum. On the one hand, the explanation is put in pure processing terms. Integration of a new locatum is boosted by the early encoding of the constituent denoting the new place, which is first and foremost a matter of optimal serialization, which happens to correspond to the marked form of a spatial relational statement. On the other hand, the explanation relies on information-structural terms, namely, that the marked word order is poset licensed. One might be tempted to deny that the information-structural aspect has an explanatory value in its own right. There is evidence that such an objection is unjustified. Hartsuiker, Kolk, and Huiskamp (1999) gathered acceptability ratings for unmarked and marked Dutch spatial relational statements. They cross-varied the definiteness of the determiners of the locatum and the relatum expression and found that unmarked statements were rated as more acceptable than marked statements unless the preverbal relatum expression was paired with a definite determiner and the postverbal locatum expression with an indefinite determiner, for example *Op de tafel ligt een bal* "On the table is-lying a ball". Assuming that indefinite determiners signal newness and definite ones signal givenness, this finding agrees with the posited default given–new partitioning of marked statements, though the acceptability judgments do not reflect integration processes. This finding also suggests that the discourse status of the locatum is important for the default given–new partitioning of spatial relational statements.

15.5.1 *Word order and the discourse status of the locatum: Experimental evidence*

In the production study that we turn to next (Hörnig and Féry, in press), participants orally described in their own words a linear layout of three toy animals on a table in front of them. The experimenter repeatedly altered the layout by removing a given animal and adding a new one, either at the place of the removed animal or at the opposite place. Occasionally, one animal was removed and one of the two remaining given animals was displaced relative to the other animal (e.g. from left to front). We expected that speakers linguistically localize added and displaced animals relative to given non-moved animals and accordingly assign the locatum role to the former and the relatum role to the latter. This is what speakers did with only few exceptions. What is central to our present concerns is whether word order responds to the discourse status of the locatum if roles are assigned as expected. Relocated animals were discourse-given; added animals were frequently discourse-new, except for occasional instances in which participants mentioned the newly added animal before they localized it. We expected word order to be marked for localizations of discourse-new animals, but unmarked for localizations of discourse-given animals. This is what we observed. The majority of relational localizations of a discourse-new locatum had a marked word order, whereas the unmarked word order predominated if the locatum was discourse-given. The production data support the assumption that the word order of spatial relational statements responds to the discourse status of the locatum. Candidate 4 with an unmarked word order therefore provides a partitioning in which both the relatum and the locatum are given.

To extend the findings from the production study to comprehension, we implemented two variants of the Hörnig *et al.* (2005) study which differed from the original experiment in that a single sentence intervened between the first and second premise of a pair. In the first variant, the intervening sentence mentioned the new entity prior to the second premise without conveying its place, for example, *Es gibt auch einen Hirsch* "There is also a deer". The intervening sentence does not affect integration requirements, though it should have information-structural consequences. If the discourse status of the locatum is indeed a criterion for contextual licensing, the processing of a marked second premise afterwards should suffer from the locatum being discourse-given. In the second variant, the intervening sentence mentioned the given entity which was picked up in the second premise, for example, *Es gibt also eine Schlange* "There is hence a snake". Again, the intervening sentence does not affect integration requirements. In addition, it has no information structural consequences. A comparison of second premises in variants 1 and 2 therefore reveals the explanatory value of contextual licensing for the original finding. The comprehension time pattern for second premises was the same as in the original study if the intervening sentence mentioned an already given entity. However, comprehension times for second premises with a given relatum were not shorter anymore if word order was marked rather than unmarked. This finding confirms two things. First, information structure even affects the processing of second premises in a difficult reasoning task. Second, the licensing criterion for marked spatial relational sentences must account for the discourse status of the locatum. A marked word order is only licensed if the locatum is discourse-new, and not if it is discourse-given.

15.5.2 *Candidate 4 and the default partitioning of marked and unmarked localizations*

We conclude that the default given–new partitionings of spatial relational statements with unmarked and marked word order have in common that the relatum expression denotes a given entity and that the locative expression denotes a new place that is poset related to the given place of the relatum. They differ, however, in just one respect, namely, the discourse status of the locatum. As claimed by Chafe, the locatum expression of a spatial relational statement with unmarked word order usually denotes a given entity. The locatum expression of a spatial relational statement with marked word order is expected to denote a new entity. We therefore propose, at least for German, that candidate 4 presented in (P2′) represents the default given–new partitioning of a spatial relational statement with unmarked word order, whereas (P4) represents the default given–new partitioning of a spatial relational statement with marked word order.

(P2')

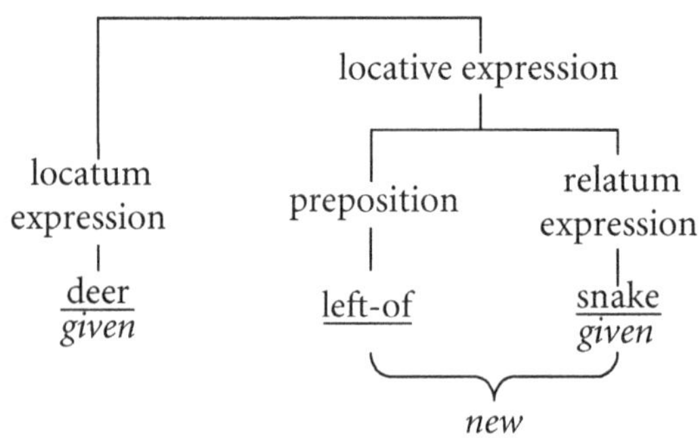

(P4)

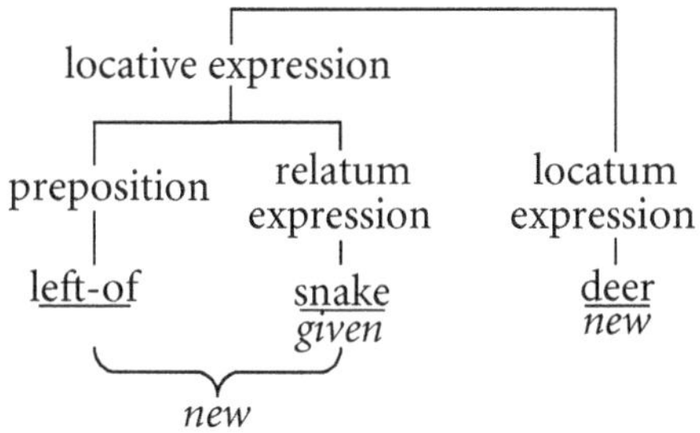

15.6 Conclusions

In this chapter we have examined the given–new partitioning of (German) spatial relational statements, for the most part by considering experimental evidence from comprehension studies and to some extent by considering production data (see Skopeteas and Fanselow, this volume, on the influence of given and new information in production studies on German and other languages; see also Hinterhölzl, this volume, on the role of given–new ordering in the development of word-order patterns in Old High German). Our first argument relied on the idea that spatial relational statements are used to linguistically localize a locatum relative to a relatum. On this account, the place of the locatum, denoted by the locative expression, is expected to be new. This argument extends to non-relational statements which are expressed, for example, by means of adverbs without an overt relatum expression. What is specific to relational statements is the additional constraint that the place of the relatum should be part of the common ground in order to help the addressee to restrict the domain of where to place or find the locatum.[4] What is left unspecified by the linguistic-localization approach is the discourse status of the locatum.

A clarification of this latter point should take word order into account. In most comprehension studies on this issue, the locatum was new whenever the relatum was given and vice versa. We found that spatial relational statements are easiest to comprehend if the relatum has been previously placed in the model and if the complete locative expression, which denotes the new place of the new locatum, is in preverbal position. We explained this finding in terms of poset licensing. Poset licensing exhibits a particularly strong effect on comprehensibility if the interpreter integrates a new entity into the model

[4] As a corollary, the pronominalization pattern of spatial relational sentences should differ from that of ordinary sentences in that the prepositional object is more likely to be pronominalized than the grammatical subject (cf. Kaiser, this volume).

constructed thus far. In order to demonstrate an influence on information structure beyond integration facilitation in terms of optimal serialization of constituents, it is necessary to identify circumstances in which a marked word order is not preferred although integration occurs. Such circumstances were evident in the production data. The marked word order, which suits integration requirements, was only preferred when the added or relocated animal figuring in the locatum role was discourse-new. When it was discourse-given, the unmarked word order was the preferred one. Hence, we propose (P4) as the default given–new partitioning of marked statements and (P2′) as the default partitioning of unmarked statements. This proposal is further supported by the comprehension study reported above in which an intervening sentence varied in whether or not it rendered the new entity discourse-given before it is localized in the second premise. In comparison to unmarked second premises with a given relatum, comprehensibility of marked second premises with a given relatum suffered from a discourse-given locatum. We conclude that the two word-order variants of spatial relational statements are associated with complementary given–new partitionings as regards the discourse status of the locatum.

There remains at least one open question: is the poset licensing criterion underspecified for spatial statements? Since we assume that a marked word order is inadequate for a spatial statement with a given locatum, the licensing criterion should address the discourse status of the locatum. What is still puzzling from a processing perspective: why should model integration be suboptimal with a preposed locative expression just because the locatum has been mentioned before (or is a salient part of the common ground for other reasons)? We speculate that the aboutness concept in the sense of Reinhart (1981) may help illuminate this issue (cf. Ehrich and Koster 1983 on aboutness and word order of spatial statements; see also Ebert and Hinterwimmer, this volume, on aboutness and topicality). On the aboutness account, an unmarked spatial statement asserts about the locatum which place it occupies; a marked statement asserts about a place which locatum it harbors. The two statements differ in how they are filed in memory (see Broadbent 1973). If we consider the possibility that the integration processes that we are concerned with lead to memory addressing in terms of places rather than objects, we come to conclude that integrating a marked statement might be easy—as long as addressing by place is adequate.

References

Abels, K. (2003). "Who Gives a Damn about Minimizers in Questions?", in R. B. Young and Y. Zhou (eds.), *Proceedings of SALT XIII*. Ithaca, NY: CLC Publications.

Aboh, E. O. (2004a). "The Morphosyntax of Complement-Head Sequences. Clause Structure and Word Order Patterns in Kwa", *Oxford Studies in Comparative Syntax*. Oxford, NY: Oxford University Press.

—— (2004b). "Left or Right? A View from the Kwa Peripheral Positions", in D. Adger, C. de Cat and G. Tsoulas (eds.), *Peripheries: Syntactic Edges and Their Effects*. Dordrecht: Kluwer, 165–89.

—— (2006). "If We See *Focus*, You Go Left and I Go Right!". Paper presented at the International Conference on Bantu Grammar. SOAS, London, April 2006.

Abusch, D. (1994). "The Scope of Indefinites", *Natural Language Semantics* 2: 83–135.

Adger, D. and Ramchand, G. (2005). "Merge and Move: *Wh*-Dependencies Revisited", *Linguistic Inquiry* 26 (2): 161–93.

Aissen, J. (1992). "Topic and Focus in Mayan", *Language* 68: 43–80.

Alexiadou, A. and Anagnostopoulou, E. (1998). "Parametrizing AGR: Word Order, V-Movement and EPP Checking", *Natural Language and Linguistic Theory* 16: 491–539.

—— —— (2000). "Greek Syntax: A Principles and Parameters Perspective", *Journal of Modern Greek Linguistics* 1: 171–222.

Alexopoulou, T. and Kolliakou, D. (2002). "On Linkhood, Topicalization, and Clitic Left Dislocation", *Journal of Linguistics* 38 (2): 193–245.

Allan, E. J. (1973). *A Grammar of Buem: The Lelemi Language, School of Oriental and African Studies*. University of London.

Almor, A. (1999). "Noun-Phrase Anaphora and Focus: The Informational Load Hypothesis", *Psychological Review* 106 (4): 748–65.

Ameka, F. (1992). "Focus Constructions in Ewe and Akan: A Comparative Perspective", in C. Collins and V. Manfredi (eds.), *Proceedings of the Kwa Comparative Syntax Workshop* 1–25. Cambridge.

Amiridze, N. (2006). *Reflexivization Strategies in Georgian*. Ph.D. dissertation, University of Utrecht. Utrecht: LOT.

Anderson, S. C. (1979). "Verb structure", in Hyman (1979b), 73–136.

Anderson, S. R. (1984). "On Representations in Morphology: Case, Agreement, and Inversion in Georgian", *Natural Language and Linguistic Theory* 2: 157–218.

Angantýsson, Á. (2001). "Skandinavísk orðaröð í íslenskum aukasetningum" [Scandinavian Word Order in Embedded Clauses in Icelandic], *Íslenskt mál* 23: 95–122.

Ariel, M. (1990). *Accessing NP Antecedents*. London: Routledge, Croom Helm.

Arnold, J. (1998). *Reference Form and Discourse Patterns*. Ph.D. dissertation, Stanford University.

Arnold, J. E. (1999). *Marking Salience: The Similarity of Topic and Focus.* Manuscript, University of Pennsylvania.

—— Eisenband, J. G., Brown-Schmidt, S., and Trueswell, J. C. (2000). "The Immediate Use of Gender Information: Eyetracking Evidence of the Time-Course of Pronoun Resolution", *Cognition* 76, B13–B26.

Aronson, H. I. (1994). "Paradigmatic and Syntagmatic Subject in Georgian", in H. I. Aronson (ed.), *Non-Slavic Languages of the USSR, Papers from the Fourth Conference.* Columbus, OH: Slavica Publishers, 13–33.

Asher, N. (1993). *Reference to Abstract Objects in Discourse.* Kluwer Academic Publishers.

—— (2003). *Logics of Conversation.* Cambridge University Press.

—— and Lascarides, A. (2001). "Indirect Speech Acts", *Synthese*, 129 (1–2): 183–228.

—— and Pustejovsky, J. (2004). *Word Meaning and Commonsense Metaphysics.* Technical report, Brandeis University, Rutgers.

—— and Reese, B. (2005). "Negative Bias in Polar Questions", in E. Maier, C. Bary and J. Huitink (eds.), *Proceedings of SuB9*, 30–43. <www.ru.nl/ncs/sub9>.

Baker, M. (2000). *The Natures of Nonconfigurationality*, in M. Baltin and C. Collins (eds.), *Handbook of Contemporary Syntactic Theories.* London: Blackwell, 407–38.

—— and Collins C. (2006). "Linkers and the Internal Structure of *v*P", *Natural Language and Linguistic Theory* 24: 307–54.

—— and Kandybowicz J. (2003). "On Directionality and the Structure of the Verb Phrase: Evidence from Nupe", *Syntax* 6 (2): 115–55.

Ballantyne, H. G. (2004). "Givenness as a Ranking Criterion in Centering Theory: Evidence from Yapese", *Oceanic Linguistics* 43 (1): 49–72.

Baltazani, M. and Jun, S.-A. (1999). "Focus and Topic Intonation in Greek", in *Proceedings of 14th International Congress of Phonetic Sciences (ICPhS)*: 1305–8.

Bar-Hillel, Y. and Carnap, R. (1953). "Semantic Information", in *Proceedings of the Symposium on Applications of Communication Theory.* London: Butterworth Scientific Publications.

Bartels, C. (2004). "Acoustic Correlates of *Second Occurrence* Focus: Towards an Experimental Investigation", in H. Kamp and B. Partee (eds.), *Context-Dependence in the Analysis of Linguistic Meaning.* Amsterdam: Elsevier, 354–61.

Barwise, J. and Cooper, R. (1981). "Generalized Quantifiers and Natural Language". *Linguistics and Philosophy* 4: 159–219.

Bates, E. and MacWhinney, B. (1989). "Functionalism and the Competition Model", in E. Bates and B. MacWhinney (eds.), *The Crosslinguistic Study of Sentence Processing.* Cambridge: Cambridge University Press.

Beaver, D. (2004). "The Optimization of Discourse Anaphora," *Linguistics and Philosophy* 27 (1): 3–56.

—— and Clark, B. (2003). "Always and Only: Why Not All Focus-Sensitive Operators are Alike", *Natural Language Semantics.*

—— and Krahmer, E. (2001). "A Partial Account of Presupposition Projection", *Journal of Logic, Language and Information* 10: 147–82.

——Clark, B., Flemming, E., Jaeger, T. F., and Wolters, M. (2007). "When Semantics Meets Phonetics: Acoustical Studies of Second Occurrence Focus", *Language* 83: 245–76.

Beck, S. (1996). "Quantified Structures as Barriers for LF Movement", *Natural Language Semantics* 4: 1–56.

——(2006). "Intervention Effects Follow from Focus Interpretation", *Natural Language Semantics* 14: 1–56.

Beckman, M. and Elam, G. A. (1997). *Guidelines for ToBI Labelling. Version* 3. Ohio State University.

——and Pierrehumbert, J. (1986). "Intonational Structure in Japanese and English", *Phonology Yearbook* 3: 255–309.

Behaghel, O. (1932). *Deutsche Syntax IV*. Heidelberg: Carl Winters.

Belletti, A. (2002). "Aspects of the Low IP Area", in Luigi Rizzi (ed.), *The Structure of IP and CP: The Cartography of Syntactic Structures* 2. Oxford: Oxford University Press.

Bentzen, K. (2005). "What's the Better Move? On Verb Placement in Standard and Northern Norwegian", *Nordic Journal of Linguistics* 28: 153–88.

——(2007). "The Degree of Verb Movement in Embedded Clauses in Three Varieties of Norwegian", in K. Bentzen and Ø. A. Vangsnes (eds.), *Scandinavian Dialect Syntax, Nordlyd* 34: 127–46.

Benz, L. (2009). *Sprachvergleich: Deutsch-Russisch, Wortstellung vs. Passivierung, Effekte der Gegebenheitsmanipulation von Satzargumenten Transitiver und Intransitiver Verben.* To appear in *Working Papers of the SFB632, Interdisciplinary Studies on Information Structure (ISIS)* 12. Potsdam: Universitätsverlag Potsdam.

Berg, R. van den, Gussenhoven C., and Rietveld, T. (1992). "Downstep in Dutch: Implications for a Model", in G. Docherty and D. R. Ladd (eds.), *Papers in Laboratory Phonology II: Gesture, Segment, Prosody*. Cambridge: Cambridge University Press, 335–67.

Berman, S. (1987). "Situation-Based Semantics for Adverbs of Quantification", in J. Blevins and A. Vainikka (eds.), *University of Massachusetts Occasional Papers* 12. Amherst: GLSA, University of Massachusetts.

Besten, H. den (1977). *On the Interaction of Root Transformations and Lexical Deletive Rules*. Manuscript, MIT and University of Amsterdam.

Bhatt, R. and Pancheva, R. (2001). "Conditionals", in M. Everaert and H. van Riemsdijk (eds.), *The Blackwell Companion to Syntax* 1. Oxford: Blackwell, 638–67.

Biberauer, T. and Roberts, I. G. (2005). "Changing EPP-Parameters in the History of English: Accounting for Variation and Change", *English Language and Linguistics* 9: 157–79.

——and Richards M. (2006). "True Optionality: When the Grammar doesn't Mind", in Cedric Boeckx (ed.), *Minimalist essays*. Amsterdam: John Benjamins, 35–67.

Birch, S. L. and Garnsey, S. (1995). "The Effect of Focus on Memory for Words in Sentences", *Journal of Memory and Language* 34: 232–67.

Birch, S. L. and Rayner, K. (1997). "Linguistic Focus Affects Eye Movements During Reading", *Memory & Cognition* 25: 653–60.
——Albrecht, J. E., and Myers, J. L. (2000). "Syntactic Focusing Structures Influence Discourse Processing", *Discourse Processes* 30: 285–304.
Birner, B. and Ward G. (2004). "Information Structure and Non-canonical Syntax", in L. R. Horn and G. Ward (eds.), *The Handbook of Pragmatics*. Oxford: Blackwell, 153–74.
Bobaljik, J. D. (2002). "Realizing Germanic Inflection: Why Morphology does Not Drive Syntax", *Journal of Comparative Germanic Linguistics* 6: 129–67.
——and Thráinsson, H. (1998). "Two Heads aren't Always Better than One", *Syntax* 1: 37–71.
Bochvar, D. A. (1981). "On a Three-Valued Logical Calculus and its Application to the Analysis of the Paradoxes of the Classical Extended Functional Calculus", *History and Philosophy of Logic* 2, 87–112.
Boeder, W. (1989). "Verbal Person Marking, Noun Phrase and Word Order in Georgian", in L. Marácz and P. Muysken (eds.), *Configurationality: The Typology of Asymmetries*. Dordrecht: Foris, 159–84.
Borkin, A. (1971). "Polarity Items in Questions", in *Papers from the 7th Regional Meeting of the Chicago Linguistic Society*. University of Chicago, 53–63.
Bošković, Ž. (2000). "Sometimes in SpecCP, Sometimes In-situ", in R. Martin, D. Michaels, and J. Uriagereka (eds.), *Step by step: Essays on Minimalism in Honor of Howard Lasnik*. Cambridge, MA: MIT Press, 53–87.
Breheny, R. (2008). "A New Look and the Semantics and Pragmatics of Numerically Modified Noun Phrases", *Journal of Semantics* 25: 93–139.
Brennan, S. E., Friedman, M. A., and Pollard, C. J. (1987). "A Centering Approach to Pronouns". *Proceedings of the 25th Annual Meeting of the Association for Computational Linguistics*. Stanford, 155–62.
Broadbent, D. E. (1973). *In Defence of Empirical Psychology*. London: Methuen.
Bródy, M. (1990). "Some Remarks on the Focus Field in Hungarian", in *UCL Working Papers in Linguistics* 2: 201–25.
——(1991). "Remarks on the Order of Elements in the Hungarian Focus Field", in I. Kenesei (ed.), *Approaches to Hungarian* III. Szeged: JATE, 95–122.
——(1995). "Focus and Checking Theory", in I. Kenesei (ed.), *Approaches to Hungarian* V. Szeged: JATE, 29–44.
Bruce, G. (1977). *Swedish Word Accents in Sentence Perspective*. Lund: Gleerup.
Buell, L. (2005). *Issues in Zulu Verbal Morphosyntax*. Ph.D. dissertation, UCLA.
——(2006). "The Zulu Conjoint/Disjoint Verb Alternation: Focus or Constituency?" *ZAS Papers in Linguistics* 43: 9–30.
Büring, D. (1997). "The Great Scope Inversion Conspiracy", *Linguistics and Philosophy* 20: 175–94.
——(2001). "Let's Phrase it! Focus, Word Order, and Prosodic Phrasing in German Double Object Constructions", in G. Müller and W. Sternefeld (eds.), *Competition in Syntax*. New York: Mouton de Gruyter, 69–105.

——(2003). "On D-trees, Beans, and B-accents", *Linguistics and Philosophy* 26: 511–45.

——(2006). "Focus Projection and Default Prominence", in V. Molnár and S. Winkler (eds.), *The Architecture of Focus*. Berlin/New York: Mouton de Gruyter, 321–46.

——(2008). *Been there, Marked that—a Tentative Theory of Second Occurrence Focus.* Revised manuscript. To appear in M. Kanazawa and C. Tancredi (eds.).

——and Gunlogson, C. (2000). *Aren't Positive and Negative Polar Questions the Same?* Manuscript, UCSC/UCLA. <www.semanticsarchive.net>

——and Gutiérrez-Bravo, R. (2001). "Focus-related Word Order Variation without the NSR: A Prosody-Based Crosslinguistic Analysis", in Séamas Mac Bhloscaidh (ed.), *Syntax at Santa Cruz* 3: 41–58.

Butler, A. and Mathieu, E. (2005). "Split-DPs, Generalized EPP and Visibility", *MIT Working Papers in Linguistics*: 49–57.

Cahill, M. C. (2007). *Aspects of the Morphology and Phonology of Konni*. Dallas: SIL International.

Carlson, L. (1983). *Dialogue Game: An Approach to Discourse Analysis*, Dordrecht: Reidel.

Cat, C. de (2002). *French Dislocation*. Ph.D. dissertation, University of York.

Chafe, W. (1970). *Meaning and the Structure of Language*. Chicago: University of Chicago Press.

——(1976). "Givenness, Contrastiveness, Definiteness, Subjects and Topics and Point of View", in C. Li (ed.), *Subject and topic*. Academic Press.

Chambers, C. G. and Smyth, R. (1998). "Structural Parallelism and Discourse Coherence: A Test of Centering Theory", *Journal of Memory and Language* 3: 593–608.

Cheng, L. and Downing, L. (2006). *Phonology and Syntax at the Right Edge in Zulu*. Paper presented at the Prosody-Syntax Interface Workshop, University College London, October 2006.

——and Rooryck J. (2001). "Licensing WH-in-situ". *Syntax* 3: 1–19.

Chierchia, G. (1995). *Dynamics of Meaning*. Chicago: University of Chicago Press.

——(1998). "Reference to Kinds Across Languages". *Natural Language Semantics* 6: 339–405.

Chomsky, N. (1981), *Lectures on government and binding*. Dordrecht: Foris.

Cinque, G. (1993). "A Null Theory of Phrase and Compound Stress", *Linguistic Inquiry* 24: 239–98.

Clark, H. H. (1972). "Difficulties People have in Answering the Question *Where is it?*", *Journal of Verbal Learning and Verbal Behavior* 11: 265–77.

——(1973). "Space, Time, Semantics, and the Child", in T. E. Moore (ed.), *Cognitive development and the acquisition of language*. New York: Academic Press, 27–63.

——and Clark, E. (1977). *Psychology and Language: An Introduction to Psycholinguistics*. New York: Harcourt Brace Jovanovich.

——and Haviland S. (1977). "Comprehension and the Given-New Contrast", in R. O. Freedle (ed.), *Discourse Production and Comprehension*. Hillsdale, NJ: Lawrence Erlbaum Associates, 1–40.

Cohen, A. (2001). "Relative Readings of *many*, *often*, and Generics". *Natural Language Semantics* 9: 41–67.

Cohen, P. R. and Levesque, H. J. (1990). "Rational Interaction as the Basis for Communication", in P. R. Cohen, J. Morgan, and M. E. Pollack (eds.), *Intentions in Communication*. MIT Press, Cambridge, MA, 221–55.

Collins, P. (1995). "The Indirect Object Construction in English: an Informational Approach", *Linguistics* 33: 35–49.

Comrie, B. (1984). "Some Formal Properties of Focus in Modern Eastern Armenian", *Annual Armenian Linguistics* 5: 1–21.

Condoravdi, C. (1990). "Sandhi Rules of Greek and Prosodic Theory", in Inkelas and Zec (1990), 63–84.

Cowles, H. W. (2003). *Processing Information Structure: Evidence from Comprehension and Production*. Ph.D. dissertation, UCSD.

—— Walenski, M., and Kluender, R. (2007). "Linguistic and Cognitive Prominence in Anaphor Resolution: Topic, Constrastive Focus and Pronouns", *Topoi* 26: 3–18.

Crawley, R. J. and Stevenson, R. J. (1990). "Reference in Single Sentences and in Texts", *Journal of Psycholinguistic Research* 19 (3): 191–210.

Creissels, D. (1996). "Conjunctive and Disjunctive Verb Forms in Setswana". *South African Journal of African Languages* 16: 109–15.

Cresswell, M. and von Stechow, A. (1982). "*De re* Belief Generalized", *Linguistics and Philosophy* 5: 503–35.

Cresti, D. (1995). *Indefinite Topics*. Ph.D. thesis, MIT.

Csirmaz, A. (2004). "Particles and Phonologically Defective Predicates", in K. É. Kiss and H. van Riemsdijk (eds.), *Verb Clusters. A Study of Hungarian, German, and Dutch*. Amsterdam: John Benjamins, 225–52.

Culicover, P. W. (1992). "English Tag Questions in Universal Grammar", *Lingua* 88, 193–226.

Cutler, A. and Fodor, J. D. (1979). "Semantic Focus and Sentence Comprehension", *Cognition* 7: 49–59.

Delin, J. (1990). *A Psychological Investigation of the Processing of Cleft Constructions*. Blue Book note no. 2. Department of Artificial Intelligence, Edinburgh.

Demuth, K. and Mmusi S. (1997). "Presentational Focus and Thematic Structure in Comparative Bantu", *Journal of African Languages and Linguistics* 18: 1–19.

Déprez, V. (1989). *On the Typology of Syntactic Positions and the Nature of Chains*. Ph.D. dissertation, Cambridge, MA: MIT.

—— (1990). "Two Ways of Moving the Verb in French", *MIT Working Papers on Linguistics* 13: 47–85.

Diesing, M. (1992). *Indefinites*. Cambridge, MA: MIT Press.

—— (1997). "Yiddish VP order and the Typology of Object Movement in Germanic", *Natural Language and Linguistic Theory* 15: 369–427.

Dik, S. (1997). *The Theory of Functional Grammar. Part I: The Structure of the Clause*. (Functional Grammar Series 20) Berlin: Mouton de Gruyter.

Dikken, M. den (1995). *Particles: On the Syntax of Verb-particle, Triadic, and Causative Constructions*. New York: Oxford University Press.

Ding, S. (1998), *Fundamentals of Prinmi (Pumi): A Tibeto-Burman Language of Northwestern Yunnan, China*. Ph.D. dissertation, Australian National University.

Dittmer, A. and Dittmer, E. (1998). *Studien zur Wortstellung—Satzgliedstellung in der althochdeutschen Tatianübersetzung*. Göttingen: Vandenhoeck & Ruprecht.

Du Bois, J. (2003). "Argument Structure: Grammar in Use", in J. Du Bois, E. Lorraine, W. Kumpf and J. Ashby (eds.), *Preferred Argument Structure: Grammar as Architecture for Function*. Amsterdam: John Benjamins, 11–60.

Durbin, M. and Ojeda, F. (1978). "Basic Word-order in Yucatec Maya", in N. C. England (ed.), *Papers in Mayan Linguistics* 2. Columbia: University of Missouri, Department of Anthropology, 69–77.

Ebert, C. and Endriss, C. (2004). "Topic Interpretation and Wide Scope Indefinites", in *Proceedings of NELS 34*. GLSA, Amherst.

—— —— (2007). "Functional Topics", in *Proceedings of the Sinn und Bedeutung XI (11th annual meeting of the Gesellschaft für Semantik)*, Barcelona.

Ehrich, V. and Koster, C. (1983). "Discourse Organization and Sentence Form: The Structure of Room Descriptions in Dutch", *Discourse Processes* 6: 169–95.

É. Kiss, K. (1987). *Configurationality in Hungarian*. Dordrecht: Foris.

—— (1995). *Discourse-Configurational Languages*. New York, Oxford University Press.

—— (1996). "A *csak* szintaxisához és szemantikájához", in I. Terts (ed.), *Nyelv, nyelvész, társadalom* II. Pécs: JPTE, 102–8.

—— (1998a). "Identificational Focus versus Information Focus", *Language* 74: 245–73.

—— (1998b). "Multiple Topic, one Focus?", *Acta Linguistica Hungarica* 45: 3–30.

—— (1998c). "Discourse-Configurationality in the Languages of Europe", in A. Siewierska (ed.), *Constituent Order in the Languages of Europe*. Berlin: Mouton de Gruyter, 681–729.

—— (2003). "Argument Scrambling, Operator Movement and Topic Movement in Hungarian", in S. Karimi (ed.), *Word Order and Scrambling*. London: Blackwell, 22–43.

—— (2006a). "Focussing as Predication", in V. Molnár and S. Winkler (ed.), *Architecture of Focus*. Berlin: Mouton de Gruyter, 169–96.

—— (2006b). "Apparent or Real? On the Complementary Distribution of Identificational Focus, and the Verbal Particle", in K. É. Kiss (ed.), *Event structure and the left periphery*. Dordrecht: Springer, 201–25.

—— (2006c). "Jól megoldottuk? Rosszul oldottuk meg? Az összefoglaló és a kirekesztő kifejezést tartalmazó mondatok szórendjének magyarázata". *Magyar Nyelv* 102: 442–58.

—— (2008). "Free Word Order, (Non-)configurationality, and Phases". *Linguistic Inquiry* 39: 441–75.

Endriss, C. (2009). "Quantificational Topics – A Scopal Treatment of Exceptional Wide Scope Phenomena", *Studies in Linguistics and Philosophy*. Springer.

Endriss, C. and Hinterwimmer, S. (2008): "Tense and Adverbial Quantification", in J. Doelling and T. Heyde-Zybatow (eds.), *Event Structures in Linguistic Form and Interpretation*. Berlin: Mouton de Gruyter.

Erguvanlı, E. E. (1984). *The Function of Word Order in Turkish Grammar*. Berkeley.

Ernst, T. (2002). *The Syntax of Adjuncts*. Cambridge: Cambridge University Press.

—— (2003). "Semantic Features and the Distribution of Adverbs", in E. Lang, C. Maienborn, and C. Fabricius-Hansen (eds.), *Modifying adjuncts*. Berlin: Mouton de Gruyter, 307–34.

Erteschik-Shir, N. (1997). *The Dynamics of Focus Structure*. Cambridge: Cambridge University Press.

—— (2007). *Information Structure: The Syntax-Discourse Interface*. Oxford: Oxford University Press.

Evans, J. St. B. T, Newstead, S. E., and Byrne, R. M. J. (1993). *Human Reasoning: The Psychology of Deduction*. Hove: Erlbaum.

Fanselow, G. (2001). "Features, θ-Roles, and Free Constituent Order", *Linguistic Inquiry* 32, 405–37.

—— (2003). "Münchhausen-Style Head Movement and the Analysis of Verb Second", in A. Mahajan (ed.), *Head Movement and Syntactic Theory* 10 of UCLA/Potsdam Working Papers in Linguistics.

Fauconnier, G. (1975). "Polarity and the Scale Principle", in *Papers from the 11th Regional Meeting of the Chicago Linguistic Society*. University of Chicago, 188–99.

Féry, C. (2006). "The Prosody of Topicalization", in K. Schwabe and S. Winkler (eds.), *On Information Structure: Meaning and Form*. Amsterdam: John Benjamins, 69–86.

—— and Herbst L. (2004). "German Sentence Accent Revisited", in S. Ishihara, M. Schmitz, and A. Schwarz (eds.), *Working Papers of the SFB632, Interdisciplinary Studies on Information Structure (ISIS)* 1. Potsdam: Universitätsverlag Potsdam, 43–75.

—— and Ishihara, S. (2009). "The Phonology of Second Occurrence Focus". *Journal of Linguistics* 45, 285–313.

—— and Kügler, F. (2008). "Pitch Accent Scaling on Given, New and Focused Constituents in German". *Journal of Phonetics* 36 (4), 680–703.

—— and Samek-Lodovici, V. (2006). "Focus Projection and Prosodic Prominence in Nested Foci", *Language* 82 (1), 131–50.

—— and Truckenbrodt, H. (2005). "Sisterhood and Tonal Scaling", *Studia Linguistica* 59: 223–43.

—— Fanselow, G. and Krifka, M. (2008). *The Notions of Information Structure*. Acta Linguistica Hungarica 55 (3–4).

Fiedler, I. (1998). "Fokus im Aja", in I. Fiedler, C. Griefenow-Mewis, and B. Reineke (eds.), *Afrikanische Sprachen im Brennpunkt der Forschung. Linguistische Beiträge zum 12. Afrikanistentag, Berlin, 3.–6. Oktober 1996*. Köln: Köppe, 75–89.

—— (2006). "Focus Expressions in Yom", *Cahiers Voltaïques / Gur Papers* 7: 112–21.

—— (2007). "Focus Expressions in Foodo", in S. Ishihara, S. Jannedy, and A. Schwarz (eds.), *Interdisciplinary Studies on Information Structure (ISIS)* 8, 97–114. Potsdam: Potsdam University.

—— (2008). "Aja". Manuscript (submitted to *The Expression of Information Structure: A Cross-linguistic Survey in Speech Production*, in S. Skopeteas, S. Hellmuth, G. Fanselow, and C. Féry (eds.), Berlin/New York: Mouton de Gruyter).

—— and Schwarz, A. (2005). "Out-of-Focus Encoding in Gur and Kwa", in S. Ishihara, M. Schmitz, and A. Schwarz (eds.), *Interdisciplinary Studies on Information Structure* 3. Potsdam: Potsdam University, 111–42.

—— —— (2007). "Focal Aspects in the Lelemi Verb System", Ms. (submitted to *Journal of West African Languages*).

Fintel, K. von (1994). *Restrictions on Quantifier Domains*. Ph.D. thesis, University of Massachusetts, Amherst.

Fodor, J. A. (1983). *The Modularity of Mind*. Cambridge, MA: MIT Press.

Foos, P. W. (1980). "Constructing Maps from Sentences", *Journal of Experimental Psychology: Human Learning and Memory* 6: 25–38.

Foraker, S. and McElree, B. (2007). "The Role of Prominence in Pronoun Resolution: Active Versus Passive Representations". *Journal of Memory and Language* 56 (3): 357–83.

Fox, D. (2006). "Free Choice and a Theory of Scalar Implicature", Manuscript, MIT.

Frascarelli, M. (2000). *The Syntax-Phonology Interface in Focus and Topic Constructions in Italian*. Dordrecht: Kluwer Academic Publishers.

—— (2007). *Narrow Focus, Clefting and Predicate Inversion*. Paper presented at GLOW XXX, 11–14 April 2007, University of Tromsø.

Fretheim, T. (1992). "The Effect of Intonation on a Type of Scalar Implicature", *Journal of Pragmatics* 18: 1–30.

Frey, W. (2000). "Über die Syntaktische Position des Satztopiks im Deutschen", in E. Lang, M. Rochon, K. Schwabe, and O. Teuber (eds.), *ZAS Papers in Linguistics* 20. Berlin: ZAS.

—— (2001). "About the Whereabouts of Indefinites". *Theoretical Linguistics* 27: 137–61.

—— (2004). "A Medial Topic Position for German". *Linguistische Berichte* 198: 153–90.

—— (2005). "Zur Syntax der linken Peripherie im Deutschen", to appear in F. d'Avis (ed.), *Deutsche Syntax: Empirie und Theorie*, Göteborger Germanistische Forschungen.

Fuchs, A. (1976). "*Normaler* und *kontrastiver* Akzent", *Lingua* 38, 293–312.

Gärtner, H.-M. (2001). "Are there V2 Relative Clauses in German?", *Journal of Comparative Germanic Linguistics* 3: 97–141.

Geenhoven, V. van (1998). *Semantic Incorporation and Indefinite Descriptions: Semantic and Syntactic Aspects of Noun Incorporation in West Greenlandic*. Stanford: CS.

Gernsbacher, M. A. and Hargreaves, D. (1988). "Accessing Sentence Participants: The Advantage of First Mention", *Journal of Memory and Language* 27: 699–717.

—— —— (1992). "The Privilege of Primacy: Experimental Data and Cognitive Explanations", in D. L. Payne (ed.), *Pragmatics of Word Order Flexibility*. Philadelphia, PA: John Benjamins, 83–116.

Geurts, B. (2006). "Take *Five*. The Meaning and Use of a Number Word". Manuscript, to appear in L. Tasmowski and S. Vogeleer (eds.), *Indefiniteness and Plurality*. Amsterdam: John Benjamins.

Giannakidou, A. and Quer, J. (1995). "Two Mechanisms for the Licensing of Negative Indefinites", in L. Gabriele, D. Hardison, and R. Westmoreland (eds.), *Formal Linguistics Society of Mid-America (FLSM)* 6, Bloomington: Indiana Linguistics Club, 103–14.

Ginzburg, J. and Sag I. (2000). *Interrogative Investigations: The Form, Meaning, and Use of English Interrogatives*. Stanford: CSLI.

Givón, T. (1983). *Topic Continuity in Discourse: A Quantitative Cross-language Study*. Amsterdam: John Benjamins.

Godjevac, S. (2000). *Intonation, Word Order and Focus Projection in Serbo-Croatian*. Ph.D. thesis, Ohio State University.

Good, J. (To appear). "Topic and Focus Fields in Naki", in I. Fiedler and A. Schwarz (eds.), *Information Structure in African Languages*.

Gordon, M. (2007). "The Intonational Realization of Contrastive Focus in Chickasaw", in C. Lee, M. Gordon, and D. Büring (eds.), *Intonational and Meaning from a Cross-linguistic Perspective*. Dordrecht: Kluwer.

Gordon, P. C., Grosz, B. J., and Gilliom, L. A. (1993). "Pronouns, Names, and the Centering of Attention in Discourse", *Cognitive Science* 17: 311–47.

Green, M. and Jaggar, P. (2003). "Ex-situ and In-situ Focus in Hausa: Syntax, Semantics and Discourse", in J. Lecarme (ed.), *Research in Afroasiatic Grammar II*. [CILT 241]. Amsterdam: John Benjamins, 187–213.

Grewendorf, G. (2002). "Left Dislocation as Movement", in S. Mauck and J. Mittelstaedt (eds.), *Georgetown University Working Papers in Theoretical Linguistics* 2: 31–81.

Grice, H. P. (1967/1975). "Logic and Conversation", in P. Cole and J. L. Morgan (eds.), *Syntax and Semantics* 3: *Speech Acts*. New York: Academic Press.

Groenendijk, J. and Stokhof, M. (1984). *Studies on the Semantics of Questions and the Pragmatics of Answers*. Ph.D. thesis, University of Amsterdam.

—— —— (1990). "Dynamic Predicate Logic". *Linguistics and Philosophy* 14: 39–100.

—— —— (1997). "Questions", in J. van Benthem and A. ter Meulen (eds.), *Handbook of Logic and Language*. Amsterdam: Elsevier, 1055–124.

Grohmann, K. (2000). *Prolific Peripheries: A Radical View from the Left*. Dissertation, University of Maryland.

Grosz, B., Joshi, A., and Weinstein, S. (1986). *Towards a Computational Theory of Discourse Interpretation*. Manuscript.

——— ——— ——— (1995). "Centering: A Framework for Modelling the Local Coherence of Discourse", *Computational Linguistics* 2 (21): 203–25.

Guerzoni, E. (2004). "Even-NPIs in yes/no questions", *Natural Language Semantics* 12, 319–43.

Güldemann, T. (2003). "Present Progressive vis-à-vis Predication Focus in Bantu: A Verbal Category between Semantics and Pragmatics", *Studies in Language* 27: 323–60.

Gundel, J. K. (1985). "Shared knowledge and topicality", *Journal of Pragmatics* 9: 83–107.

—— (1988). "Universals of Topic-Comment Structure", in M. Hammond, E. A. Moravcsik, and J. Wirth (eds.), *Studies in Syntactic Typology*. Amsterdam: John Benjamins, 209–39.

—— Hedberg, N., and Zacharski, R. (1993). "Cognitive Status and the Form of Referring Expressions in Discourse", *Language* 69: 274–307.

Gunlogson, C. (2003). *True to Form: Rising and Falling Declaratives as Questions in English*. Outstanding Dissertations in Linguistics. New York: Routledge.

Gussenhoven, C. (1983). "Focus, Mode and the Nucleus", *Journal of Linguistics* 19: 377–417.

—— (1984). *On the Grammar and Semantics of Sentence Accents*. Dordrecht: Foris.

—— (1992). "Sentence Accents and Argument Structure", in I. Roca (ed.), *Thematic Structure: Its Role in Grammar*. Berlin/New York: Foris, 79–106.

Gutiérrez, B. R. and Monforte y Madera, J. (2007). *La alternancia sujeto inicial/verbo inicial y la teoría de la optimidad*. Manuscript.

Hagstrom, P. (1998). *Decomposing Questions*. Ph.D. dissertation, MIT.

Haider, H. (2000). "Adverb Placement-Convergence of Structure and Licensing", *Theoretical Linguistics* 26: 95–134.

—— (2004). "Pre- and Postverbal Adverbials in OV and VO", *Lingua* 114: 779–807.

—— and Rosengren I. (2003). "Scrambling: Nontriggered Chain Formation in OV Languages", *Journal of German Linguistics* 15 (3): 203–67.

Halle, M. and Vergnaud, J. (1980). "Three Dimensional Phonology", *Journal of Linguistic Research* 1: 83–105.

Hamblin, C. L. (1958). "Questions". *Australasian Journal of Philosophy* 36 (3): 159–68.

—— (1973). "Questions in Montague English". *Foundations of Language* 10: 41–53.

Han, C. (2002). "Interpreting Interrogatives as Rhetorical Questions". *Lingua* 112: 201–29.

Hara, Y. (2006). "Grammar of Knowledge Representation: Japanese Discourse Items at Interfaces". Doctoral dissertation, University of Delaware.

—— and Rooij R. van (2007). "Contrastive Topics Revisited: a Simpler Set of Topic-Alternatives". Talk presented at NELS, University of Ottawa.

Harris, A. C. (1981). *Georgian Syntax: A Study in Relational Grammar*. Cambridge: Cambridge University Press.

—— (1982). "Towards the Universals of Q-Word Formation", in *Papers from the Parasession on Nondeclaratives*. Chicago: Chicago Linguistic Society, 67–75.

Harris, L. J. (1975). "Spatial Direction and Grammatical Form of Instructions Affect the Solution of Spatial Problems", *Memory & Cognition* 3: 329–34.

Hartmann, K. and Zimmermann, M. (2007a). "In Place – Out of Place: Focus in Hausa", in K. Schwabe, S. Winkler (eds.), *On Information Structure: Meaning and Form*. Amsterdam: John Benjamins, 365–403.

——— (2007b). "Focus Strategies in Chadic – The Case of Tangale Revisited", *Studia Linguistica* 61 (2): 95–129.

——— (2007c). "Exhaustivity Marking in Hausa: A Reanalysis of the Particle *nee/cee*", in E. O. Aboh, K. Hartmann, and M. Zimmermann (eds.), *Focus Strategies in African Languages. The Interaction of Focus and Grammar in Niger-Congo and Afro-Asiatic*. Berlin: Mouton de Gruyter, 241–63.

——— (in press). "Morphological Focus Marking in Guruntum", *Lingua* (2009), doi: 10.1016/j.lingua.2009.02.002.

Hartsuiker, R. J., Kolk, H. H. J., and Huiskamp, P. (1999). "Priming Word Order in Sentence Production", *The Quarterly Journal of Experimental Psychology* 52A: 129–47.

Hayes, B. (1995). *Metrical Stress Theory: Principles and Case Theory*. Chicago: University of Chicago Press.

—— and Lahiri, A. (1991). "Bengali Intonational Phonology", *Natural Language & Linguistic Theory* 9: 47–96.

Heck, F. (2004). *A Theory of Pied-Piping*. Ph.D. dissertation, University of Tübingen.

Heim, I. (1982): *The Semantics of Definite and Indefinite Noun Phrases*. Ph.D. thesis, University of Massachusetts, Amherst.

—— (1984). "A Note on Negative Polarity and DE-ness", in C. Jones and P. Sells (eds.), NELS *14 Proceedings*. UMASS Amherst: GLSA, 98–107.

—— and Kratzer, A. (1998): *Semantics in Generative Grammar*. Blackwell, Oxford.

Herburger, E. (2000): *What Counts. Focus and Quantification*. Cambridge, MA: MIT Press.

Herring, S. C. and Polillo, J. C. (1995). "Focus Position in SOV languages", in P. Downing and M. Noonan (eds.), *Word Order in Discourse*. Amsterdam: John Benjamins, 163–98.

Herskovits, A. (1986). *Language and Spatial Cognition. An Interdisciplinary Study of the Prepositions in English*. Cambridge: Cambridge University Press.

Hewitt, B. G. (1995). *Georgian: A Structural Reference Grammar*. Amsterdam: John Benjamins.

Heycock, C. (2008). "Japanese *-wa*, *-ga*, and Information Structure", in S. Miyagawa and M. Saito (eds.), *Handbook of Japanese Linguistics*. Oxford: Oxford University Press.

Higgins, F. Roger (1973). *The Pseudo-cleft Construction in English*. Ph.D. dissertation, MIT.

Hinterhölzl, R. (2001). "Event-related Adjuncts and the OV/VO Distinction", in K. Megerdoomian and L. Bar-el (eds.), *WCCFL 20 Proceedings*. Sommerville: Cascadilla Press.

—— (2002). "Parametric Variation and Scrambling in English", in W. Abraham and J. W. Zwart (eds.), *Studies in Comparative Germanic Syntax* (Proceedings of CGSW 15 in Groningen, May 2000). Amsterdam: John Benjamins.

—— (2004). "Language Change versus Grammar Change: What Diachronic Data Reveal about the Distinction between Core Grammar and Periphery", in E. Fuss and C. Trips (eds.), *Diachronic Clues to Synchronic Grammar*. Linguistik Aktuell 72. Amsterdam: John Benjamins.

—— (2006). *Scrambling, Remnant Movement and Restructuring*. Oxford: Oxford University Press.

—— (to appear). "A Phase-based Comparative Approach to Modification and Word Order in Germanic", *Syntax* (accepted April 2008).

Hinterwimmer, S. (2006): "The Interpretation of Universally Quantified DPs and Singular Definites in Adverbially Quantified Sentences", in D. Baumer, D. Montero, and M. Scanlon (eds.), *Proceedings of the 25th West Coast Conference on Formal Linguistics*. Cascadilla Proceedings Project, Somerville, MA.

—— (2008). *Q-Adverbs as Selective Binders: The Quantificational Variability of Free Relatives and Definite DPs*. Berlin: Mouton de Gruyter.

Hirschberg, J. (1985). *A Theory of Scalar Implicature*. Ph.D. dissertation. Philadelphia: University of Pennsylvania.

Hobbs, J. R. (1979). "Coherence and Coreference", *Cognitive Science* 3: 67–90.

Hockett, C. F. (1958): *A Course in Modern Linguistics*. New York: McMillan.

Höftmann, H. (1993). *Grammatik des Fon*. Leipzig: Langenscheidt.

Holmberg, A. (1986). *Word Order and Syntactic Feature in the Scandinavian Languages and English*. Doctoral dissertation, University of Stockholm.

—— (1997). "Scandinavian Stylistic Fronting: Movement of Phonological Features in the Syntax", *Working Papers in Scandinavian Syntax* 60: 81–124.

—— (2000). "Scandinavian Stylistic Fronting: How any Category can become an Expletive", *Linguistic Inquiry* 31 (3): 445–83.

—— and Platzack C. (1995). *The Role of Inflection in Scandinavian Syntax*. New York/Oxford: Oxford University Press.

Horn, L. R. (1972). *On the Semantic Properties of Logical Operators in English*. Ph.D. dissertation, UCLA.

—— (1981). "Exhaustiveness and the Semantics of Clefts", in V. Burke and J. Pustejovsky (eds.), *Papers from the 11th Annual Meeting of NELS*. Amherst: GLSA, 124–42.

—— (1989). *A Natural History of Negation*. Chicago and London: The University of Chicago Press.

—— (1992). "The Said and the Unsaid", *Ohio State University Working Papers in Linguistics (SALT II Proceedings)* 40: 163–92.

—— (1996). "Presupposition and Implicature", in S. Lappin (ed.), *Handbook of Contemporary Linguistic Theory*. Oxford: Blackwell, 299–319.

Hornby, P. A. (1974). "Surface Structure and Presupposition", *Journal of Verbal Learning and Verbal Behavior* 13: 530–38.

Hörnig, R. and Féry, C. (in press). "Linguistic Markers of Discourse Status in Describing Altered Spatial Layouts".

—— Oberauer, K., and Weidenfeld, A. (2005). "Two Principles of Premise Integration in Spatial Reasoning", *Memory & Cognition* 33: 131–9.

—— Weskott, T., Kliegl, R., and Fanselow, G. (2006). "Varying Word Order in Spatial Descriptions with Adverbs", *Memory & Cognition* 34: 1183–92.

Horváth, J. (1986). *FOCUS in the theory of Grammar and the Syntax of Hungarian.* Dordrecht: Foris.

—— (2005). "Is *Focus-Movement* Driven by Stress?", in C. Piñón and P. Siptár, (eds.), *Approaches to Hungarian* 9. Budapest: Akadémiai Kiadó, 131–58.

Howell, J. (2007). "Second Occurrence Focus and the Acoustics of Prominence", in *WCCFL 26*. Cascadilla Press Proceedings Project, 114–25.

Hrafnbjargarson, G. H. (2004). "Stylistic fronting", *Studia Linguistica* 58 (2): 88–134.

Hróarsdóttir, Þ. (1996). "The Decline of OV Word Order in the Icelandic VP: A diachronic study". *Working Papers in Scandinavian Syntax* 57: 92–141.

—— (2000). *Word Order Change in Icelandic. From OV to VO.* [*Linguistik Aktuell/Linguistics Today* 35]. Amsterdam: John Benjamins.

—— Hrafnbjargarson G. H., Wiklund A.-L. and Bentzen K. (2006). "The Tromsø Guide to Scandinavian Verb Movement", *Working Papers in Scandinavian Syntax* 78: 1–36.

—— Wiklund A.-L., Bentzen K., and Hrafnbjargarson, G. H. (2007). "The Afterglow of Verb Movement", *Working Papers in Scandinavian Syntax* 80: 45–75.

Huang, C.-T. (1982). *Logical Relations in Chinese and the Theory of Grammar.* Ph.D. dissertation, MIT.

Huber, S. (2000). *Es-Clefts und det-Clefts. Zur Syntax, Semantik und Informationsstruktur von Spaltsätzen im Deutschen und Schwedischen.* Stockholm: Almquist and Wiksell International.

Huddleston, R. and Pullum, G. K. (2002). *The Cambridge Grammar of the English Language.* Cambridge University Press.

Huttenlocher, J. (1968). "Constructing Spatial Images: A Strategy in Reasoning", *Psychological Review* 75: 550–60.

—— Eisenberg, K., and Strauss, S. (1968). "Comprehension: Relation between Perceived Actor and Logical Subject", *Journal of Verbal Learning and Verbal Behavior* 7: 527–30.

—— and Strauss, S. (1968). "Comprehension and a Statement's Relation to the Situation it Describes", *Journal of Verbal Learning and Verbal Behavior* 7: 300–4.

—— and Weiner, S. L. (1971). "Comprehension of Instructions in Varying Contexts", *Cognitive Psychology* 2: 369–85.

Hyman, L. M. (1979a). "Phonology and Noun Structure", in Hyman (1979b), 1–72.

—— (1979b). *Aghem Grammatical Structure.* (SCOPIL 7.) Los Angeles: USC Dept of Linguistics Publications.

—— (1985). "Dependency Relations in Aghem Syntax: The Mysterious Case of the Empty Determiner in Aghem", *Studies in African Linguistics, Supplement* 9: 151–56.

—— (to appear). "Focus Marking in Aghem: Syntax or Semantics?", in I. Fiedler and A. Schwarz (eds.), *Information Structure in African Languages*. Amsterdam: John Benjamins.

—— and Watters J. R. (1984). "Auxiliary focus". *Studies in African Linguistics* 15: 233–73.

Hyun-Oak, K. A. (1988). "Preverbal Focusing and Type XXIII Languages", in E. Moravcsik, M. Hammond, and J. Wirth (eds.), *Studies in Syntactic Typology*. Amsterdam: John Benjamins, 149–71.

Iatridou, S. (1993). "On the Contribution of Conditional Then", *Natural Language Semantics* 2 (3): 171–99.

—— (1995). "Clitics and Island Effects", *Penn Working Papers in Linguistics* 2, 11–31.

Indriðason, Þ. (1987). *Skýrsla um orðaröð í sagnlið*. Manuscript. Reykjavík: University of Iceland.

Inkelas, S. and Zec, D. (1990). *The Phonology-Syntax Connection*. Chicago: University of Chicago Press.

Ishihara, S. (2003). *Intonation and Interface Conditions*. Unpublished Ph.D. dissertation, MIT.

—— (2007). "Major Phrase, Focus Intonation and Multiple Spell-out (MaP, FI, MSO)". *The Linguistic Review* 24: 137–67.

—— (2008). *Japanese downstep revisited*. Manuscript, University of Potsdam.

—— and Féry, C. (in prep). *Syntactic and Prosodic Realization of Information Structure in German, Hungarian and Japanese*.

Israel, M. (2001). "Minimizers, Maximizers and the Rhetoric of Scalar Reasoning", *Journal of Semantics* 18, 297–331.

Ito, J. and Mester, A. (2006). "Prosodic Adjunction in Japanese Compounds", *Formal Approaches to Japanese Linguistics 4 MITWPL #55*. Cambridge, MA: MITWPL.

—— —— (2007). *Prosodic Categories and Recursion*. Paper presented at the Workshop on Prosody, Syntax, and Information Structure 3 (WPSI3).

Ivanishvili, M. and Ether, S. (1999). "A Morphological Structure and Semantics of the Georgian so-called Passive Forms", *Proceedings of the Fourth International Tbilisi Symposium on Language, Logic, and Computation, Batumi 1999*.

Jackendoff, R. S. (1972). *Semantic Interpretation in Generative Grammar*. Cambridge, MA: MIT Press.

Jacobs, J. (1984). "Funktionale Satzperspektive und Illokutionssemantik", *Linguistische Berichte* 91: 25–58.

—— (1988). "Fokus-Hintergrund-Gliederung und Grammatik", in H. Altmann (ed.), *Intonationsforschungen*. Tübingen: Niemeyer, 183–216.

—— (1991/2a). *Informationsstruktur und Grammatik*. Opladen: Westdeutscher Verlag. (= Linguistische Berichte Sonderheft 4). 32.

—— (1991/2b). *Neutral Stress and the Position of Heads*, in Jacobs (1991/2a), 220–244.

Jäger, G. (1996). *Topics in Dynamic Semantics*. Ph.D. thesis. Berlin: Humboldt University.

Jaggar, P. (2001). *Hausa*. Amsterdam: John Benjamins.

Johnson-Laird, P. N. (1983). *Mental Models*. Cambridge, MA: Harvard University Press.

Jónsson, J. G. (1991). "Stylistic Fronting in Icelandic", *Working Papers in Scandinavian Syntax* 48: 1–43.

—— (1996). *Clausal Architecture and Case in Icelandic*, Doctoral dissertation, University of Massachusetts, Amherst.

Joppen-Hellwig, S. (2001). *Verbklassen und Argumentlinking: Nicht-kanonische Argumente, Expletiva und Vierstellige Kausativa in Ergativ- versus Akkusativsprachen*. Tübingen: Niemeyer.

Jun, S.-A. (1996). *The Phonetics and Phonology of Korean Prosody: Intonational Phonology and Prosodic Structure*. New York: Garland.

—— and Fougeron C. (2000). "A Phonological Model of French Intonation", in A. Botinis (ed.), *Intonation: Analysis, Modeling and Technology*. Kluwer Academic Publishers, 209–42.

Kadmon, N. (1985). "The Discourse Representation of Noun Phrases with Numeral Determiners", in J. Choe, S. Bereman, and J. McDonough (eds.), *Proceedings of NELS 15*.

—— (2001). *Formal Pragmatics*. Oxford: Blackwell.

—— and Landman, F. (1993). "Any", *Linguistics and Philosophy* 16, 353–422.

Kahnemuyipour, A. (2004). *The Syntax of Sentential Stress*. Unpublished Ph.D dissertation, University of Toronto.

Kaiser, E. (2003). *The Quest for a Referent: A Crosslinguistic Look at Reference Resolution*. Ph.D. dissertation, University of Pennsylvania.

—— and Trueswell, J. C. (2008). "Interpreting Pronouns and Demonstratives in Finnish: Evidence for a Form-specific Approach to Reference Resolution", *Language and Cognitive Processes* 23 (5): 709–48.

Kameyama, M. (1996). "Indefeasible Semantics and Defeasible Pragmatics", in M. Kanazawa, C. Pinon, and H. de Swart (eds.), *Quantifiers, Deduction and Context*. Stanford, CA: CSLI, 111–38.

Kamp, H. (1981): "A Theory of Truth and Semantic Representation", in J. Groenendijk, T. Janssen, and M. Stokhof (eds.), *Formal Methods in the Study of Language*. Amsterdam: Mathematisch Centrum, University of Amsterdam, 277–322.

Kandybowicz, J. (2006). *Conditions on Multiple Copy Spell-Out and the Syntax-Phonology Interface*. Ph.D. dissertation, UCLA.

Kanerva J. M. (1990). "Focusing on Phonological Phrases in Chichewa", in S. Inkelas and D. Zec (eds.), *The Phonology-Syntax Connection*. Chicago: University of Chicago Press, 145–61.

Karttunen, L. (1977). "Syntax and Semantics of Questions", *Linguistics and Philosophy* 1: 3–44.

Kay, P. (2002). "English Subjectless Tagged Sentences", *Language* 78 (3), 453–81.

Kayne, R. (1994). "The Antisymmetry of Syntax", *Linguistic Inquiry Monograph* 25. Cambridge, MA: MIT Press.

—— and Pollock J.-Y. (1978). "Stylistic Inversion, Successive Cyclicity, and Move NP in French", *Linguistic Inquiry* 9: 595–621.

—— —— (2001). "New Thoughts on Stylistic Inversion", in A. Hulk and J.-Y. Pollock (eds.), *Subject Inversion in Romance and the Theory of Universal Grammar*. New York: Oxford University Press, 107–61.

Kehler, A. (1997). "Current Theories of Centering for Pronoun Interpretation: A Critical Evaluation", *Computational Linguistics* 23 (3): 467–75.

—— (2002). *Coherence, Reference and the Theory of Grammar*. Stanford: CSLI

—— Kertz, L., Rohde, H., and Elman, J. (2008). "Coherence and Coreference Revisited", *Journal of Semantics (Special Issue on Processing Meaning)*, 25 (1): 1–44.

Kemenade, A. van (1987). *Syntactic Case and Morphological Case in the History of English*. Dordrecht: Foris.

Kempen, G. and Harbusch, K. (2005). "The Relationship between Grammaticality Ratings and Corpus Frequencies: A Case Study into Word Order Variability in the Midfield of German Clauses", in S. Kepser and M. Reis (eds.), *Linguistic Evidence—Empirical, Theoretical, and Computational Perspectives*. Berlin: Mouton de Gruyter, 327–47.

Kenesei, I. (1986). "On the Logic of Word Order in Hungarian", in W. Abraham and S. de Meij (eds.), *Topic, Focus, and Configurationality*. Amsterdam: John Benjamins, 143–59.

—— and Vogel I. (1990). *Focus and Phonological Structure*. Manuscript, University of Cambridge and University of Budapest.

Kenstowicz, M. (1985). "The Phonology and Syntax of WH-Expressions in Tangale", *Studies in the Linguistic Sciences* 15: 79–91.

—— and Sohn H.-S. (1996). "Phrasing and Focus in Northern Kyungsang Korean", in Pier Marco Bertinetto (ed.), *Certamen Phonologicum III*. Rosenberg & Sellier, Torino, 137–56.

Khan, Sameer ud Dowla (2007). "The Distribution and Realization of H-Tones in Bengali", *Working Papers in Phonetics* 106. Los Angeles: UCLA, 34–40.

Kidda, M. (1993). *Tangale Phonology*. Berlin: Dietrich Reimer Verlag.

Kidwai, A. (2000). *XP-adjunction in Universal Grammar: Scrambling and Binding in Hindi-Urdu*. New York: Oxford University Press.

Kishimoto, H. (2005). "Wh-in-situ and Movement in Sinhala Questions". *Natural Language and Linguistic Theory* 23: 1–51.

Koeneman, O. (2000). *The Flexible Nature of Verb Movement*. Doctoral dissertation, University of Utrecht.

Koopman, H. (2000). *The Syntax of Specifiers and Heads*. London: Routledge.

Kosmeijer, W. (1986). "The Status of the Finite Inflection in Icelandic and Swedish", *Working Papers in Scandinavian Syntax* 26.

Kratzer, A. (1991). "Representation of Focus", in A. von Stechow and D. Wunderlich (eds.), *Handbook of Semantics*. Berlin: de Gruyter, 825–34.

—— (1995): "Stage-Level Predicates and Individual-Level Predicates", in G. Carlson and F. J. Pelletier (eds.), *The Generic Book*. Chicago: University of Chicago Press, 125–75.

Kratzer, A. (1999). "Beyond *ouch* and *oops*: How Descriptive and Expressive Meaning Interact". Handout, Cornell Conference on Theories of Context Dependence. <www.semanticsarchive.net>.

—— and Selkirk, E. (2007). "Phase Theory and Prosodic Spellout: The Case of Verbs", *The Linguistic Review* 24: 93–135.

Krifka, M. (1991). "A Compositional Semantics for Multiple Focus Constructions", in J. Jacobs (ed.), *Informationsstruktur und Grammatik (Sonderheft 4, Linguistische Berichte)*. Opladen.

—— (1995). "The Semantics and Pragmatics of Polarity Items", *Linguistic Analysis* 25: 209–57.

—— (1998). "Scope Inversion under Rise-fall Contour in German", *Linguistic Inquiry* 29 (1): 75–112.

—— (2001). "Non-Novel Indefinites in Adverbial Quantification", in C. Condoravdi and G. Renardel de Lavelette (eds.), *Logical Perspectives on Language and Information*. CSLI Publications, Stanford.

—— (2002). "Embedded Speech Acts". Handout for the talk given at Workshop *In the Mood*, Graduiertenkolleg *Satzarten: Variation und Interpretation*, Universität Frankfurt am Main.

—— (2003). "Polarity Items in Questions", *Slides from talk at Korean Society of Logic and Information*, Inha University, Incheon, Korea.

—— (2004a). Focus and/or Context: "A Second Look at Second Occurrence Expressions", in H. Kamp and B. Partee (eds.), *Context-dependence in the Analysis of Linguistic Meaning*. Amsterdam: Elsevier, 187–207.

—— (2004b). "Semantics below and above Speech Acts". Handout for the talk given at Stanford University.

—— (2006). "Association with Focus Phrases", in V. Molnar and S. Winkler (eds.), *Architecture of Focus*. Berlin: Mouton de Gruyter, 105–36.

—— (2007). "Basic Notions of Information Structure", in C. Féry, G. Fanselow, and M. Krifka (eds.), *Working Papers of the SFB 632* 6, Interdisciplinary Studies on Information Structure (ISIS). Potsdam: Universitätsverlag Potsdam, 13–56.

—— (2008). "Basic Notions of Information Structure". *Acta Linguistica Hungarica* 55 (3–4): 243–76.

—— Pelletier, F. J., Carlson, G., Meulen, A. ter, Chierchia, G., and Link, G. (1995): "Genericity: An Introduction", in G. Carlson and F. J. Pelletier (eds.), *The Generic Book*. Chicago: University of Chicago Press, 1–124.

Kroch, A. and Taylor, A. (1997). "Verb Movement in Old and Middle English: Dialect Variation and Language Contact", in A. van Kemenade and N. Vincent (eds.), *Parameters of Morphosyntactic Change*. Cambridge: CUP.

—— —— (2000). *Verb-Object Order in Early Middle English*. Manuscript, University of Pennsylvania.

Kubozono, H. (1993). *The Organization of Japanese Prosody*. Tokyo: Kurosio.

—— (2007). "Focus and Intonation in Japanese: Does Focus Trigger Pitch Reset?", in S. Ishihara (ed.), *Proceedings of the 2nd Workshop on Prosody, Syntax and Information*

Structure (WPSI2), Interdisciplinary Studies on Information Structure 9. Potsdam: University of Potsdam, 1–27.

Kučerová, I. (2007). *The Syntax of Givenness.* Ph.D. dissertation, MIT.

Kuno, S. (1972): "Functional Sentence Perspective: A Case Study from Japanese and English", *Linguistic Inquiry* 3: 269–336.

—— (1973). *The Structure of the Japanese Language*, Cambridge, MA: MIT Press.

Kuppevelt, J. van (1996). "Inferring from topics", *Linguistics and Philosophy* 19: 393–443.

Kuroda, S. (1972): "The Categorical and the Thetic Judgement: Evidence from Japanese Syntax", *Foundations of Language*, 9 (2): 153–85.

Ladd, D. R. (1980) *The Structure of Intonational Meaning: Evidence from English.* Distributed (1980) by the Indiana University Linguistic Club, Bloomington.

—— (1981). "A First Look at the Semantics and Pragmatics of Negative Questions and Tag Questions", in *Papers from the 17th Regional Meeting of the Chicago Linguistic Society.* University of Chicago: 164–71.

—— (1990). "Metrical Representation of Pitch Register", in J. Kingston and M. Beckman (eds.), *Papers in Laboratory Phonology I: Between the Grammar and Physics of Speech.* Cambridge: Cambridge University Press, 35–57.

—— (1996). *Intonational Phonology.* Cambridge: Cambridge University Press.

Ladusaw, W. A. (1979). *Polarity Sensitivity as Inherent Scope Relations.* Ph.D. thesis, University of Texas at Austin.

Lambrecht, K. (1994). *Information Structure and Sentence Form. Topic, Focus, and the Mental Representations of Discourse Referents.* Cambridge: Cambridge University Press.

—— (1996). "On the Formal and Functional Relationship between Topics and Vocatives. Evidence from French", in A. Goldberg (ed.), *Conceptual Structure, Discourse and Language.* Stanford: CSLI, 267–88.

Landau, B. and Jackendoff, R. (1993). "*What* and *Where* in Spatial Language and Spatial Cognition", *Behavioral & Brain Sciences* 16: 217–65.

Lasnik, H. and Saito M. (1992). *Move [alpha]: Conditions on Its Applications and Output.* Cambridge, MA: MIT Press.

Lee, C. (1999). "Contrastive Topic: A Locus of the Interface", in K. Turner (ed.), *The Semantics/Pragmatics Interface from Different Points of View* 1. London: Elsevier, 317–41.

—— (2006). "Contrastive Topics/Focus and Polarity in Discourse", in K. von Heusinger and K. Turner (eds.), *Where Semantics Meets Pragmatics*, CRiSPI 16. Elsevier, 381–420.

Lee-Schoenfeld, V. (2005). *Beyond Coherence: The Syntax of Opacity in German.* Ph.D. dissertation, UC Santa Cruz.

Lefebvre, C. and Brousseau, A.-M. (2002). *A Grammar of Fongbe.* Berlin/New York: Mouton de Gruyter.

Lenerz, J. (1984). *Syntaktischer Wandel und Grammatiktheorie.* Tübingen: Niemeyer.

Levelt, W. J. M. (1984). "Some Perceptual Limitations on Talking about Space", in A. J. van Doorn, W. A. van der Grind, and J. J. Koenerdink (eds.), *Limits in perception*. Utrecht: VNU Science Press, 323–58.

Levinson, S. C. (1996). "Frames of Reference and Molyneux's Question: Crosslinguistic Evidence", P. Bloom, M. A. Peterson, L. Nadal, and M. F. Garrett (eds.), *Language and Space*. Cambridge, MA: MIT Press, 109–69.

—— (2001). *Presumptive Meanings*. Cambridge, MA: MIT Press.

Lewis, D. (1975): "Adverbs of Quantification" in E. L. Keenan (ed.), *Formal Semantics of Natural Language*. Cambridge: Cambridge University Press, 3–15.

Li, C. N. and Thompson, S. A. (1976). "Subject and Topic: A New Typology of Language", in C. N. Li (ed.), *Subject and topic*. New York/San Francisco/London: Academic Press, 457–89.

Liberman, M. and Prince, A. (1977). "On Stress and Linguistic Rhythm", *Linguistic Inquiry* 8: 249–336.

—— and Sag, I. (1974). "Prosodic Form and Discourse Function", in *Papers from the 10th Regional Meeting of the Chicago Linguistic Society*. University of Chicago: 41–127.

Lightfoot, D. W. (1999). *The Development of Language: Acquisition, Change, and Evolution*. Oxford and Malden, MA: Blackwell Publishers.

Link, G. (1983). "The Logical Analysis of Plurals and Mass Terms: A Lattice-Theoretical Approach", in R. Bäuerle, C. Schwarze, and A. von Stechow (eds.), *Meaning, Use and Interpretation of Language*. Berlin: de Gruyter, 303–23.

Löbner, S. (2000). "Polarity in Natural Language Predication, Quantification and Negation in Particular and Characterizing Sentences", *Linguistics and Philosophy* 23: 213–308.

Lohnstein, H. (2000). *Satzmodus – kompositionell. Zur Parametrisierung der Modusphrase im Deutschen*. Berlin: Akademie-Verlag.

MacDonald, M. C., Pearlmutter N. J. and Seidenberg, M. S. (1994). "The Lexical Nature of Syntactic Ambiguity Resolution", *Psychological Review* 101: 676–703.

Mahajan, A. K. (1990). *The A/A-bar distinction and Movement Theory*. Ph.D. dissertation, Massachusetts Institute of Technology, Cambridge, Massachusetts.

—— (2003). "Word Order and (Remnant) VP Movement", in S. Karimi (ed.), *Word Order and Scrambling*. Oxford: Blackwell, 217–37.

Maling, J. (1990). "Inversion in Embedded Clauses in Modern Icelandic", in J. Maling and A. Zaenen (eds.), *Modern Icelandic Syntax* [*Syntax and Semantics* 24]. San Diego: Academic Press, 71–91.

Marandin, J.-M., Beyssade, C., Delais-Roussarie E., Rialland E. A., and de Fornel, M. (2005). *The Meaning of Final Contours in French*. Manuscript.

Massam, D. (2000). "VSO and VOS: Aspects of Niuean Word Order", in A. Carnie and E. Guilfoyle (eds.), *The Syntax of Verb-initial Languages*. Oxford: Oxford University Press, 97–117.

Mathesius, V. (1975). *A Functional Analysis of Present Day English on a General Linguistic Basis*. The Hague, Paris: Mouton.

May, R. (1985): *Logical Form: Its Structure and Derivation*. Cambridge, MA: MIT Press.

McCarthy, J. and Prince, A. (1993). "Generalized Alignment", in Geert Booij & Jaap van Marle (eds.), *Yearbook of Morphology* 1993. Dordrecht: Kluwer, 79–153.

McCawley, J. D. (1988). *The Syntactic Phenomena of English* 2. Chicago: University of Chicago Press.

McGinnis, M. (2004). "Lethal Ambiguity", *Linguistic Inquiry* 35 (1): 47–95.

Michaels, J. M. and Nelson, C. E. (2004). *A Preliminary Investigation of Intonation in East Bengali.* Manuscript, UCLA.

Millar, M. and Brown, K. (1979). "Tag Questions in Edinburgh Speech", *Linguistische Berichte* 60: 24–45.

Miller, G. A. and Johnson-Laird, P. N. (1976). *Language and Perception.* Cambridge, MA: Harvard University Press.

Molnár, V. (1993): "Zur Pragmatik und Grammatik des TOPIK-Begriffes", in M. Reis (ed.), *Wortstellung und Informationsstruktur* 306 in *Linguistische Arbeiten.* Tübingen: 155–202.

Morimoto, Y. (2000). *Discourse Configurationality in Bantu Morphosyntax.* Ph.D. dissertation, Stanford University.

Müller, G. (2004). "Verb-second as vP-first", *Journal of Comparative Germanic Linguistics* 7 (3), 179–234.

—— and Sternefeld W. (1994). "Scrambling as A-bar Movement", in N. Corver and H. van Riemsdijk (eds.), *Studies on Scrambling.* Berlin: Mouton de Gruyter, 331–85.

Munro, P. and Willmond, C. (1994). *Chickasaw: an Analytical Dictionary.* Norman: University of Oklahoma Press.

Nagahara, H. (1994). *Phonological Phrasing in Japanese.* Doctoral dissertation. Los Angeles: University of California.

Navarretta, C. (2002). "Combining Information Structure and Centering-based Models of Salience for Resolving Intersential Pronominal Anaphora", in A. Branco, T. McEnery, and R. Mitkov (eds.), *Proceedings of the 4th Discourse Anaphora and Anaphor Resolution Colloquium.* Lisbon: Edições Colibri, 135–40.

Neeleman, A. and Koot, H. van de (2008). "Dutch Scrambling and the Nature of Discourse Templates", *The Journal of Comparative Germanic Linguistics* 11: 137–89.

—— and Reinhart, T. (1998). "Scrambling and the PF Interface", in M. Butt and W. Geuder (eds.), *The Projection of Arguments.* Stanford: CSLI Publications, 309–53.

Nespor, M. and Vogel, I. (1986). *Prosodic Phonology.* Dordrecht: Foris Publications.

—— Guasti, M. T., and Christophe, A. (1996). "Selecting Word Order: The Rhythmic Activation Principle", in U. Kleinhenz (ed.), *Interfaces in Phonology.* Berlin: Akademie Verlag, 1–26.

Newman, P. (2000). *The Hausa Language.* New Haven and London: Yale University Press.

Nilsenová, M. (2006). *Rises and Falls: Studies in the Semantics and Pragmatics of Intonation.* Ph.D. thesis, Institute For Logic, Language and Computation, Universiteit van Amsterdam.

Nkemnji, M. A. (1995). *Heavy pied-piping in Nkweh.* Ph.D. dissertation, UCLA.

Oda, K. (2005). "V1 and Wh-questions: A Typology", in A. Carney, H. Harley, and S. Dooley (eds.), *Verb First: On the Syntax of Verb-initial Languages*. Amsterdam: John Benjamins: 107–34.

Oehrle, R. T. (1976). *The Grammatical Status of the English Dative Alternation*. Ph.D. dissertation, MIT.

—— (1987). "Multi-dimensional Categorial Grammars and Linguistic Analysis", in E. Klein and J. van Bentham (eds.), *Categories, Polymorphism and Unification*. Center for Cognitive Science, Edinburgh: University of Edinburgh, 231–60.

Olsvay, C. (2000). *A Syntactic Analysis of Negative Universal Quantifiers in Hungarian*. MA thesis, Eötvös Loránd University, Budapest.

Partee, B. (1991). "Topic, Focus and Quantification", in S. Moore and A. Z. Wyner (eds.), *Proceedings from Semantics and Linguistic Theory*, Vol. 1 of *Cornell Working Papers in Linguistics*, 159–188. Cornell.

—— (1995). "Quantificational Structures and Compositionality", in A. Kratzer, E. Bach, E. Jelinek, and B. Partee (eds.), *Quantification in Natural Languages*. Dordrecht: Kluwer, 541–602.

Partee, B. H. and Borschev, V. (2003). "Genitives, Relational Nouns, and Argument-modifier Ambiguity", in E. Lang, C. Maienborn, C. Fabricius-Hansen (eds.), *Modifying adjuncts*. Berlin: Mouton de Gruyter, 67–112.

Paulson, L. (1996). *ML for the Working Programmer*. 2nd edn. Cambridge: Cambridge University Press.

Pierrehumbert, J. (1980). *The Phonology and Phonetics of English Intonation*. Ph.D. dissertation, MIT.

—— and Beckmann, M. (1988). *Japanese Tone Structure*. Cambridge, MA: MIT Press.

—— and Hirschberg, J. (1990). "The Meaning of Intonational Contours in the Interpretation of Discourse", in P. R. Cohen, J. Morgan, and M. E. Pollack (eds.), *Intentions in Communication*. Cambridge, MA: MIT Press, 271–311.

Pinker, S. (2007). *The Stuff of Thought*. Viking.

Pintzuk, S. (1999). *Phrase Structures in Competition: Variation and Change in Old English Word Order*. New York: Garland.

Poesio, M., Stevenson, R., di Eugenio, B., and Hitzeman, J. (2004). "Centering: A Parametric Theory and its Instantiations", *Computational Linguistics* 30 (3): 309–63.

Portner, P. and Yabushita, K. (1998). "The Semantics and Pragmatics of Topic Phrases." *Linguistics and Philosophy* 21: 117–57.

Poser, W. (1984). *The Phonetics and Phonology of Tone and Intonation in Japanese*. Ph.D. dissertation, MIT.

Potts, C. (2005). *The Logic of Conventional Implicatures*. Oxford University Press.

Prat Sala, M. (1997). *The Production of Different Word Orders: A Psycholinguistic and Developmental Approach*. Ph.D. dissertation, University of Edinburgh.

Prince, A. (1983). "Relating to the Grid", *Linguistic Inquiry*: 19–100.

Prince, E. F. (1981). "Toward a Taxonomy of Given-New Information", in P. Cole (ed.), *Radical Pragmatics*. New York: Academic Press, 223–56.

—— (1997). "On the Functions of Left-dislocation in English Discourse", in A. Kamio (ed.), *Directions in Functional Linguistics*. Amsterdam: John Benjamins, 117–43.

—— (1999). "How Not to Mark Topics. *Topicalization* in English and Yiddish", in *Texas Linguistic Forum*. Austin: University of Texas. <www.ling.upenn. edu/~ellen/ home.html>.

—— (2003). *The Yiddish Impersonal Pronoun Men "one" in discourse*. Presented at the Annual Meeting of the Linguistic Society of America, Atlanta, January 2003. <www.ling.upenn.edu/~ellen>.

Quesada, J. D. (2000). *Teribe*. München, etc.: Lincom Europa.

—— and Skopeteas, S. (2008). *The Discourse Functions of Inverse Voice: an Experimental Study in Teribe (Chibchan)*. Manuscript, University of Potsdam and Universidad Nacional de Costa Rica.

Quirk, R., Greenbaum, S., Leech, G., and Svartvik, J. (1985). *A Comprehensive Grammar of the English Language*. London: Longman Publishers.

Rando, E. (1980). "Intonation in Discourse", in L. Waugh and C. van Schooneveld (ed.), *The Melody of Language*. Baltimore: University Park Press, 243–78.

Reese, B. (2006a). "Negative Polar Interrogatives and Bias", in T. Washio, A. Sakurai, K. Nakajima, H. Takeda, S. Tojo, and M. Yokoo (eds.), *New Frontiers in Artificial Intelligence: Joint JSAI 2005 Workshop Post-Proceedings*, Lecture Notes in Computer Science. Berlin: Springer-Verlag, 85–92.

—— (2006b). "The Meaning and Use of Negative Polar Interrogatives", in O. Bonami and P. Cabredo Hofherr (eds.), *Empirical Issues in Syntax and Semantics* 6: 331–54.

—— (2007). *Bias in Questions*. Ph.D. thesis, University of Texas at Austin.

—— and Asher, N. (2007). "Prosody and the Interpretation of Tag Questions", in E. Puig-Waldmüller (ed.), *Proceedings of Sinn und Bedeutung* 11. Barcelona: Universitat Pompeu Fabra, 433–47.

Reineke, B. (2006a). "Focus et topique en tant que deux phénomènes pragmatiques dans les langues Oti-Volta orientales", *Cahïers Voltaïques / Gur Papers* 7: 100–11.

—— (2006b). "Verb- und Prädikationsfokus im Ditammari und Byali", in K. Winkelmann and D. Ibriszimow (eds.), *Zwischen Bantu und Burkina. Festschrift für Gudrun Miehe zum 65. Geburtstag*. Köln: Rüdiger Köppe Verlag, 163–80.

—— (2007). "Identificational Operation as Focus Strategy in Byali", in E. O. Aboh, K. Hartmann, and M. Zimmermann (eds.), *Focus Strategies in African Languages*. Berlin: Mouton de Gruyter, 223–40.

Reinhart, T. (1981). "Pragmatics and Linguistics. An Analysis of Sentence Topics", *Philosophica* 27: 53–94.

—— (1982). *Pragmatics and Linguistics: an Analysis of Sentence Topics*. University of Indiana Linguistics Club.

—— (1995). *Interface Strategies*. OTS Working Papers (TL-95-002).

—— (1996). "Interface Economy: Focus and Markedness", in C. Wilder, H.-M. Gaertner, and M. Bierwisch (eds.), *The Role of Economy Principles in Linguistic Theory*. Berlin: Akademische Verlag.

Reinhart, T. (1998). "Wh-in-situ in the Framework of the Minimalist Program", *Natural Language Semantics* 6: 29–56.

—— (2006). *Interface Strategies: Optimal and Costly Computations*. Cambridge, MA: MIT Press.

Rizzi, L. (1997). "The Fine Structure of the Left Periphery", in L. Haegeman (ed.), *Elements of Grammar*. Dordrecht: Kluwer, 281–337.

Roberts, C. (1996). "Information Structure in Discourse: Towards an Integrated Formal Theory of Pragmatics", in J. Yoon and A. Kathol (eds.), *OSU Working Papers in Linguistics* 49. The Ohio State University Department of Linguistics, 91–136.

Roberts, I. G. (1995). "Object Movement and Verb Movement in Early Modern English", in H. Haider, S. Olsen, and S. Vikner (eds.), *Studies in Comparative Germanic Syntax*. Dordrecht: Kluwer, 269–84.

—— (1997). "Directionality and Word Order Change in the History of English", in A. Kemenade and N. Vincent (eds.), *Parameters of Morphosyntactic Change*. Cambridge: Cambridge University Press, 396–426.

—— (2001). "Language Change and Learnability", in S. Bertolo (ed.), *Language Acquisition and Learnability*. Cambridge: Cambridge University Press, 81–125.

Rochemont, M. (1986). *Focus in Generative Grammar*. Amsterdam: John Benjamins.

Rögnvaldsson, E. (1996). "Word Order Variation in the VP in Old Icelandic", *Working Papers in Scandinavian Syntax* 58: 55–86.

Rohrbacher, B. W. (1994). *The Germanic VO Languages and the Full Paradigm: A Theory of V-to-I raising*. Doctoral dissertation, University of Massachusetts, Amherst.

—— (1999). *Morphology-Driven Syntax: A Theory of V to I Raising and Pro-drop*. Amsterdam: John Benjamins.

Romero, M. (2005). "Two Approaches to Biased yes/no Questions", in J. Alderete, C.-H. Han, and A. Kochetov (eds.), *Proceedings of the 24th West Coast Conference on Formal Linguistics*. Somerville, MA: Cascadilla Press, 352–60.

—— and Han, C.-H.(2004). On Negative *yes/no* Questions. *Linguistics and Philosophy* 27: 609–658.

Rooij R. van (2003). "Negative Polarity Items in Questions: Strength as Relevance", *Journal of Semantics* 20: 239–73.

—— (2004). "Utility, Informativity and Protocols", *Journal of Philosophical Logic* 33: 389–419.

—— and Šafářová, M. (2003). "On Polar Questions", in R. Young and Y. Zhou (eds.), *Proceedings of SALT XIII*. Ithaca, NY: CLC Publications.

—— and Schulz, K. (2004). "Exhaustive Interpretation of Complex Sentences", *Journal of Logic, Language and Information*: 491–519.

Rooth, M. (1985): *Association with Focus*. Ph.D. thesis, University of Massachusetts, Amherst.

—— (1992): "A Theory of Focus Interpretation", *Natural Language Semantics* 1: 75–116.

—— (1993). *A hybrid architecture for focus*. Handout for talk given at the Amsterdam colloquium.
—— (1995). "Indefinites, Adverbs of Quantification and Focus Semantics", in G. Carlson and F. J. Pelletier (eds.), *The Generic Book*. Chicago: University of Chicago Press, 265–99.
—— (1996a). "Focus", in S. Lappin (ed.), *The Handbook of Contemporary Semantic Theory*. Oxford: Blackwell, 271–97.
—— (1996b). "On the Interface Principles for Intonational Focus", in T. Galloway and J. Spence (eds.), *Proceedings of SALT VI*. Ithaca, NY: CLC Publications.
—— (2004). "Comments on Krifka's Paper", in H. Kamp and B. Partee (eds.), *Context-Dependence in the Analysis of Linguistic Meaning*. Amsterdam: Elsevier, 475–87.
Sabel, J. and Zeller, J. (2006). "Wh-question Formation in Nguni", *Selected Proceedings of the 35th Annual Conference on African Linguistics*: 271–283.
Sadock, J. M. (1971). "Queclaratives", in *Papers from the 7th Regional Meeting of the Chicago Linguistic Society*. University of Chicago, 223–31.
—— (1974). *Towards a Linguistic Theory of Speech Acts*. New York: Academic Press.
—— and Zwicky, A. M. (1985). "Speech Act Distinctions in Syntax", in T. Shopen (ed.), *Language Typology and Syntactic Description* 1. Cambridge: Cambridge University Press, 155–96.
Šafářová, M. (2005). "The Semantics of Rising Intonation in Interrogatives and Declaratives", in E. Maier, C. Bary, and J. Huitink (eds.), *Proceedings of SuB9*: 355–69. <www.ru.nl/ncs/sub9>.
Saito, M. (1992). "Long-distance Scrambling in Japanese", *Journal of East Asian Linguistics* 1: 69–118.
Saltarelli, M. (1988). *Basque*. London: Routledge.
Samek-Lodovici, V. (2005). "Prosody-Syntax Interaction in the Expression of Focus". *Natural Language and Linguistic Theory* 23: 687–755.
Sasse, H.-J. (1987). "The Thetic/categorical Distinction Revisited". *Linguistics* 25: 511–80.
—— (1995). "Theticity and VS order: A case study", in Y. Matras and H.-J. Sasse (eds.), *Verb-subject order and theticity in European languages (Sprachtypologie und Universalienforschung* 48 (1/2)). Berlin: Akademie, 3–31.
Sauerland, U. (2004). "Scalar Implicatures in Complex Sentences", *Linguistics and Philosophy* 27 (3): 363–91.
Schmerling, S. (1976). *Aspects of English Sentence Stress*. Austin: University of Texas Press.
Schuh, R. (1982). "Questioned and Focused Subjects and Objects in Bade/Ngizim", in H. Jungraithmayr (ed.), *The Chad Languages in the Hamitosemitic-Negritic Border Area*. Berlin: Reimer, 160–74.
—— (2005). *Yobe State, Nigeria, as a Linguistic Area*. Paper presented at the 31st Annual Meeting of the Berkeley Linguistics Society, February 18–20, 2005.
Schulz, K. and Rooij, R. van (2006) "Pragmatic Meaning and Non monotonic Reasoning: The Case of Exhaustive Interpretation", *Linguistics and Philosophy* 29: 205–50.

Schwarz, A. (2005). *Aspekte der Morphosyntax und Tonologie im Buli. Mit Schwerpunkt auf dem Buli von Wiaga.* Ph.D. dissertation, Humboldt-Universität zu Berlin.

—— (2007). "The Particles *lé* and *lá* in the Grammar of Konkomba", in S. Ishihara, S. Jannedy and A. Schwarz (eds.), *Interdisciplinary Studies on Information Structure* 8. Potsdam: Potsdam University, 115–39.

—— (2008a). "Buli". Manuscript (submitted to *The expression of information structure: A cross-linguistic survey in speech production*, in S. Skopeteas, S. Hellmuth, G. Fanselow, and C. Féry. Berlin, New York: Mouton de Gruyter.)

—— (2008b). "The Syntax of Lelemi Predicates". Manuscript (submitted to *Afrika und Übersee*).

—— (2009a). "Tonal Focus Reflections in Buli and Some Gur Relatives", in *Tone and intonation in a Typological Perspective. Lingua* 119: 950–72.

—— (2009b). "How Many Focus Markers are there in Konkomba?", in M. Matondo, F. McLaughlin and E. Potsdam (eds.), *Selected Proceedings of the 38th Annual Conference on African Linguistics.* Somerville: Cascadilla Press, 182–92.

—— (to appear). "Verb and Predication Focus Marking in Gur", in I. Fiedler and A. Schwarz (eds.), *Information Structure in African Languages.* Amsterdam: John Benjamins.

—— and Fiedler I. (2007). "Narrative Focus Strategies in Gur and Kwa", in E. O. Aboh, K. Hartmann, and M. Zimmermann (eds.), *Focus Strategies in African Languages.* Berlin: Mouton de Gruyter, 267–86.

Schwarzschild, R. (1999). "GIVENness, AvoidF and Other Constraints on the Placement of Accent", *Natural Language Semantics* 7: 141–77.

—— (2004). "Focus Interpretations: Comments on Geurts and van der Sandt", *Theoretical Linguistics* 30(1): 137–47.

Searle, J. R. (1969). *Speech Acts: An Essay in the Philosophy of Language.* Cambridge University Press.

—— (1975). "Indirect Speech Acts", in P. Cole and J. Morgan (eds.), *Syntax and Semantics* 3: *Speech Acts.* New York: Academic Press, 59–82.

—— and Vanderveken, D. (1985). *Foundations of Illocutionary Logic.* Cambridge University Press.

Selkirk, E. (1984). *Phonology and Syntax: The Relation Between Sound and Structure.* Cambridge, MA: MIT Press.

—— (1986). "On Derived Domains in Sentence Phonology", *Phonology Yearbook* 3: 371–405.

—— (1995). "Sentence Prosody: Intonation, Stress, and Phrasing", in J. A. Goldsmith (ed.), *The Handbook of Phonological Theory.* London: Blackwell, 550–69.

—— (2002). "Contrastive FOCUS versus Presentational Focus: Prosodic Evidence from Right Node Raising in English", in Speech Prosody 2002. Aix-en-Provence, France.

—— (2006a). "Bengali Intonation Revisited: An Optimality Theoretic Analysis in which FOCUS Stress Prominence Drives FOCUS phrasing", in C. Lee and D. Büring (eds.), *Topic and Focus: Cross-linguistic Perspectives on Meaning and Intonation.* Dordrecht: Springer, 215–44.

—— (2006b). Contrastive Focus, Givenness, and Phrase Stress. Unpublished manuscript 20.

—— (2007). "*Bengali Intonation Revisited: An Optimality Theoretic Analysis in which FOCUS Stress Prominence Drives FOCUS Phrasing*", in Lee *et al.* (2007), 215–44.

—— (2008). "Contrastive Focus, Givenness and Unmarked Status of *Discourse New*", in C. Féry, G. Fanselow, and M. Krifka (eds.), *The Notions of Information Structure, Acta Linguistica Hungarica*: 125–45.

—— and Tateishi, K. (1991). "Syntax and Downstep in Japanese", in C. Georgopoulos and R. Ishihara (eds.), *Interdisciplinary Approaches to Language: Essays in Honor of S.-Y. Kuroda.* Dordrecht: Kluwer Academic Publishers, 519–43.

Sestoft, P. (1996). *Standard ML on the Web Server: Visualizing Lambda Calculus Reduction.*

Shaer, B. (2003). "*Manner* Adverbs and the Association Theory: Some Problems and Solutions", in E. Lang, C. Maienborn, and C. Fabricius-Hansen (eds.), *Modifying adjuncts.* Berlin: Mouton de Gruyter: 211–60.

—— Cook, P., Frey, W., and Maienborn, C. (eds., 2008). *Dislocated Elements in Discourse. Syntactic, Semantic and Pragmatic Perspectives.* London: Routledge.

Shan, C. (2004). "Binding Alongside Hamblin Alternatives Calls for Variable-free Semantics", is *Proceedings of SALT XIV.* Ithaca, NY: CLC Publications.

Shinya, T. (1999). "Eigo to nihongo ni okeru fookasu ni yoru daunsteppu no sosi to tyooon-undoo no tyoogoo" [The blocking of downstep by focus and articulatory overlap in English and Japanese]. In *Proceedings of Sophia Linguistics Society* 14, 35–51.

Siewierska, A. (1993). "On the Interplay of Factors in the Determination of Word Order", in J. Jacobs *et al.* (eds.), *Syntax: An International Handbook of Contemporary Research.* Berlin: Walter de Gruyter, 826–46.

—— (1998). "Variation in Major Constituent Order: A Global and a European Perspective", in A. Siewierska (ed.). *Constituent Order in the Languages of Europe.* Berlin: Mouton de Gruyter, 475–551.

Sigurðsson, H. Á. (1988). "From OV to VO: Evidence from Old Icelandic", *Working Papers in Scandinavian Syntax* 34.

Singer, M. (1976). "Thematic Structure and the Integration of Linguistic Information", *Journal of Verbal Learning and Verbal Behavior* 15: 549–58.

Skopeteas, S. and Fanselow, G. (2009). *Focus in Georgian and the Expression of Contrast.* Manuscript, University of Potsdam (submitted to *Lingua*).

—— and Verhoeven, E. (2005). "Postverbal Argument Order in Yucatec Maya", *Sprachtypologie und Universalienforschung* 53 (1): 71–9.

—— Féry, C., and Asatiani, R. (2009). "Word Order and Intonation in Georgian", *Lingua* 119: 102–27.

—— Fiedler, I., Hellmuth, S., Schwarz, A., Stoel, R., Fanselow, G., Féry, C., and Krifka, M. (2006). "Questionnaire on Information Structure (QUIS)",

Interdisciplinary Studies on Information Structure (ISIS) 4. Universitätsverlag Potsdam, Germany.

Smyth, R. (1994). "Grammatical Determinants of Ambiguous Pronoun Resolution", *Journal of Psycholinguistic Research*, 23: 197–229.

Sperber, D. and Wilson, D. (1986/1995). *Relevance: Communication and Cognition.* Oxford: Blackwell.

Speyer, A. (2007). "Die Bedeutung der Centering Theory für Fragen der Vorfeldbesetzung im Deutschen", *Zeitschrift für Sprachwissenschaft* 26: 83–115.

Stechow, A. von (1982). *Structured Propositions.* Manuscript, Universität Konstanz.

—— (1985/89). Focusing and backgrounding operators. Technical Report 6, Fachgruppe Sprachwissenschaft, Universität Konstanz.

—— (1990). *Current Issues in the Theory of Focus.* Fachgruppe Sprachwissenschaft Universität Konstanz: Arbeitspapier Nr.24.

—— and Uhmann S. (1986). "On the Focus-pitch Accent Relation", *Groninger Arbeiten zur Germanistischen Linguistik* 25, 228–63.

Steedman, M. (2000). "Information Structure and the Syntax-Phonology Interface", *Linguistic Inquiry* 31(4): 649–89.

—— (2003). "Information-structural Semantics for English Intonation". *Paper to LSA Summer Institute Workshop on Topic and Focus.* Santa Barbara.

Stevenson, R. J., Crawley, R. J., and Kleinman, D. (1994). "Thematic Roles, Focus and the Representation of Events", *Language and Cognitive Processes* 9: 519–48.

Strawson, P. F. (1964). "Identifying Reference and Truth-values", *Theoria* 30: 96–118. Reprinted in D. Steinberg and L. Jakobovits (eds.), *Semantics: An Interdisciplinary Reader in Linguistics, Philosophy and Psychology.* Cambridge: Cambridge University Press.

Strube, M. and Hahn, U. (1996). "Functional Centering", in *Proceedings of ACL '96*, 270–7.

—— —— (1999). "Functional Centering: Grounding Referential Coherence in Information Structure", *Computational Linguistics* 25 (3): 309–44.

Sturgeon, A.-M. (2006). *The Syntax and Pragmatics of Contrastive Topic in Czech.* Ph.D. dissertation, UC Santa Cruz.

Sugahara, M. (2002). "Conditions on Post-FOCUS Dephrasing in Tokyo Japanese", in B. Bell and I. Marlien (eds.), *Proceedings of the First International Conference on Speech Prosody.* Aix-En-Provence, France: 655–8.

—— (2003): *Downtrends and Post-FOCUS Intonation in Japanese.* Ph.D. dissertation, University of Massachusetts, Amherst.

Swart, H. de (1992). "Intervention Effects, Monotonicity and Scope", *Proceedings of SALT* 2: 387–406.

—— (1993): *Adverbs of Quantification: A Generalized Quantifier Approach.* New York: Garland.

Swerts, M. and Krahmer, E. (2008). "Meaning, Intonation and Negation", in H. Bunt and R. Muskens (eds.), *Computing Meaning* Volume 3. Dordrecht: Springer.

——— and Avesani, C. (2002). "Prosodic Marking of Information Status in Dutch and Italian: A Comparative Analysis", *Journal of Phonetics* 4(30): 629–54.

Szabolcsi, A. (1980). "Az aktuális mondattagolás szemantikájához", *Nyelvtudományi Közlemények* 82: 59–82.

—— (1981a). "The Semantics of Topic-Focus Articulation" in J. Groenendijk *et al.* (eds.), *Formal Methods in the Study of Language*. Amsterdam: Matematisch Centrum, 513–40.

—— (1981b). "Compositionality in Focus", *Folia Linguistica* 15: 141–62.

—— (1983). "Focusing Properties, or the Trap of First Order", *Theoretical Linguistics* 10: 125–45.

—— (1985). "Puszta főnév a fókuszban, avagy az elsőrendű predikátumkalkulus csapdája", *Általános Nyelvészeti Tanulmányok* 16: 229–50.

—— (1994). "All Quantifiers are Not Equal: the Case of Focus", *Acta Linguistica Hungarica* 42: 171–87.

—— (1997). "Strategies for Scope Taking" in A. Szabolcsi (ed.), *Ways of Scope Taking*. Dordrecht: Kluwer, 109–55.

—— and Zwarts, F. (1992–3). "Weak Islands and an Algebraic Semantics for Scope Taking", *Natural Language Semantics* 1: 235–84.

Szendrői, K. (no year). *Stress-Focus Correspondence in Italian*. Manuscript University College London.

—— (2001). "Focus and the Syntax-Phonology Interface". Ph.D. thesis, University College London.

—— (2003). "A Stress-based Approach to the Syntax of Hungarian Focus", *The Linguistic Review* 20: 37–78.

Talmy, L. (1983). "How Language Structures Space", in H. L. Pick Jr. and L. P. Acredolo (eds.), *Spatial Orientation. Theory, Research, and Application*. New York: Plenum Press, 225–82.

Tancredi, C. (2005). *Association with Focus and Discourse Licensing*. Presented at National Institute of Informatics Workshop on Focus.

Tanenhaus, M. K. and Trueswell, J. C. (1995). "Sentence Comprehension", in J. L. Miller and P. D. Eimas (eds.), *Handbook of Perception and Cognition* 11. San Diego: Academic Press, 217–62.

Taraldsen, K. T. (2000). "V-movement and VP-movement in Derivations Leading to VO Order", in P. Svenonius (ed.), *The Derivation of VO and OV*. Amsterdam: John Benjamins: 97–122.

Temürcü, C. (2001). *Word Order Variations in Turkish: Evidence from Binding and Scope*. Ph.D. dissertation, Middle East Technical University, Ankara.

Tomioka, S. (2007). "Intervention Effects in Focus: From a Japanese Point of View", in S. Ishihara (ed.), *Potsdam Working Papers in Linguistics: Proceedings from WPSI 2*, Universität Potsdam.

—— (2009). "A Scope Theory of Contrastive Topics", in *Current Issues in Unity and Diversity of Languages: Collection of the Papers Selected from the CIL 18*, The Linguistic Society of Korea, Seoul, 2282–96.

Tomlin, R. (1995). "Focal Attention, Voice, and Word Order: an Experimental, Cross-linguistic Study", in P. Downing and M. Noonan (eds.), *Word order in discourse*. Amsterdam: John Benjamins, 517–54.

Trips, C. (2002). *From OV to VO in Early Middle English*. Amsterdam: John Benjamins.

Truckenbrodt, H. (1995). "Phonological Phrases: Their Relation to Syntax, Focus, and Prominence". Ph.D. thesis, MIT, Cambridge, MA.

—— (2002). "Upstep and Embedded Register Levels", *Phonology* 19: 77–120.

—— (2004). "Final Lowering in Non-final Position", *Journal of Phonetics* 32: 313–48.

—— (2006). "Phrasal Stress", in K. Brown (ed.), *The Encyclopedia of Languages and Linguistics* 9, 2nd edition. Oxford: Elsevier, 572–79.

—— and Féry, C. (2004). *More on Hierarchical Organization and Tonal Scaling*. Manuscript, Univeristy of Tübingen/University of Potsdam.

Trueswell, J. C., Tanenhaus, M. K., and Garnsey, S. (1994). "Semantic Influences on Parsing: Use of Thematic Role Information in Syntactic Ambiguity Resolution", *Journal of Memory and Language* 33: 285–318.

Tsai, W.-T. (1995). *On Economizing the Theory of A-bar Dependencies*. Ph.D. dissertation, MIT.

Tuller, L. (1986). *Bijective Relations in Universal Grammar and the Syntax of Hausa*. Ph.D. dissertation, UCLA, Los Angeles.

—— (1992). "The Syntax of Postverbal Focus Constructions in Chadic", *Natural Language and Linguistic Theory* 10: 303–34.

Vallduví, E. (1992). *The Informational Component*. Ph.D. dissertation, University of Pennsylvania, Garland.

Vermeulen, R. (2009). "Topics, Contrast and Contrastive Topics in Japanese", in R. Vermeulen and R. Shibagaki (eds.), *MIT Working Papers in Linguistics 58: Proceedings of the 5th Workshop of Altaic Formal Linguistics (WAFL 5)*, 361–72.

Vikner, S. (1995). *Verb Movement and Expletive Subjects in the Germanic Languages*. New York: Oxford University Press.

—— (1997). "V-to-I Movement and Inflection for Person in All Tenses", in L. Hageman (ed.), *The New Comparative Syntax*. London: Longman, 189–213.

—— (2001). *Verb Movement Variation in Germanic and Optimality Theory*. Habilitation thesis, University of Tübingen.

Vilkuna, M. (1989). *Free Word Order in Finnish. Its Syntax and Discourse Functions*. Ph.D. thesis, Helsinki.

Wagner, M. (2005). *Prosody and recursion*. Unpublished Ph.D. dissertation, MIT.

—— (2006). "Givenness and Locality", in *Proceedings of SALT XVI*. CLC Publications.

—— (2008). "A Compositional Analysis of Contrastive Topics", to appear in the *Proceedings of NELS 38*.

Walker, M. A., Joshi, A. K., and Prince, E. F. (1998). *Centering Theory in Discourse*. Oxford: Clarendon Press.

Warburton, I. (1975). "The Passive in English and Greek", in W. Abraham (ed.), *Foundations of Language, International Journal of Language and Philosophy*. Boston: Reidel, 563–78.

Ward, G. (1988). *The Semantics and Pragmatics of Preposing*. New York: Garland.

—— and Hirschberg, J. (1985). "Implicating uncertainty: The Pragmatics of Fall–Rise Intonation", *Language* 61(4): 747–76.

Watters, J. (1979). "Focus in Aghem: A Study of its Formal Correlates and Typology", in Hyman (1979b), 137–97.

Weber, A. and Müller, K. (2004). "Word Order Variation in German Main Clauses: A Corpus Analysis", *Proceedings of the 20th International Conference on Computational Linguistics*. Geneva, Switzerland, 71–7.

Wedgwood, D. (2005). *Shifting the Focus. From Static Structures to the Dynamics of Interpretation*. Amsterdam: Elsevier.

Westergaard, M. R. (2004a). *What Norwegian Child Language can Tell us about the History of English: The Interaction of Syntax and Information Structure in the Development of Word Order*. Paper presented at Workshop on the Syntax of English and the Nordic Languages, Ninth Nordic Conference for English Studies. Århus, May 27–29, 2004.

—— (2004b). "The Interaction of Input and UG in the Acquisition of Verb Movement in a Dialect of Norwegian", *Nordlyd* 32 (1): 110–34.

Wiklund, A.-L., Hrafnbjargarson, G. H., Bentzen, K., and Hróarsdóttir, Þ. (2007). "Rethinking Scandinavian Verb Movement", *Journal of Comparative Germanic Linguistics* 10 (3): 203–33.

Wiltschko, M. (1997). "Parasitic Operators in German Left-Dislocation", in E. Anagnostopoulou, H. van Riemsdijk, and F. Zwarts (eds.), *Materials on Left Dislocation*. Amsterdam: John Benjamins, 307–39.

Winograd, T. (1972). *Understanding Natural Language*. New York, Academic Press.

Wold, D. E. (1996). "Long Distance Selective Binding: The Case of Focus", in T. Galloway, and J. Spence (eds.), *Proceedings from Semantics and Linguistic Theory (SALT)* 6. Cornell University, 311–28.

Xu, L. (2006). "Topicalization in Asian languages", in M. Everaert and H. van Riemsdijk (eds.), *The Blackwell Companion to Syntax*, Vol. V. Oxford: Blackwell, 137–74.

Yabushita, K. (2008). "A New Approach to Contrastive Topic: Partition Semantics and Pragmatics", to appear in T. Friedman and S. Ito (eds.), *The Proceedings of Semantics and Linguistic Theory XVIII*. Ithaca, NY: Cornell University.

Yekovich, F. R., Walker, C. H., and Blackman, H. S. (1979). "The Role of Presupposed and Focal Information in Integrating Sentences", *Journal of Verbal Learning and Verbal Behavior* 18: 535–48.

Zimmer, H. D. and Engelkamp, J. (1981). "The Given-New Structure of Cleft Sentences and their Influence on Picture Viewing", *Psychological Research* 43: 375–89.

Zimmermann, M. (2005). "Focus in Western Chadic: A Unified OT-Account", in C. Davis *et al.* (eds.), *Proceedings of NELS 36*. UMass, Amherst.

——(2006). "Focus Realization in West Chadic", Manuscript, Humboldt University, Berlin.

——(2008). "Contrastive Focus and Emphasis", *Acta Linguistica Hungarica* 55: 347–60.

Zubizarreta, M. L. (1998). *Prosody, Focus and Word Order*. Cambridge, MA: MIT Press.

Author Index

Subject Index

The manufacturer's authorised representative in the EU for product safety is Oxford University Press España S.A. of el Parque Empresarial San Fernando de Henares, Avenida de Castilla, 2 – 28830 Madrid (www.oup.es/en or product.safety@oup.com). OUP España S.A. also acts as importer into Spain of products made by the manufacturer.

www.ingramcontent.com/pod-product-compliance
Lightning Source LLC
LaVergne TN
LVHW011904200926
848484LV00002B/525

9780199570959